Catalytic Nanomaterials: Energy and Environment

Catalytic Nanomaterials: Energy and Environment

Editors

Hongda Li
Mohammed Baalousha
Victor A. Nadtochenko

MDPI • Basel • Beijing • Wuhan • Barcelona • Belgrade • Manchester • Tokyo • Cluj • Tianjin

Editors

Hongda Li
School of Electronic Engineering
Guangxi University of Science and Technology
Liuzhou
China

Mohammed Baalousha
Department of Environmental Health Sciences
University of South Carolina
Columbia
United States

Victor A. Nadtochenko
N.N. Semenov Federal Research Center for Chemical Physics
Russian Academy of Sciences
Moscow
Russia

Editorial Office
MDPI
St. Alban-Anlage 66
4052 Basel, Switzerland

This is a reprint of articles from the Special Issue published online in the open access journal *Molecules* (ISSN 1420-3049) (available at: www.mdpi.com/journal/molecules/special_issues/Photocatalytic_Nanomaterials).

For citation purposes, cite each article independently as indicated on the article page online and as indicated below:

LastName, A.A.; LastName, B.B.; LastName, C.C. Article Title. *Journal Name* **Year**, *Volume Number*, Page Range.

ISBN 978-3-0365-7705-0 (Hbk)
ISBN 978-3-0365-7704-3 (PDF)

© 2023 by the authors. Articles in this book are Open Access and distributed under the Creative Commons Attribution (CC BY) license, which allows users to download, copy and build upon published articles, as long as the author and publisher are properly credited, which ensures maximum dissemination and a wider impact of our publications.

The book as a whole is distributed by MDPI under the terms and conditions of the Creative Commons license CC BY-NC-ND.

Contents

About the Editors vii

Preface to "Catalytic Nanomaterials: Energy and Environment" ix

Fangzhi Wang, Xiaoyan Zhou, Jing Li, Qiuyue He, Ling Zheng and Qing Liu et al.
Rationally Designed $Bi_2M_2O_9$ (M = Mo/W) Photocatalysts with Significantly Enhanced Photocatalytic Activity
Reprinted from: *Molecules* **2021**, *26*, 7334, doi:10.3390/molecules26237334 1

Lanqin Tang, Yin Jia, Zhishang Zhu, Yue Hua, Jun Wu and Zhigang Zou et al.
Effects of Co Doping on the Growth and Photocatalytic Properties of ZnO Particles
Reprinted from: *Molecules* **2022**, *27*, 833, doi:10.3390/molecules27030833 17

Quanhui Li, Liang Jiang, Yuan Li, Xiangrong Wang, Lixia Zhao and Pizhen Huang et al.
Enhancement of Visible-Light Photocatalytic Degradation of Tetracycline by Co-Doped TiO_2 Templated by Waste Tobacco Stem Silk
Reprinted from: *Molecules* **2023**, *28*, 386, doi:10.3390/molecules28010386 29

Hongxia Fan, Xiaohui Ma, Xinyang Li, Li Yang, Yongzhong Bian and Wenjun Li
Fabrication of Novel g-C_3N_4@Bi/$Bi_2O_2CO_3$ Z-Scheme Heterojunction with Meliorated Light Absorption and Efficient Charge Separation for Superior Photocatalytic Performance
Reprinted from: *Molecules* **2022**, *27*, 8336, doi:10.3390/molecules27238336 43

Yayang Wang, Xiaojie Yang, Jiahui Lou, Yaqiong Huang, Jian Peng and Yuesheng Li et al.
Enhance ZnO Photocatalytic Performance via Radiation Modified g-C_3N_4
Reprinted from: *Molecules* **2022**, *27*, 8476, doi:10.3390/molecules27238476 55

Xiaomin Guo, Guotao Pan, Lining Fang, Yan Liu and Zebao Rui
Z-Scheme CuO_x/Ag/TiO_2 Heterojunction as Promising Photoinduced Anticorrosion and Antifouling Integrated Coating in Seawater
Reprinted from: *Molecules* **2023**, *28*, 456, doi:10.3390/molecules28010456 67

Guowei Wang and Hefa Cheng
Application of Photocatalysis and Sonocatalysis for Treatment of Organic Dye Wastewater and the Synergistic Effect of Ultrasound and Light
Reprinted from: *Molecules* **2023**, *28*, 3706, doi:10.3390/molecules28093706 81

Sung-Hyun Kim and Hee-Joon Kim
Photocatalytic Hydrogen Production by the Sensitization of Sn(IV)-Porphyrin Embedded in a Nafion Matrix Coated on TiO_2
Reprinted from: *Molecules* **2022**, *27*, 3770, doi:10.3390/molecules27123770 99

Rongyang Yin, Pengfei Sun, Lujun Cheng, Tingting Liu, Baocheng Zhou and Xiaoping Dong
A Three-Dimensional Melamine Sponge Modified with MnOx Mixed Graphitic Carbon Nitride for Photothermal Catalysis of Formaldehyde
Reprinted from: *Molecules* **2022**, *27*, 5216, doi:10.3390/molecules27165216 109

Ewelina Wierzyńska, Marcin Pisarek, Tomasz Łecki and Magdalena Skompska
Comparative Studies of g-C_3N_4 and $C_3N_3S_3$ Organic Semiconductors—Synthesis, Properties, and Application in the Catalytic Oxygen Reduction
Reprinted from: *Molecules* **2023**, *28*, 2469, doi:10.3390/molecules28062469 121

Dongli Zhang, Yujun Shen, Jingtao Ding, Haibin Zhou, Yuehong Zhang and Qikun Feng et al.
A Combined Experimental and Computational Study on the Adsorption Sites of Zinc-Based MOFs for Efficient Ammonia Capture
Reprinted from: *Molecules* **2022**, *27*, 5615, doi:10.3390/molecules27175615 137

Dongli Zhang, Yujun Shen, Jingtao Ding, Haibin Zhou, Yuehong Zhang and Qikun Feng et al.
Tunable Ammonia Adsorption within Metal–Organic Frameworks with Different Unsaturated Metal Sites
Reprinted from: *Molecules* **2022**, *27*, 7847, doi:10.3390/molecules27227847 147

Hongda Li, Chenpu Li, Hao Zhao, Boran Tao and Guofu Wang
Two-Dimensional Black Phosphorus: Preparation, Passivation and Lithium-Ion Battery Applications
Reprinted from: *Molecules* **2022**, *27*, 5845, doi:10.3390/molecules27185845 157

About the Editors

Hongda Li

Hongda Li is an associate professor at the Guangxi University of Science and Technology. He got his Ph.D. in Chemistry in 2019 from the University of Science & Technology Beijing. He is a visiting scholar at Huazhong University of Science and Technology to research new energy materials. His research focus is on photocatalytic materials, 2D electronics/optoelectronics, and new energy materials.

Mohammed Baalousha

Dr. Mohammed Baalousha obtained his Ph.D. in 2006 from the University of Bordeaux, France. Between 2006 and 2013, he worked as a postdoctoral fellow at the University of Birmingham, United Kingdom. He joined the University of South Carolina in 2014. He teaches and conducts research in the broad field of Environmental Health Sciences, with projects related to environmental fate and effects of engineered nanomaterials and nanoplastics and the applications of engineered nanomaterials. He has raised over $3M in research funding and has published over 80 peer-reviewed journal papers. He has served as the president of the Carolinas Society of Environmental Toxicology and Chemistry Chapter.

Victor A. Nadtochenko

Victor A. Nadtochenko is a professor at the Russian Academy of Sciences. His research focuses on femtosecond spectroscopy, ultrafast reactions in natural photosystems (sight, photosynthesis), laser microsurgery of cells, and nanoparticles.

Preface to "Catalytic Nanomaterials: Energy and Environment"

Catalysts, substances that increase the rate of reaction without undergoing permanent chemical changes, play important roles in the chemical industry and the environment. Catalysts have been used to treat fuels such as oil, natural gas, and coal, and to purify wastewater and industrial waste gases. Catalysts can be classified as homogeneous and heterogeneous catalysts. The former refers to catalysts that exist in the same phase (gas or liquid) as the reactants, whereas the latter refers to catalysts that are not in the same phase as the reactants. Heterogenous catalysis occurs at the interface of two phases, usually with a porous solid catalyst and a liquid or gas reactant. For chemicals to react, their chemical bonds must be changed, and certain energy is needed to change or break the chemical bonds. The minimum energy threshold required to change the chemical bonds is called the activation-free energy, and the catalyst affects the reaction rate by changing the activation-free energy of the chemical reactants. Compared with homogeneous catalysts, heterogeneous catalysts possess higher selectivity and better yield. Therefore, heterogeneous catalysts are receiving much more attention than homogeneous catalysts. In order to improve the efficiency, yield, and purity of catalysts, it is very important to study new catalytic materials or optimize the existing catalyst system.

At present, the research in this field mainly focuses on nano-structured catalysts with special physical and chemical properties. Nanoscale catalysts possess higher surface area and surface energy than corresponding non-nanoscale catalysts, ultimately resulting in higher catalytic activity. Nanocatalysts improve reaction selectivity by reducing the activation energy, reducing the occurrence of side reactions, and improving the recovery of energy consumption. Therefore, nanocatalysts are widely used in green chemistry, environmental remediation, efficient biomass conversion, renewable energy development, and other fields. This special issue focuses on the application of nanocatalysts in the fields of energy and environment.

Hongda Li, Mohammed Baalousha, and Victor A. Nadtochenko
Editors

Article

Rationally Designed $Bi_2M_2O_9$ (M = Mo/W) Photocatalysts with Significantly Enhanced Photocatalytic Activity

Fangzhi Wang [1], Xiaoyan Zhou [1], Jing Li [1], Qiuyue He [1], Ling Zheng [1], Qing Liu [1], Yan Chen [1], Guizhai Zhang [1], Xintong Liu [2,*] and Hongda Li [3,*]

[1] School of Resources and Environmental Engineering, Shandong Agriculture and Engineering University, Jinan 250100, China; wfz0814@126.com (F.W.); yaling110@163.com (X.Z.); ripplelj@126.com (J.L.); ajheqiuyue@163.com (Q.H.); zhengl1988@126.com (L.Z.); liuqingliuqing2021@163.com (Q.L.); ychen0612@163.com (Y.C.); zgzok2005@163.com (G.Z.)
[2] School of Light Industry, Beijing Technology and Business University, Beijing 100048, China
[3] School of Microelectronics and Materials Engineering, Guangxi University of Science and Technology, Liuzhou 545006, China
* Correspondence: liuxt@btbu.edu.cn (X.L.); hdli@gxust.edu.cn (H.L.)

Abstract: Novel $Bi_2W_2O_9$ and $Bi_2Mo_2O_9$ with irregular polyhedron structure were successfully synthesized by a hydrothermal method. Compared to ordinary Bi_2WO_6 and Bi_2MoO_6, the modified structure of $Bi_2W_2O_9$ and $Bi_2Mo_2O_9$ were observed, which led to an enhancement of photocatalytic performance. To investigate the possible mechanism of enhancing photocatalytic efficiency, the crystal structure, morphology, elemental composition, and optical properties of Bi_2WO_6, Bi_2MO_6, $Bi_2W_2O_9$, and $Bi_2Mo_2O_9$ were examined. UV-Vis diffuse reflectance spectroscopy revealed the visible-light absorption ability of Bi_2WO_6, Bi_2MO_6, $Bi_2W_2O_9$, and $Bi_2Mo_2O_9$. Photoluminescence (PL) and photocurrent indicated that $Bi_2W_2O_9$ and $Bi_2Mo_2O_9$ pose an enhanced ability of photogenerated electron–hole pairs separation. Radical trapping experiments revealed that photogenerated holes and superoxide radicals were the main active species. It can be conjectured that the promoted photocatalytic performance related to the modified structure, and a possible mechanism was discussed in detail.

Keywords: $Bi_2W_2O_9$; $Bi_2Mo_2O_9$; modified structure; visible light; photocatalysis

1. Introduction

Semiconductor photocatalysis technology has received increasing attention as a green approach since it can be widely applied in the areas of carbon dioxide reduction, water splitting, and organic pollutants degradation [1–4]. It is no doubt that TiO_2 is a popular photocatalyst. However, the band gap of TiO_2 is so wide that it can only respond to ultraviolet (UV) light. Therefore, tremendous efforts have been made to develop new visible-light-driven (VLD) photocatalysts, such as Bi_2WO_6, Bi_2MoO_6, MoS_2, $g-C_3N_4$, $BiOBr$, $BiVO_4$, and so on [5–8].

Both Bi_2WO_6 and Bi_2MoO_6 are active members of the Aurivillius oxide family with a special layer structure [9–12]. Bi_2WO_6 and Bi_2MoO_6 (Bi_2MO_6, M = W/Mo) possess similar crystal structure. Bi_2WO_6 consists of WO_6 layers and $[Bi_2O_2]^{2+}$ layers, and Bi_2MoO_6 consists of $[MoO_2]^{2+}$ layers and $[Bi_2O_2]^{2+}$ layers. Such a layered structure is favorable to the separation and transfer of photogenerated carriers [13]. Moreover, Bi_2MO_6 can absorb visible light and has good stability against photocorrosion. Thus, they have displayed potential photocatalytic performance for the decontamination of contaminants [14,15]. However, their practical application remains limited because of the high recombination rate of photogenerated electron–hole pairs in photocatalytic processes, and the visible-light use efficiency is still limited, which only responds to the light under 500 nm [16,17]. To solve these issues, composite materials with a heterojunction structure, doping of other ions, and loading of noble metal co-catalysts have been extensively investigated [18–22]. The results indicate that these methods essentially changed the structure of Bi_2MO_6 crystal to effectively inhibit the recombination of photogenerated electron–hole pairs under charge transmission, resulting in high photocatalytic performance. In this case, finding a kind of bismuthate with appropriate crystal structure is a potential approach to promote photocatalytic performance. Based on the special layer structure of Bi_2MO_6, a train of thought to modify the layer structure can be attempted.

The designation of $Bi_2M_2O_9$ (M = W/Mo) was adopted because of the similar crystal structure between $Bi_2W_2O_9$ and $Bi_2Mo_2O_9$. There are similarities and differences between $Bi_2M_2O_9$ and Bi_2MO_6. Both $Bi_2M_2O_9$ and Bi_2MO_6 consist of $(Bi_2O_2)^{2+}$ and $(M_xO_y)^{2-}$ (M = W/Mo) layers, while the difference is that the $(M_xO_y)^{2-}$ (M = W/Mo) layer is $(M_2O_7)^{2-}$ in $Bi_2M_2O_9$ and $(MO_4)^{2-}$ in Bi_2MO_6. Given this kind of difference, a sort of structure modification phenomenon can take place in $Bi_2M_2O_9$ crystal, and certainly lead to chemical bond changes.

In this study, novel morphology $Bi_2M_2O_9$ photocatalysts and ordinary Bi_2MO_6 were synthesized by a hydrothermal process. The modified structure of $Bi_2M_2O_9$ can facilitate charge separation to promote photocatalytic performance. Moreover, the modification of structure and chemical bond changes were studied, and the relationship between modified structure and promoted photocatalytic performance was investigated. We hope to explore a potential strategy to obtain a highly efficient visible-light-driven photocatalyst.

2. Result and Discussion

2.1. XRD Analysis

The typical diffraction patterns of the as-prepared samples can be observed in Figure 1, which indicates the successful synthesis of samples using the hydrothermal method. It also reveals the crystal style and major diffraction peaks of Bi_2WO_6, Bi_2MO_6, $Bi_2W_2O_9$, and $Bi_2Mo_2O_9$ in panel A to D, respectively. The major diffraction peaks at 2θ values of 27.5°, 33.4°, and 47.1° were indexed to (0 0 2), (6 0 0) and (0 2 0) of Bi_2WO_6 in Figure 1A, 2θ values of 27.7°, 29.9°, and 55.7° were indexed to (1 1 4), (1 1 5), and (1 3 4) of $Bi_2W_2O_9$ in Figure 1B, 2θ values of 10.9°, 28.3°, 33.1°, 47.2°, and 56.2° were indexed to (0 2 0), (1 3 1), (0 6 0), (0 6 2), and (1 9 1) of Bi_2MoO_6 in Figure 1C, and 2θ values of 25.8°, 31.8°, 36.9°, and 54.3° were indexed to (031), (330), (332) and (361) of $Bi_2Mo_2O_9$ in Figure 1D, respectively [23–25]. No signal for any crystalline phase of bismuth oxides were observed in the as-prepared photocatalysts.

It was reported that the crystal structure of Bi_2MO_6 and $Bi_2M_2O_9$ photocatalysts can be described as $(Bi_2O_2)^{2+}$ layer and $(MO_4)^{2-}$ or $(M_2O_7)^{2-}$ layer alternately connect to each other [26,27]. The polyhedron style model of Bi_2WO_6 and $Bi_2W_2O_9$ were exhibited in Figure 2. As shown in Figure 2A, there is a double W-O layer between two $(Bi_2O_2)^{2+}$ layers. The W atom and six surrounding oxygen atoms formed a WO_6 octahedra, which connects with other similar octahedra by axial O3 in a lengthways direction and equatorial O4, O5, O7 and O8 oxygen atoms in a crosswise direction, forming double $(W_2O_7)^{2-}$ layers. Top and bottom oxygen atoms O6 and O9 are located close to $(Bi_2O_2)^{2+}$ layers above and below WO_6 octahedra respectively, which eventually formed $Bi_2W_2O_9$ consisting of $(W_2O_7)^{2-}$ and $(Bi_2O_2)^{2+}$ layers. In addition, as-prepared $Bi_2W_2O_9$ structure can be regarded as a modification of the tetragonal structure. The formation principle of Bi_2WO_6 (Figure 2B) is similar to that of $Bi_2W_2O_9$. The WO_6 octahedra join another octahedra by axial O3 in a lengthways direction forming a single $(WO_4)^{2-}$ layer, and equatorial O5, O6, and O2 no longer join other WO_6 octahedra but join $(Bi_2O_2)^{2+}$ instead. Bi_2WO_6 with a single $(WO_4)^{2-}$ layer is formed in this way.

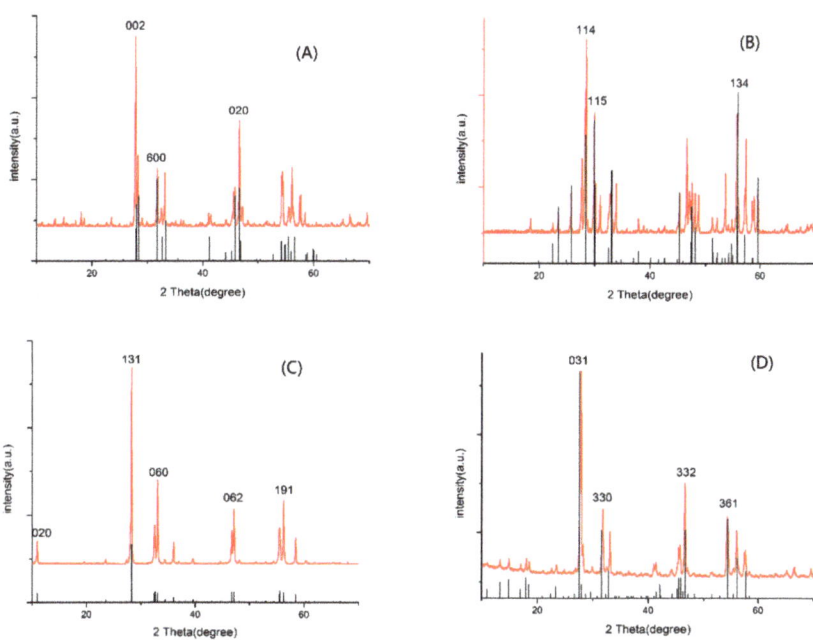

Figure 1. X-ray diffraction patterns of prepared samples (**A**) Bi_2WO_6, (**B**) $Bi_2W_2O_9$, (**C**) Bi_2MoO_6, and (**D**) $Bi_2Mo_2O_9$.

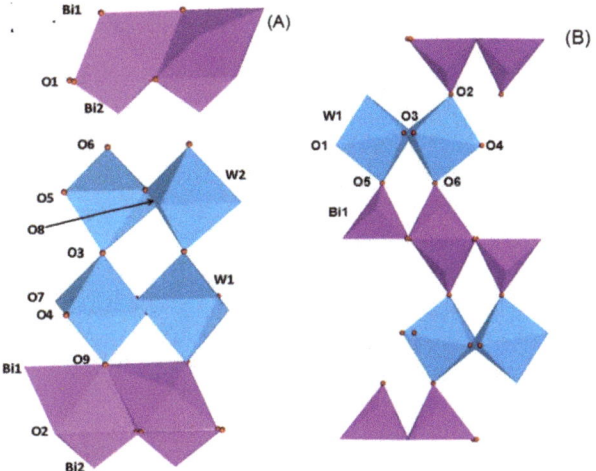

Figure 2. Models of $Bi_2W_2O_9$ (**A**) and Bi_2WO_6 (**B**) consist of $(Bi_2O_2)^{2+}$ and $(W_2O_7)^{2-}$ or $(WO_4)^{2-}$ layers.

As a result of the layer structure of Bi_2WO_6 and $Bi_2W_2O_9$, some translational motions of layered crystal can be regarded as a Rigid Unit (RU) layer motion. There are two shared RU modes in $Bi_2W_2O_9$, and the adjacent layers move parallel to the layer planes, which leads to modified compressional RU modes between adjacent layers in the perpendicular direction of the layer planes. As for Bi_2WO_6 structure, there is only one RU mode and the effect of compression between adjacent layers is weaker than that of $Bi_2W_2O_9$. As a result, compression of $Bi_2W_2O_9$ layer structure changes the chemical bond properties.

2.2. Morphology Characterization

SEM images of Bi_2WO_6, $Bi_2W_2O_9$, Bi_2MoO_6, and $Bi_2Mo_2O_9$ are illustrated in Figure 3 panels a to d, respectively. The morphology of Bi_2WO_6 in Figure 3A exhibits fastener-like nanoparticles with the radius of no more than 100 nm. The dimerization bismuthate $Bi_2W_2O_9$ certainly kept a similar fastener-like morphology as that of Bi_2WO_6 (Figure 3B). However, it can be observed that a volume increase phenomenon exists in $Bi_2W_2O_9$ crystal, and the average radius rose to about 400 nm at the same plotting scale compared to Bi_2WO_6. In terms of Bi_2MoO_6 and $Bi_2Mo_2O_9$, Figure 3C,D represent the panoramic SEM images of Bi_2MoO_6 and $Bi_2Mo_2O_9$. Both Bi_2MoO_6 and $Bi_2Mo_2O_9$ are of irregular polyhedron morphology and the variation is that there is also an increase of radius in $Bi_2Mo_2O_9$ compared to Bi_2MoO_6, but the increasing rate is less than that of $Bi_2W_2O_9$ and Bi_2WO_6. Double RU of $Bi_2M_2O_9$ increase of the distance of adjacent $(Bi_2O_2)^{2+}$ layers may be the possible reason for the volume augment. The novel morphology observed in SEM images may be attributed to the introduction of AOT surfactant during the synthesis process.

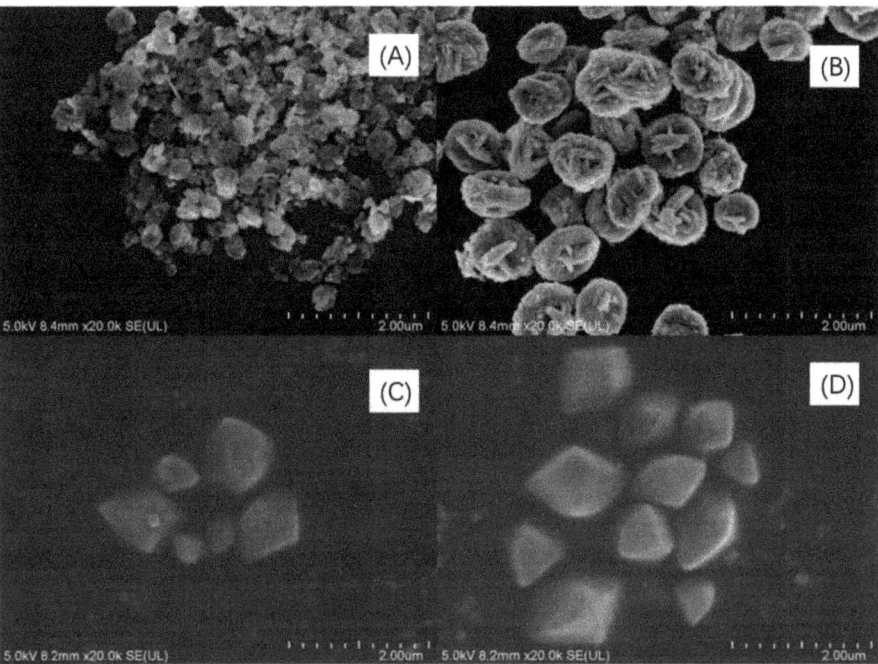

Figure 3. SEM images of (**A**) Bi_2WO_6, (**B**) $Bi_2W_2O_9$, (**C**) Bi_2MoO_6 and (**D**) $Bi_2Mo_2O_9$.

The possible synthesis mechanism is as follows. Raw materials have aggregated together to form some spheres during the hydrothermal process and finally the products have grown into clear-cut fastener-like or irregular polyhedron microspheres. At the beginning of the reaction, Bi^{3+} with $(M_xO_y)^{2-}$ ions precipitated quickly under the driving force of low-solubility products of $Bi_2M_xO_y$. Subsequently, AOT molecules are absorbed on the surface of nanoparticles through intermolecular interaction, and the newly formed nanoparticles are aggregated into loose microspheres, forming the $Bi_2M_xO_y$-AOT composite systems. These $Bi_2M_xO_y$-AOT composite systems connected with each other in various shapes. Finally, these amorphous nanoparticles underwent Ostwald ripening from the inside out as their surfaces come into contact with the surrounding solution. As a result, the internal nanoparticles tend to dissolve, which provides the driving force for spontaneous inside-out Ostwald ripening. This dissolution process could initiate at regions either near the surface or around the center of the microspheres. Redundant AOT was washed by the solution. Ostwald ripening occurred during the synthesis process of Bi_2WO_6, $Bi_2W_2O_9$, Bi_2MoO_6, and $Bi_2Mo_2O_9$ presumably depending on the packing of primary nanoparticles and ripening characteristics of AOT. A simple schematic illustration for the formation of the process is given in Figure 4.

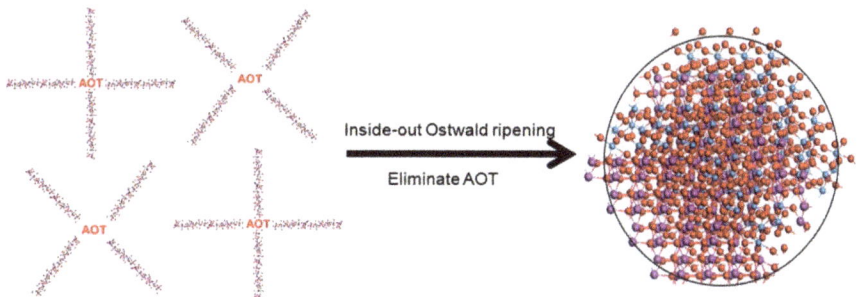

Figure 4. Possible formation process of novel morphology samples with AOT surfactant.

To confirm the atomic composition of Bi_2WO_6, $Bi_2W_2O_9$, Bi_2MoO_6, and $Bi_2Mo_2O_9$ samples conform to the theoretical value, EDX measurement was carried out and the data are listed in Table 1.

Table 1. EDX data of Bi_2WO_6, $Bi_2W_2O_9$, Bi_2MoO_6 and $Bi_2Mo_2O_9$.

Sample	Element	Wt%	At%
Bi_2WO_6	Bi	58.82	22.28
	W	27.90	12.02
	O	13.28	65.70
$Bi_2W_2O_9$	Bi	24.48	11.73
	W	24.32	13.25
	O	11.98	75.02
Bi_2MoO_6	Bi	67.66	21.56
	Mo	16.20	11.25
	O	16.14	67.19
$Bi_2Mo_2O_9$	Bi	54.13	14.62
	Mo	26.01	15.30
	O	19.86	70.07

The EDX data show that although there is deviation compared to the experimental data with theoretical value, the deviation is within acceptable limits.

2.3. Chemical State Analysis

X-ray photoelectron spectroscopy (XPS) analysis is used to further investigate the chemical state and surface chemical composition of Bi_2WO_6, $Bi_2W_2O_9$, Bi_2MoO_6, and $Bi_2Mo_2O_9$, especially to understand the structure modification effect of $Bi_2W_2O_9$ and $Bi_2Mo_2O_9$ on the binding energy, which has a great influence on photocatalytic performance [28].

The overall XPS spectra of Bi_2WO_6, $Bi_2W_2O_9$, Bi_2MoO_6, and $Bi_2Mo_2O_9$ are shown in Figure 5. The characteristic peaks of the Bi, W, Mo, and O elements were detected. Before the analysis, all peaks of the other elements were calibrated according to the deviation between the C 1s peak and the standard signal of C 1s at 284.8 eV [29]. No XPS characteristic peaks of N 1s were detected at around 400 eV, although raw material contained nitrogen, which indicated no nitrogen was doped in Bi_2WO_6, $Bi_2W_2O_9$, Bi_2MoO_6, and $Bi_2Mo_2O_9$ samples. The signals of Bi_2WO_6 in Figure 5B were attributed to Bi $4f_{7/2}$ and Bi $4f_{5/2}$ states respectively at 159.2 and 164.5 eV [30]. The binding energy of W $4f_{7/2}$ and W $4f_{5/2}$ were observed at 35.4 and 37.6 eV (Figure 5C) that can be attributed to W^{6+} [31]. The binding energy peak in Figure 5D located at 530.0 eV corresponds to O 1s state in Bi_2WO_6 [32]. The same examined element spectra in $Bi_2W_2O_9$ were exhibited together with those in Bi_2WO_6 in the same panels. It can be obviously observed that all Bi, W, and O have similar peak patterns, and the difference is that characteristic peaks of $Bi_2W_2O_9$ shift towards higher binding energy, which illustrates that the chemical environment has changed, and a higher binding energy indicates the existence of the electron-drawing group. The same phenomenon can be observed in Bi_2MoO_6 and $Bi_2Mo_2O_9$ in panels E to H, where Bi, Mo, and O elements in $Bi_2Mo_2O_9$ pose a higher binding energy.

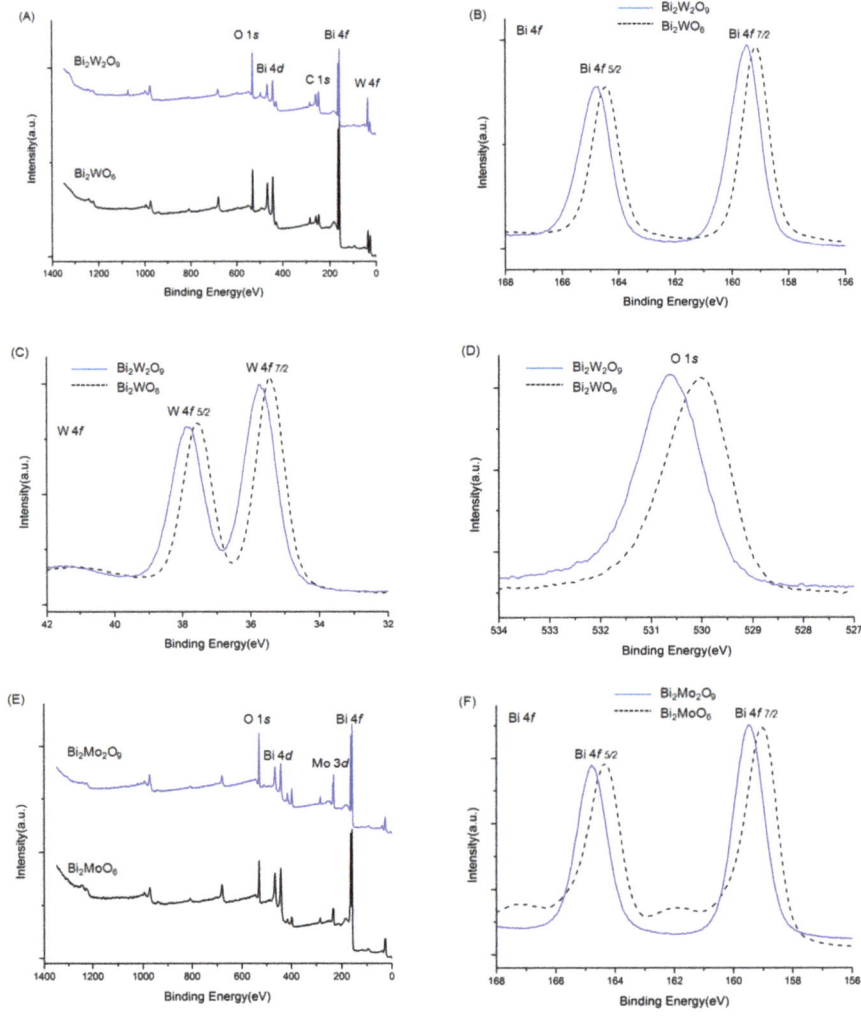

Figure 5. *Cont.*

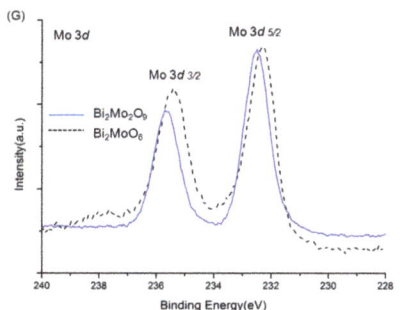
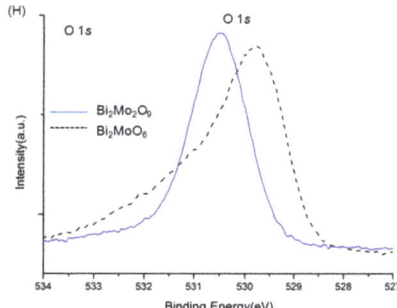

Figure 5. XPS spectra of Bi_2WO_6, $Bi_2W_2O_9$, Bi_2MoO_6 and $Bi_2Mo_2O_9$: ((**A**): Bi_2WO_6 and $Bi_2W_2O_9$) ((**E**): $Bi_2W_2O_9$) overall spectra; (**B**,**F**) Bi 4f; (**C**) W f; (**D**,**H**) O 1s; (**G**) Mo 3d.

2.4. Optical Properties

The optical properties of Bi_2WO_6, $Bi_2W_2O_9$, Bi_2MoO_6, and $Bi_2Mo_2O_9$ were characterized through a UV-Vis diffuse reflectance spectrometer in the wavelength range of 250–800 nm [33]. Compared to Bi_2WO_6 and Bi_2MoO_6, it can be observed obviously that there is a red shift of the light absorption edge from $Bi_2W_2O_9$ and $Bi_2Mo_2O_9$ samples, respectively. The optical band gap of the as-prepared photocatalysts was calculated using the following Equation (1) [34].

$$Ah\nu = \alpha(h\nu - E_g)^{n/2} \quad (1)$$

in which h, α, A, ν, and E_g represent Planck's constant with the unit of eV, a constant, the absorption coefficient near the absorption edge, light frequency, the absorption band gap energy, respectively, and n is equal to 1 or 4, depending on whether the optical transition type is direct or indirect. Bi_2WO_6, $Bi_2W_2O_9$, Bi_2MoO_6, and $Bi_2Mo_2O_9$ have a direct band gap, and n is 1 herein. The inset shows the curve of $(\alpha h\nu)^2$ versus hν for the as-prepared samples. It can be observed that an evidential red shift exists in $Bi_2M_2O_9$ samples compared to Bi_2MO_6 and $Bi_2Mo_2O_9$ has the greatest absorption range (Figure 6). The band gap energy (E_g) of Bi_2WO_6, $Bi_2W_2O_9$, Bi_2MoO_6 and $Bi_2Mo_2O_9$ were computed to be 2.78, 2.76, 2.72, and 2.70 eV respectively and exhibit in the inset of Figure 6. The result indicated that $Bi_2M_2O_9$ samples presented an enhanced absorbance ability compared to Bi_2MO_6.

2.5. Photocatalytic Properties

The photocatalytic performance of the as-synthesized photocatalysts were evaluated by examining the photodegradation of MB solution under visible-light irradiation (Figure 7). Generally, the different concentrations of MB adsorbed on the catalyst surface will have a great influence on the photocatalytic performance, so the adsorption ratio was collected when adsorption–desorption equilibrium was achieved before irradiation [35]. Bi_2WO_6, $Bi_2W_2O_9$, Bi_2MoO_6, and $Bi_2Mo_2O_9$ samples presented a similar capacity for MB absorption, and it can be obviously observed that $Bi_2M_2O_9$ samples exhibit a higher photocatalytic activity than that of Bi_2MO_6. $Bi_2Mo_2O_9$ displayed the best photocatalytic activity among the four test samples, and the photodegradation rate reached up to 75% within 4 h.

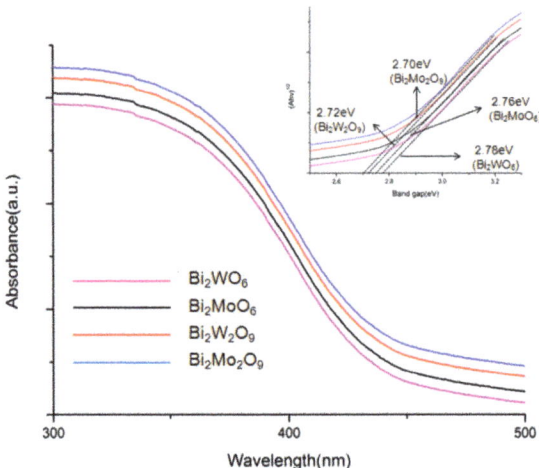

Figure 6. DRS spectra of Bi_2WO_6, $Bi_2W_2O_9$, Bi_2MoO_6 and $Bi_2Mo_2O_9$. The inset shows the band gap energies.

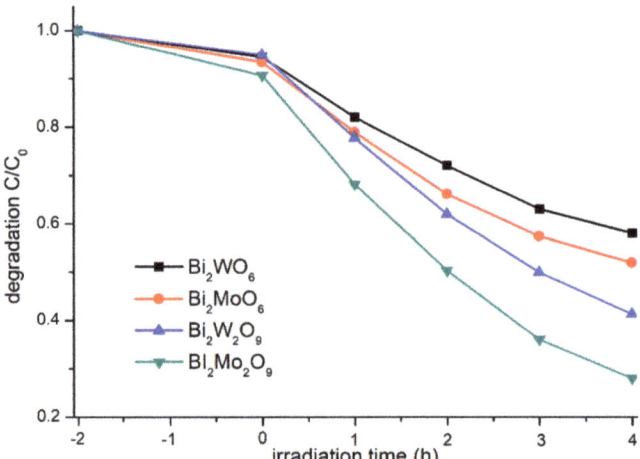

Figure 7. Comparison of absorption and degradation rate of MB using Bi_2WO_6, $Bi_2W_2O_9$, Bi_2MoO_6 and $Bi_2Mo_2O_9$.

In our work, the cycling experiment for MB photocatalytic degradation was performed under visible-light irradiation to evaluate the stability of the best photocatalyst ($Bi_2Mo_2O_9$). As shown in Figure 8, the photocatalytic performance of $Bi_2Mo_2O_9$ did not display any significant reduction for MB degradation. This result confirms that $Bi_2M_2O_9$ photocatalysts are not easily photo-corroded during the photodegradation of the pollutant molecules, which is important for their application.

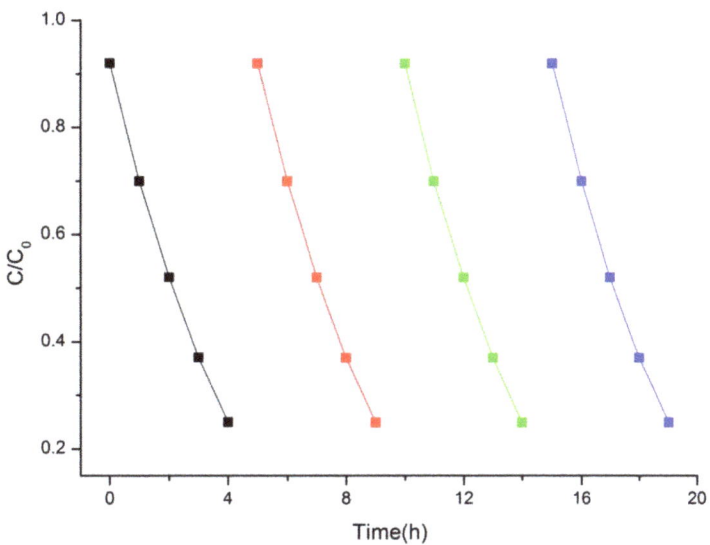

Figure 8. Cyclic photocatalytic activity of MB by $Bi_2Mo_2O_9$ photocatalyst.

2.6. Photocatalytic Mechanism

Overall, the photocatalytic activity of $Bi_2M_2O_9$ was highly improved compared to ordinary Bi_2MO_6. As noted above, such enhancement may partially come from structure modification. To understand the structure modification effect on $Bi_2M_2O_9$ and the charge behavior during the photocatalyst process, trapping experiments with different scavengers were performed with $Bi_2Mo_2O_9$ (Figure 9). In this way, active species could be determined, including holes (h^+), superoxide radicals ($·O_2^-$) and hydroxyl radicals ($·OH$) with effective oxidation and reduction potentials [36–38]. In the present study, isopropyl alcohol (IPA), ethylenediaminetetraacetic acid disodium salts (EDTA), and 1,4-benzoquinone (BQ) were used as scavengers of $·OH$, h^+, and $·O_2^-$, respectively. Remarkably, MB degradation was halted as we added the scavenger EDTA (1 mM) for h^+ to the reaction system. Meanwhile, the photodegradation rate of MB evidently declined with the addition of the scavenger BQ (1 mM) for $·O_2^-$. However, there was no clear reduction in the degradation rate of MB when the scavenger IPA (1 mM) for $·OH$ was added. These results indicate that h^+ and $·O_2^-$ were the main reactive species, while $·OH$ had little influence on the MB degradation process.

The electron–hole separation condition was tested by PL in the range of 400–700 nm. As shown in Figure 10, compared to Bi_2MO_6, the PL intensities of $Bi_2M_2O_9$ were perceptibly weaker, and $Bi_2Mo_2O_9$ has the lowest peak intensity. Usually, a lower PL intensity shows stronger photogenerated charge separation, which leads to excellent photocatalytic efficiency of the photocatalyst [39]. The experiment results reveal that $Bi_2M_2O_9$ has an elevated ability of electron–hole separation.

The photocurrent responses directly related to the generation and transfer of the photogenerated electrons and holes [40,41]. Figure 11 shows the photocurrent response of Bi_2WO_6, $Bi_2W_2O_9$, Bi_2MoO_6 and $Bi_2Mo_2O_9$ samples. Obviously, the current abruptly increased and decreased through on–off cycles under visible-light irradiation. $Bi_2M_2O_9$ samples showed a significantly improved photocurrent response compared to that of the Bi_2MO_6 samples. This suggests that more efficient separation of the photogenerated charge carriers occurred in $Bi_2M_2O_9$ samples. In addition, it should be noted that $Bi_2M_2O_9$ samples hold obvious residual currents when the light source is switched off. This is probably attributed to the modified structure; it may lead to some remnant electron–hole pairs when the visible-light source is removed, which could release the trapped electrons or holes because of the self-thermal motion.

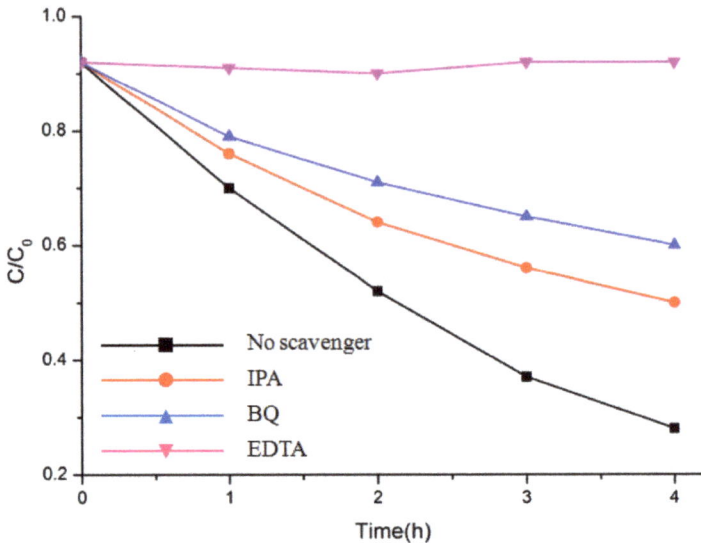

Figure 9. Photocatalytic degradation of MB over $Bi_2Mo_2O_9$ photocatalyst with the addition of scavengers EDTA, BQ, and IPA.

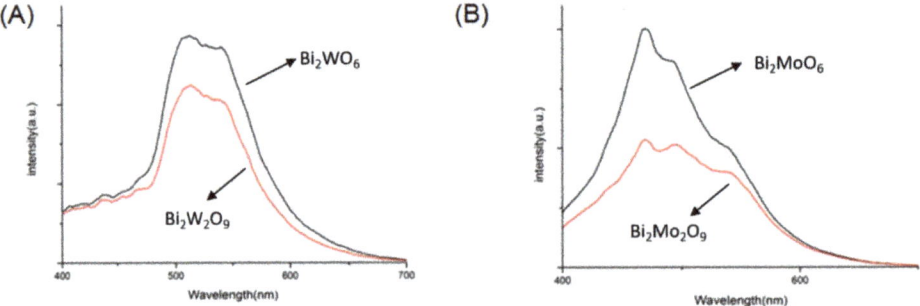

Figure 10. Photoluminescence (PL) spectra of (**A**) Bi_2WO_6 and $Bi_2W_2O_9$; (**B**) Bi_2MoO_6 and $Bi_2Mo_2O_9$.

Based on the characterization methods above, a possible photocatalytic mechanism of the $Bi_2M_2O_9$ photocatalyst under visible-light irradiation is therefore proposed, and the $Bi_2M_2O_9$ photocatalyst has a stable double Rigid Unit (RU), which consists of two octahedral layers. Double RU mode can be regarded as a strong electron-drawing group, and $Bi_2M_2O_9$ show a stable reduced bond distance of octahedral RU. The influence of modified structure is reflected in PL test and photocurrent measurement. The reduction of PL intensity and the enhancement of photocurrent express promoted photogenerated electron–hole pairs separation ability; the strong electron-drawing group traps the photogenerated electrons and leads to the increase of active species to promote the photocatalytic performance.

As shown in Figure 12, the photogenerated electrons could migrate to the conduction band (CB) from the valence band (VB) and the photogenerated holes formed in the valence band when the semiconductor was irradiated with visible light. The shortened bond distance of octahedral RU generates a strong electron-drawing efficiency that traps the electron firmly and reduces the recombination rate. Evidentially, the trapped electrons could easily transfer to the oxygen molecules (O_2) adsorbed on the surface of the $Bi_2M_2O_9$

catalysts. Subsequently, the released electrons react with O_2 to form the active superoxide radical anion species (O_2^-). The electron capture and release process enhances the charge transfer and separation efficiency of photogenerated electrons and holes, which contributes to organic contaminant photodegradation by the h^+ and $·O_2^-$ species.

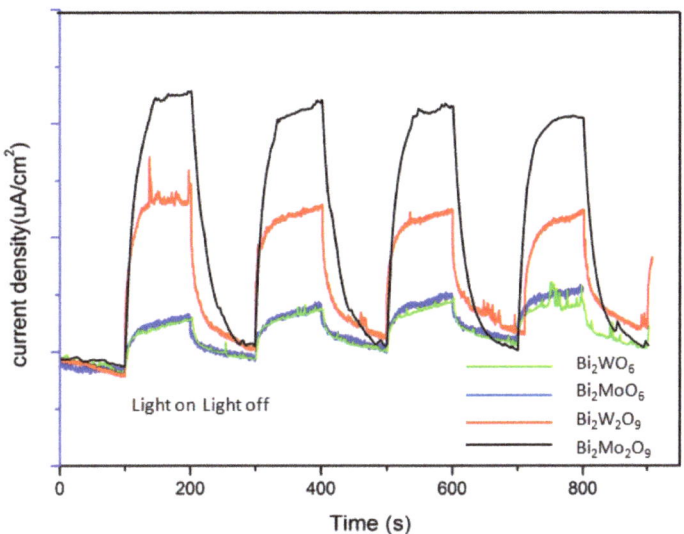

Figure 11. Photocurrent responses of Bi_2WO_6, $Bi_2W_2O_9$, Bi_2MoO_6 and $Bi_2Mo_2O_9$ samples under visible-light irradiation.

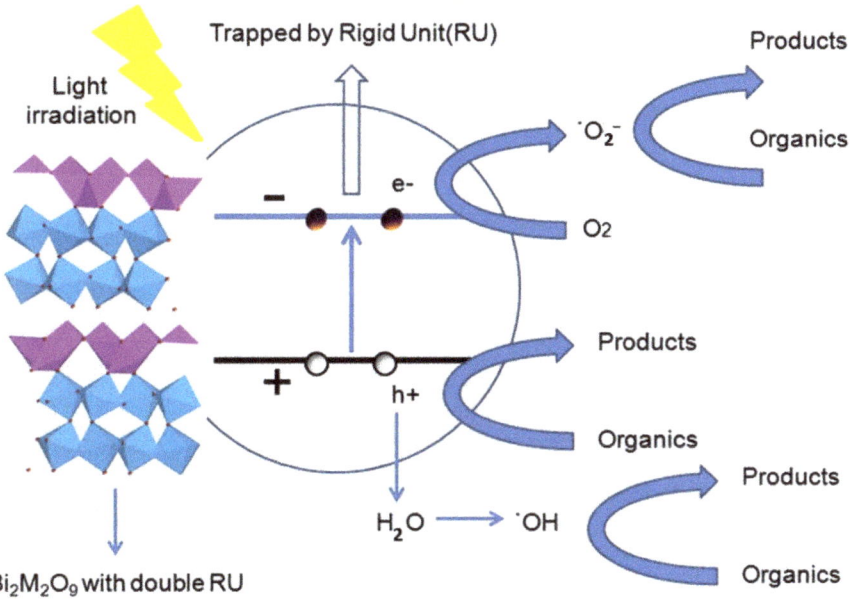

Figure 12. Schematic illustration of the mechanism of $Bi_2M_2O_9$ photocatalyst under visible-light irradiation.

Therefore, the above process indicates that an appropriate increase the number of $(Bi_2O_2)^{2+}$ layers and $(W_2O_7)^{2-}$ layers to form a stable Rigid Unit (RU) mode could significantly improve photocatalytic activity.

3. Experimental

3.1. Synthesis of the Photocatalysts

All the reagents used in this study were of analytical purity (Sinopharm Chemical Reagents Co., Ltd., Shanghai, China) and used without any further purification. Bi_2WO_6, $Bi_2W_2O_9$, Bi_2MoO_6 and $Bi_2Mo_2O_9$ photocatalysts were prepared through a hydrothermal method and the general synthesis processes were as follows: proportionate amounts of sodium tungstate ($Na_2WO_4 \cdot H_2O$) or sodium molybdate ($Na_2MoO_4 \cdot H_2O$) were dissolved in Milli-Q water. Then proportional bismuth (III) nitrate pentahydrate ($Bi(NO_3)_3 \cdot 5H_2O$) was dissolved in nitric acid and anionic surfactant AOT was introduced into the solution sequence under stirring to form a transparent solution and ensure Bi:Mo and Bi:W keeps 2:1 during the preparation process of Bi_2WO_6 and Bi_2MoO_6, similarly, a value of the ratio of Bi and W or Mo stays at 1 when preparing $Bi_2W_2O_9$ and $Bi_2Mo_2O_9$. We adjust the pH value of the solution to ca. 7 using NaOH solution. The slurry was stirred for another 30 min and then poured into Teflon-lined stainless steel autoclaves. The sealed reactors were then heated at 180 °C for 12 h. Then, the products were cooled to room temperature naturally and collected via centrifugation and washed in Milli-Q water and ethanol several times to make sure that the residual impurities were all removed, and then the product was dried at 80 °C for 8 h.

3.2. Characterization of the Photocatalysts

The crystalline phases of the as-prepared samples were analyzed using X-ray diffraction (XRD) (D/MAX-RB, Rigaku, Tokyo, Japan). The diffraction patterns were recorded in the 2θ range from 10 to 70° with a Cu Kα source (λ = 1.5418 Å) running at 40 KV and 30 mA. The morphology images of the as-prepared samples were captured using scanning electron microscopy (SEM) equipped with an energy-dispersive X-ray spectrometer (EDX) on a SUPRA 55 SAP-PHIRE instrument operating at 20 KV. High-resolution transmission electron microscopy (HRTEM) images were acquired with a transmission electron microscopy (F-20, FEI, Hillsboro, OR, USA) at an accelerating voltage of 200 KV. The UV-Vis diffuse reflectance spectra of the as-prepared samples were examined at room temperature using a UV-Vis spectrophotometer (T9s; Persee, Beijing, China) equipped with an integrating sphere. $BaSO_4$ was used as the blank reference. Photoluminescence (PL) spectra were recorded using a fluorescence spectrophotometer (F-4500; Hitachi, Tokyo, Japan) with a Xe lamp as the excitation light source. X-ray photoelectron spectroscopy (XPS) was examined on an X-ray photoelectron spectrometer (ESCALAB 250Xi, Thermo Scientific, Waltham, MA, USA) using an Al Kα radiation.

3.3. Photocatalytic Experiment

The photocatalytic activities of Bi_2WO_6, $Bi_2W_2O_9$, Bi_2MoO_6, and $Bi_2Mo_2O_9$ photocatalysts under visible light were assessed by degrading 10 mmol·L^{-1} methylene blue (MB). A 400 W Xe lamp with a UV-cut-off filter (λ > 420 nm) was used as a light source and set about 10 cm apart from the reactor. The experiments were as follows: 40 mg of the photocatalyst was dispersed in 40 mL of MB solution. It was then stirred for 120 min in the dark to achieve an adsorption–desorption equilibrium before light irradiation. During the irradiation, the reaction samples were collected at 60 min intervals and centrifuged to separate out the photocatalyst particles. The ratios (C/C_0) of the MB were adopted to evaluate the degradation efficiency (i.e., C_0 was the initial concentration, where C was the concentration at a certain time) by checking the absorbance spectrum at 664 nm for MB using a UV-Vis spectrophotometer (T9s; Persee, Beijing, China).

3.4. Measurement of Photocurrent

The measurement of the photocurrent was carried out with an electrochemical workstation (5060F, RST, Zhengzhou, China) in a standard three-electrode system including the samples, an Ag/AgCl electrode (saturated KCl), and a Pt filament used as the working electrode, reference electrode, and counter electrode, respectively. In addition, a pre-made 0.5 mol L^{-1} Na$_2$SO$_4$ aqueous solution was introduced as the electrolyte. A 100 W incandescent lamp with a 420 nm cut-off filter was used as the light source. The working electrode was manufactured as follows: 5 mg samples were appended to 2 mL of ethanol and Nafion mixture solution (v/v = 30:1), followed by spreading on the middle of an ITO glass in a rounded hole with a diameter of 6 mm.

4. Conclusions

In this study, ordinary Bi$_2$WO$_6$ and Bi$_2$MoO$_6$ samples were successfully synthesized; furthermore, a novel Bi$_2$Mo$_2$O$_9$ fastener sphere and a Bi$_2$W$_2$O$_9$ irregular polyhedron were prepared by a hydrothermal method with AOT introduced. The results revealed that structure-modified Bi$_2$W$_2$O$_9$ and Bi$_2$Mo$_2$O$_9$ exhibited enhanced photocatalytic performance. The Rigid Unit (RU) and modified bond structure influence the trapping–release process of electrons to promote the separation efficiency of photogenerated electron–hole pairs. Therefore, an appropriate increase of the layer number of (W$_2$O$_7$)$^{2-}$ or (Mo$_2$O$_7$)$^{2-}$ between (Bi$_2$O$_2$)$^{2+}$ layers to form an RU strong electron-drawing group as a method of structure modification can be a potential strategy to improve visible-light photocatalytic activity by affecting the charge behavior.

Author Contributions: Conceptualization, F.W., X.L. and H.L.; methodology, G.Z.; validation, Y.C., H.L.; formal analysis, J.L.; investigation, L.Z., Q.L.; data curation, X.Z., Q.H.; writing—review and editing, F.W., X.L. All authors have read and agreed to the published version of the manuscript.

Funding: This research was funded by the Inspiration Fund of Shandong Agriculture and Engineering University (Grant No. BSQJ201811), Qingchuang Science and Technology Program of Shandong Province in China (grant no. 2019KJF029), and the National Natural Science Foundation of China (Grant No. 61801274).

Institutional Review Board Statement: Not applicable.

Informed Consent Statement: Not applicable.

Data Availability Statement: The data presented in the study are available from the corresponding author.

Conflicts of Interest: The authors declare no conflict of interest.

Sample Availability: Samples of the compounds are not available from the authors.

References

1. Xu, C.; Anusuyadevi, P.R.; Aymonier, C.; Luque, R.; Marre, S. Nanostructured materials for photocatalysis. *Chem. Soc. Rev.* **2019**, *48*, 3868–3902. [CrossRef] [PubMed]
2. Wei, Z.; Liu, J.; Shangguan, W. A review on photocatalysis in antibiotic wastewater: Pollutant degradation and hydrogen production. *Chin. J. Catal.* **2020**, *41*, 1440–1450. [CrossRef]
3. Asadzadeh-Khaneghah, S.; Habibi-Yangjeh, A. g-C$_3$N$_4$/carbon dot-based nanocomposites serve as efficacious photocatalysts for environmental purification and energy generation: A review. *J. Clean. Prod.* **2020**, *276*, 124319. [CrossRef]
4. Bie, C.; Yu, H.; Cheng, B.; Ho, W.; Fan, J.; Yu, J. Design, Fabrication, and Mechanism of Nitrogen-Doped Graphene-Based Photocatalyst. *Adv. Mater.* **2021**, *33*, 2003521. [CrossRef]
5. Liu, X.; Gu, S.; Zhao, Y.; Zhou, G.; Li, W. BiVO$_4$, Bi$_2$WO$_6$ and Bi$_2$MoO$_6$ photocatalysis: A brief review. *J. Mater. Sci. Technol.* **2020**, *56*, 45–68. [CrossRef]
6. Wang, F.; Li, W.; Gu, S.; Li, H.; Wu, X.; Liu, X. Samarium and Nitrogen Co-Doped Bi$_2$WO$_6$ Photocatalysts: Synergistic Effect of Sm^{3+}/Sm^{2+} Redox Centers and N-Doped Level for Enhancing Visible-Light Photocatalytic Activity. *Chem.-Eur. J.* **2016**, *22*, 12859–12867. [CrossRef]
7. Zhang, J.; Zhu, Z.; Jiang, J.; Li, H. Synthesis of Novel Ternary Dual Z-scheme AgBr/LaNiO$_3$/g-C$_3$N$_4$ Composite with Boosted Visible-Light Photodegradation of Norfloxacin. *Molecules* **2020**, *25*, 3706. [CrossRef] [PubMed]

8. Mishra, A.; Mehta, A.; Basu, S.; Shetti, N.P.; Reddy, K.R.; Aminabhavi, T.M. Graphitic carbon nitride (g–C_3N_4)–based metal-free photocatalysts for water splitting: A review. *Carbon* **2019**, *149*, 693–721. [CrossRef]
9. Li, H.; Li, W.; Gu, S.; Wang, F.; Zhou, H. In-built Tb^{4+}/Tb^{3+} redox centers in terbium-doped bismuth molybdate nanograss for enhanced photocatalytic activity. *Catal. Sci. Technol.* **2016**, *6*, 3510–3519. [CrossRef]
10. Yuan, Y.J.; Shen, Z.; Wu, S.; Su, Y.; Pei, L.; Ji, Z.; Ding, M.; Bai, W.; Chen, Y.; Yu, Z.T.; et al. Liquid exfoliation of g-C_3N_4 nanosheets to construct 2D-2D MoS_2/g-C_3N_4 photocatalyst for enhanced photocatalytic H_2 production activity. *Appl. Catal. B-Environ.* **2019**, *246*, 120–128. [CrossRef]
11. Li, J.; Yin, Y.; Liu, E.; Ma, Y.; Wan, J.; Fan, J.; Hu, X. In situ growing Bi_2MoO_6 on g-C_3N_4 nanosheets with enhanced photocatalytic hydrogen evolution and disinfection of bacteria under visible light irradiation. *J. Hazard. Mater.* **2017**, *321*, 183–192. [CrossRef] [PubMed]
12. Zhou, K.; Lu, J.; Yan, Y.; Zhang, C.; Qiu, Y.; Li, W. Highly efficient photocatalytic performance of BiI/Bi_2WO_6 for degradation of tetracycline hydrochloride in an aqueous phase. *RSC Adv.* **2020**, *10*, 12068–12077. [CrossRef]
13. Fu, H.; Pan, C.; Yao, W.; Zhu, Y. Visible-light-induced degradation of rhodamine B by nanosized Bi_2WO_6. *J. Phys. Chem. B* **2005**, *109*, 22432–22439. [CrossRef] [PubMed]
14. Zhang, H.; Yu, D.; Wang, W.; Gao, P.; Bu, K.; Zhang, L.; Zhong, S.; Liu, B. Multiple heterojunction system of Bi_2MoO_6/WO_3/Ag_3PO_4 with enhanced visible-light photocatalytic performance towards dye degradation. *Adv. Powder Technol.* **2019**, *30*, 1910–1919. [CrossRef]
15. Huang, D.; Li, J.; Zeng, G.; Xue, W.; Chen, S.; Li, Z.; Deng, R.; Yang, Y.; Cheng, M. Facile construction of hierarchical flower-like Z-scheme AgBr/Bi_2WO_6 photocatalysts for effective removal of tetracycline: Degradation pathways and mechanism. *Chem. Eng. J.* **2019**, *375*, 121991. [CrossRef]
16. Vesali-Kermani, E.; Habibi-Yangjeh, A.; Diarmand-Khalilabad, H.; Ghosh, S. Nitrogen photofixation ability of g-C_3N_4 nanosheets/Bi_2MoO_6 heterojunction photocatalyst under visible-light illumination. *J. Colloid Interf. Sci.* **2020**, *563*, 81–91. [CrossRef] [PubMed]
17. Li, J.; Zhao, Y.; Xia, M.; An, H.; Bai, H.; Wei, J.; Yang, B.; Yang, G. Highly efficient charge transfer at 2D/2D layered P-$La_2Ti_2O_7$/Bi_2WO_6 contact heterojunctions for upgraded visible-light-driven photocatalysis. *Appl. Catal. B-Environ.* **2020**, *261*, 118244. [CrossRef]
18. Zhang, G.; Chen, D.; Li, N.; Xu, Q.; Li, H.; He, J.; Lu, J. Fabrication of Bi_2MoO_6/ZnO hierarchical heterostructures with enhanced visible-light photocatalytic activity. *Appl. Catal. B-Environ.* **2019**, *250*, 313–324. [CrossRef]
19. Zhen, Y.; Yang, C.; Shen, H.; Xue, W.; Gu, C.; Feng, J.; Zhang, Y.; Fu, F.; Liang, Y. Photocatalytic performance and mechanism insights of a S-scheme g-C_3N_4/Bi_2MoO_6 heterostructure in phenol degradation and hydrogen evolution reactions under visible light. *Phys. Chem. Chem. Phys.* **2020**, *22*, 26278–26288. [CrossRef] [PubMed]
20. Zhang, L.; Yang, C.; Lv, K.; Lu, Y.; Li, Q.; Wu, X.; Li, Y.; Li, X.; Fan, J.; Li, M. SPR effect of bismuth enhanced visible photoreactivity of Bi_2WO_6 for NO abatement. *Chin. J. Catal.* **2019**, *40*, 755–764. [CrossRef]
21. Ji, L.; Liu, B.; Qian, Y.; Yang, Q.; Gao, P. Enhanced visible-light-induced photocatalytic disinfection of Escherichia coli by ternary Bi_2WO_6/TiO_2/reduced graphene oxide composite materials: Insight into the underlying mechanism. *Adv. Powder Technol.* **2020**, *31*, 128–138. [CrossRef]
22. Jiang, Y.; Huang, K.; Ling, W.; Wei, X.; Wang, Y.; Wang, J. Investigation of the Kinetics and Reaction Mechanism for Photodegradation Tetracycline Antibiotics over Sulfur-Doped Bi_2WO_{6-x}/$ZnIn_2S_4$ Direct Z-Scheme Heterojunction. *Nanomaterials* **2021**, *11*, 2123. [CrossRef] [PubMed]
23. Alfaro, S.O.; la Cruz, A.M. Synthesis, characterization and visible-light photocatalytic properties of Bi_2WO_6 and $Bi_2W_2O_9$ obtained by co-precipitation method. *Appl. Catal. A Gen.* **2010**, *383*, 128–133. [CrossRef]
24. Wang, D.W.; Siame, B.; Zhang, S.Y.; Wang, G.; Ju, X.S.; Li, J.L.; Lu, Z.L.; Vardaxoglou, Y.; Whittow, W.; Cadman, D.; et al. Direct integration of cold sintered, temperature-stable $Bi_2Mo_2O_9$-K_2MoO_4 ceramics on printed circuit boards for satellite navigation antennas. *J. Eur. Ceram. Soc.* **2020**, *40*, 4029–4034. [CrossRef]
25. Zhang, L.; Xu, T.; Zhao, X.; Zhu, Y. Controllable synthesis of Bi_2MoO_6 and effect of morphology and variation in local structure on photocatalytic activities. *Appl. Catal. B Environ.* **2010**, *98*, 138–146. [CrossRef]
26. Huang, J.; Tan, G.; Ren, H.; Yang, W.; Xu, C.; Zhao, C.; Xia, A. Photoelectric Activity of a Bi_2O_3/$Bi_2WO_{6-x}F_{2x}$ Heterojunction Prepared by a Simple One-Step Microwave Hydrothermal Method. *ACS Appl. Mater. Interfaces* **2014**, *6*, 21041–21050. [CrossRef]
27. Song, J.; Zhang, L.; Yang, J.; Huang, X.; Hu, J. Facile hydrothermal synthesis of Fe^{3+} doped $Bi_2Mo_2O_9$ ultrathin nanosheet with improved photocatalytic performance. *Ceram. Int.* **2017**, *43*, 9214–9219. [CrossRef]
28. Liu, Y.; Chen, J.; Zhang, J.; Tang, Z.; Li, H.; Yuan, J. Z-scheme $BiVO_4$/Ag/Ag_2S composites with enhanced photocatalytic efficiency under visible light. *RSC Adv.* **2020**, *10*, 30245–30253. [CrossRef]
29. He, G.; Zhang, J.; Hu, Y.; Bai, Z.; Wei, C. Dual-template synthesis of mesoporous TiO_2 nanotubes with structure-enhanced functional photocatalytic performance. *Appl. Catal. B-Environ.* **2019**, *250*, 301–312. [CrossRef]
30. Zhang, H.; He, J.; Zhai, C.; Zhu, M. 2D Bi_2WO_6/MoS_2 as a new photo-activated carrier for boosting electrocatalytic methanol oxidation with visible light illumination. *Chin. Chem. Lett.* **2019**, *30*, 2338–2342. [CrossRef]
31. Wang, F.; Li, W.; Gu, S.; Li, H.; Wu, X.; Ren, C.; Liu, X. Facile fabrication of direct Z-scheme MoS2/Bi_2WO_6 heterojunction photocatalyst with superior photocatalytic performance under visible light irradiation. *J. Photoch. Photobiol. A* **2017**, *335*, 140–148. [CrossRef]

32. Ren, J.; Wang, W.Z.; Sun, S.M.; Zhang, L.; Chang, J. Enhanced photocatalytic activity of Bi_2WO_6 loaded with Ag nanoparticles under visible light irradiation. *Appl. Catal. B-Environ.* **2009**, *92*, 50–55. [CrossRef]
33. Zhang, M.; Du, H.; Ji, J.; Li, F.; Lin, Y.C.; Qin, C.; Zhang, Z.; Shen, Y. Highly Efficient $Ag_3PO_4/g-C_3N_4$ Z-Scheme Photocatalyst for its Enhanced Photocatalytic Performance in Degradation of Rhodamine B and Phenol. *Molecules* **2021**, *26*, 2062. [CrossRef] [PubMed]
34. Lu, C.; Guo, F.; Yan, Q.; Zhang, Z.; Li, D.; Wang, L.; Zhou, Y. Hydrothermal synthesis of type II $ZnIn_2S_4/BiPO_4$ heterojunction photocatalyst with dandelion-like microflower structure for enhanced photocatalytic degradation of tetracycline under simulated solar light. *J. Alloys Compd.* **2019**, *811*, 151976. [CrossRef]
35. Singla, S.; Sharma, S.; Basu, S. MoS_2/WO_3 heterojunction with the intensified photocatalytic performance for decomposition of organic pollutants under the broad array of solar light. *J. Clean. Prod.* **2021**, *324*, 129290. [CrossRef]
36. Feizpoor, S.; Habibi-Yangjeh, A.; Ahadzadeh, I.; Yubuta, K. Oxygen-rich TiO_2 decorated with C-Dots: Highly efficient visible-light-responsive photocatalysts in degradations of different contaminants. *Adv. Powder Technol.* **2019**, *30*, 1183–1196. [CrossRef]
37. Zhu, Q.; Sun, Y.; Xu, S.; Li, Y.; Lin, X.; Qin, Y. Rational design of 3D/2D In_2O_3 nanocube/$ZnIn_2S_4$ nanosheet heterojunction photocatalyst with large-area "high-speed channels" for photocatalytic oxidation of 2, 4-dichlorophenol under visible light. *J. Hazard. Mater.* **2020**, *382*, 121098. [CrossRef]
38. Yu, B.; Meng, F.; Khan, M.W.; Qin, R.; Liu, X. Facile synthesis of AgNPs modified $TiO_2@$ $g-C_3N_4$ heterojunction composites with enhanced photocatalytic activity under simulated sunlight. *Mater. Res. Bull.* **2020**, *121*, 110641. [CrossRef]
39. Wang, F.; Gu, S.; Shang, R.; Jing, P.; Wang, Y.; Li, W. Fabrication of $AgBr/La_2Ti_2O_7$ hierarchical heterojunctions: Boosted interfacial charge transfer and high efficiency visible-light photocatalytic activity. *Sep. Purif. Technol.* **2019**, *229*, 115798. [CrossRef]
40. Wang, F.; Li, W.; Gu, S.; Li, H.; Liu, X.; Wang, M. Fabrication of $FeWO_4@ZnWO_4/ZnO$ heterojunction photocatalyst: Synergistic effect of $ZnWO_4/ZnO$ and $FeWO_4@ZnWO_4/ZnO$ heterojunction structure on the enhancement of visible-light photocatalytic activity. *ACS Sustain. Chem. Eng.* **2016**, *4*, 6288–6298. [CrossRef]
41. Sun, X.; Li, H.J.; Ou, N.; Lyu, B.; Gui, B.; Tian, S.; Qian, D.; Wang, X.; Yang, J. Visible-light driven TiO_2 photocatalyst coated with graphene quantum dots of tunable nitrogen doping. *Molecules* **2019**, *24*, 344. [CrossRef] [PubMed]

Article

Effects of Co Doping on the Growth and Photocatalytic Properties of ZnO Particles

Lanqin Tang [1,2,3,*], Yin Jia [1], Zhishang Zhu [1], Yue Hua [1], Jun Wu [1], Zhigang Zou [2,3] and Yong Zhou [2,3,*]

1. College of Chemistry and Chemical Engineering, Yancheng Institute of Technology, 9 Yingbin Avenue, Yancheng 224051, China; xiaoyu2898@sina.com (Y.J.); DouglasM2000@163.com (Z.Z.); JoannaVu22@163.com (Y.H.); swujun@163.com (J.W.)
2. National Laboratory of Solid State Microstructures, Collaborative Innovation Center of Advanced Microstructures, School of Physics, Nanjing University, Nanjing 210093, China; zgzou@nju.edu.cn
3. Eco-Materials and Renewable Energy Research Center (ERERC), Nanjing University, Nanjing 210093, China
* Correspondence: lanqin_tang@163.com (L.T.); zhouyong1999@nju.edu.cn (Y.Z.)

Abstract: The present work reports on the synthesis of ZnO photocatalysts with different Co-doping levels via a facile one-step solution route. The structural and optical properties were characterized by powder X-ray diffraction (XRD), field emission scanning electron microscopy (FESEM), transmission electron microscopy (TEM), energy dispersive spectroscopy (EDS), and UV-Vis diffuse reflectance spectra. The morphology of Co-doped ZnO depends on the reaction temperature and the amount of Co and counter-ions in the solution. Changes with the c-axis lattice constant and room temperature redshift show the replacement of Zn with Co ions without changing the wurtzite structure. Photocatalytic activities of Co-doped ZnO on the evolution of H_2 and the degradation of methylene blue (MB) reduce with the doping of Co ions. As the close ionic radii of Co and Zn, the reducing photocatalytic activity is not due to the physical defects but the formation of deep bandgap energy levels. Photocurrent response experiments further prove the formation of the recombination centers. Mechanistic insights into Co-ZnO formation and performance regulation are essential for their structural adaptation for application in catalysis, energy storage, etc.

Keywords: ZnO; Co-doped; flower-like; photocatalytic properties; chemical method

Citation: Tang, L.; Jia, Y.; Zhu, Z.; Hua, Y.; Wu, J.; Zou, Z.; Zhou, Y. Effects of Co Doping on the Growth and Photocatalytic Properties of ZnO Particles. *Molecules* **2022**, *27*, 833. https://doi.org/10.3390/molecules 27030833

Academic Editor: Hongda Li

Received: 13 January 2022
Accepted: 25 January 2022
Published: 27 January 2022

Publisher's Note: MDPI stays neutral with regard to jurisdictional claims in published maps and institutional affiliations.

Copyright: © 2022 by the authors. Licensee MDPI, Basel, Switzerland. This article is an open access article distributed under the terms and conditions of the Creative Commons Attribution (CC BY) license (https:// creativecommons.org/licenses/by/ 4.0/).

1. Introduction

The photocatalytic transformation has received continuous attention with the depletion of fossil fuels and the intensification of environmental pollution [1]. Traditional semiconductors, such as TiO_2, ZnO, etc., are usually applied as photocatalysts [2]. Controlling the morphologies of catalysts with specific exposed facets will result in efficient separation of the electron–hole pairs [3]. Doping ions is an effective way to reduce the bandgap of catalysts and improve the ability of visible-light harvesting [4,5]. Consequently, considerable interest has been focused on synthesizing photocatalysts with controllable morphologies and regulating their photocatalytic properties by a doping-based method.

Zinc oxide (ZnO) has been actively investigated due to its unique optical, electronic, and piezoelectric properties, such as solar cells [6], piezotronics [7], UV detectors [8], gas sensors [9], and light-emitting diodes (LEDs) [10]. Additionally, ZnO is a promising semiconductor with the richest range of morphologies and has been found to display good photoconductivity and high transparency in the visible region [11]. There are several methods available for controlling morphologies at present, such as the sputtering deposition technique [12], chemical vapor deposition [13], pulsed laser deposition [14], and the coprecipitation process [15]. Among these methods, the aqueous solution method appears more favorable in an economical and large-scale production [16–18]. Particles with specific crystal facets and complex corrugated particles with tips also have good photocatalytic activity and dispersion properties [19–22]. Based on previous studies, the microstructure, morphology, and luminescence performance of particles are extremely sensitive to the conditions of their preparation.

The Co-doping element is of much interest among the transition metal ions. Substituting the non-magnetic element Zn ion with the Co ion can introduce ferromagnetic behavior while retaining its semiconducting properties. There were also some reports on Co-doped ZnO nanoparticles, such as nano-crystalline powders [23], nanowires [24], etc. Different methods were applied to prepare Co-doped ZnO particles. For example, Chang et al. use the solvothermal method and calcination to dope Co^{2+}/Co^{3+} ions [25]. A similar process conducted at 150 °C is also reported by Fang and Wang et al. [26]. Asif et al. reported bimetallic Co-Ni/ZnO cubes through the hydrothermal process [27]. It is suggested that structural, optical, and photocatalytic properties largely depended on the synthesis method. The aqueous solution method is more facile for regulating the structures of catalysts, which influence their properties. Furthermore, previous applications mainly focus on sensing, Raman, magnetism, and antibacterial activity. Therefore, it seems necessary to study the growth parameters and Co-dopant effects on the photocatalytic activity of the Co-doped ZnO.

A simple one-step aqueous solution route is reported for synthesizing Co-doped ZnO nanoparticles. No organic solvents or high-temperature treatments are used. The presented doping process has been achieved at an attractive low temperature (50 °C). Well-defined flower-like ZnO and Co-ZnO particles have been directly obtained. The present work investigates the formation process and photocatalytic activities of pure and cobalt-doped ZnO nanoparticles. It is worth noting that we used the Co-doped ZnO structure as a model system to understand the growth and doping-induced photocatalytic property mechanism. Effects of growth parameters such as doping content, reaction temperature, and counter-ions on the formation of particles are studied. The influences of Co-doping on structural, optical, morphological, and photocatalytic activities, including methylene blue degradation and H_2 evolution, are also investigated in detail. Due to the similar ionic radii of Co and Zn, the replacement of Zn with Co creates deep bandgap levels and acts as the recombination centers, and eventually leads to the reduction of the photocatalytic activity. Therefore, doping species should be paid more attention. This method opens a new avenue for regulating the microstructure and properties of photocatalytic inorganic compounds.

2. Experimental

2.1. Materials

The starting chemical reagents used in this work were: zinc acetate dihydrate ($Zn(CH_3COO)_2 \cdot 2H_2O$), cobaltous chloride hexahydrate ($CoCl_2 \cdot 6H_2O$), cobalt acetate tetrahydrate ($Co(CH_3COO)_2 \cdot 4H_2O$), and sodium hydroxide (NaOH), all of which were from Sinopharm Chemical Reagent Co., Ltd., Shanghai, China, and were used as analytic grade without further purification.

2.2. Experimental Procedure

In a typical synthesis, 50 mL of a mixed solution containing 0.05 mol of $Zn(CH_3COO)_2 \cdot 2H_2O$ and a certain amount of 1.0 M $CoCl_2 \cdot 6H_2O$ solution were prepared in a three-necked flask under stirring. Then, 50 mL of 2.0 M NaOH solution was added dropwise to the flask at a rate of 1.0 mL/min at 50 °C, resulting in the formation of precipitates. After adding the above-mixed solution, the obtained precipitates were washed with distilled water and dried at 70 °C in the air for 24 h. The molar ratio of Co^{2+}/Zn^{2+}, R, was varied from 0.2:100 to 0.8:100. Experimental conditions for typical Co-doped ZnO samples are listed in Table 1.

2.3. Characterization of Materials

The crystalline structures were characterized using a Rigaku D/Max-RA Cu Kα diffractometer employing a scanning rate of $0.01° \ S^{-1}$ in the 2θ ranges from 30 to 70°. Zinc oxide crystallites were quantitatively characterized and compared with the calculated texture coefficient, $T_{c(hkl)}$. The texture coefficient, $T_{c(hkl)}$, is defined as follows [28]: $T_{c(hkl)} =$

$(I_{(hkl)}/I_{r(hkl)})/[1/n\Sigma(I_{(hkl)}/I_{r(hkl)})]$, where $T_{c(hkl)}$ is the texture coefficient, n is the number of peaks considered, $I_{(hkl)}$ are the intensities of the peaks of obtained zinc oxide samples, and $I_{r(hkl)}$ is the peak intensities indicated in the JCPDS 36-1451 corresponding to the randomly oriented crystallites. The lattice parameter, c, of the samples is calculated using the formula: $\sin^2\theta = \lambda^2/4[4(h^2 + hk + k^2)/3a^2 + l^2/c^2]$, where θ is the diffraction angle, λ is the incident wavelength, and h, k, and l are all Miller's indices. The morphologies of the samples were studied by field emission scanning electron microscope (FESEM, JEOL JSM-6700F, Tokyo, Japan) and a transmission electron microscope (TEM, JEOL-1230, Tokyo, Japan). Energy dispersive spectroscopy (EDS) measurements were applied to determine the dopant content of cobalt ions in the Co-ZnO particles. The response curves were measured with the light on/off cycles at 0 V versus SEC (saturated calomel electrode).

Table 1. Preparation parameters of various Co-doped ZnO samples.

Samples	Morphology	Zn^{2+} (mL)	Co^{2+} (mL)	Co/Zn (Theoretical Zn, mol %)	Reaction Temperature (°C)
S_0	OLP	50	0.00	0	50
$S_{0.2}$	OLP	50	0.10	0.2:100	50
$S_{0.4}$	OLP	50	0.20	0.4:100	50
$S_{0.6}$	FLP	50	0.30	0.6:100	50
$S_{0.8}$	FLP	50	0.40	0.8:100	50
S_0^*	NR	50	0.00	0	70
$S_{0.2}^*$	NR	50	0.10	0.2:100	70
$S_{0.8}^*$	NR	50	0.40	0.8:100	70

OLP, NR, and FLP are abbreviations describing product morphology and refer to olive-like particles, nanorods, and flower-like particles, respectively.

2.4. Measurement of Photocatalytic Activity

Methylene Blue (MB) dye was used to evaluate the photocatalytic activity of doped and pure ZnO particles in response to UV light. A 6 W UV lamp with a wavelength of 365 nm was employed as the light source. Then, 50 mg of photocatalysts was dispersed in 100 mL of 10 mg/L of the MB aqueous solution in a 100 mL beaker. The suspension was stirred in the dark for 30 min to ensure the adsorption and desorption equilibrium of MB on the particle surface. Subsequently, the suspension was irradiated with simulated UV light. The distance between the light source and the surface of the solution was 6 cm. After every 30 min of irradiation, 5 mL of the suspension was extracted and then centrifuged to separate particles from the supernatant. The obtained solution was analyzed by recording variations in the absorption band (663 nm) in the UV-Vis spectra of MB using a UV-2100 spectrophotometer.

Photocatalytic activity in H_2 production was examined using a 100 mL Pyrex flask. The flask was sealed with a silicone rubber septum. A 300 W Xe arc lamp was used as the light source. In a typical experiment, 50.0 mg of photocatalyst was mixed in 100 mL of an aqueous solution containing 0.25 M NaS_2 and 0.25 M $NaSO_3$. Before irradiation, the suspension was bubbled with nitrogen for 30 min to remove any dissolved oxygen and ensure the system was under anaerobic conditions. Magnetic stirring was used throughout the reaction to keep the photocatalyst particles in a suspension state. In addition, 0.4 mL of gas was sampled intermittently through the septum and analyzed using a gas chromatograph equipped with a thermal conductivity detector (TCD).

3. Results and Discussion

3.1. Structural and Morphological Characteristics

Cobalt was selected for its expected ease in doping due to the similar ionic radius (0.58 Å) to Zn (0.60 Å). Figure 1 displays XRD patterns of pure ZnO powder (sample-S_0), Co-doped sample-S0.2, and sample-S0.8 photocatalysts. In each case, the characteristic (100), (002), (101), (102), (110), (103), (200), (112), and (201) reflections of simple wurtzite phase

(ZnO, JCPDS No.36-1451) were observed, and no XRD patterns arising from other phases appeared. The XRD patterns of Co-doped ZnO samples show the same characteristics as undoped ZnO, consisting of only peaks corresponding to ZnO. This is because XRD has a relatively poor detection limit of around 1% by volume. The sharp and intense peaks in these structures indicate that all samples are highly crystallized. There is a reduction of the c-axis lattice constants with the increase of Co-dopant level, to 5.2091 Å (sample-S_0), 5.2086 Å (sample-S0.2), and 5.2079 Å (sample-S0.8). Since the difference in radii of Co and Zn is minimal, the changes in cell parameters with cobalt substitution in the lattice are rather small [28]. Therefore, the difference of lattice parameters could be attributed to Co incorporation, which indicates defect evolution in the lattice.

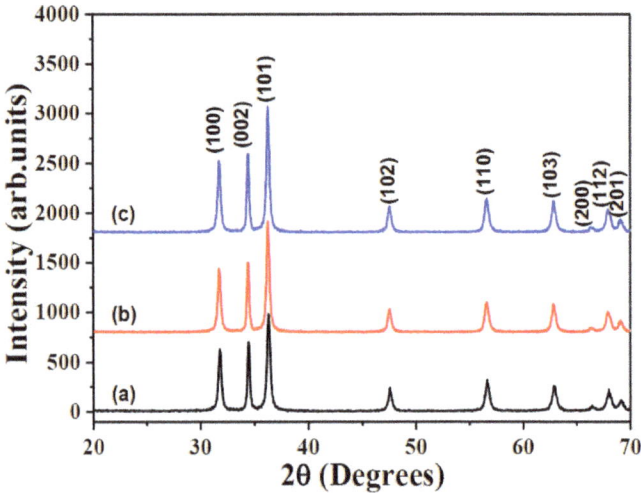

Figure 1. The powder XRD patterns of pure ZnO sample-S_0 (**a**), sample-S0.2 (**b**), and sample-S0.8 (**c**) photocatalysts.

The texture coefficient values of these samples produced in the presence or the absence of Co were also calculated according to their three main corresponding X-ray diffraction peaks (Table 2). The (002) orientation with a high texture coefficient, $T_{c(002)}$, is stronger than the (100) orientation and the (101) orientation. $T_{c(hkl)}$ of 1 presents a sample with randomly oriented crystallites, where a larger value indicates an abundance of crystallites oriented to the (*hkl*) plane [29]. These results confirmed that the addition of Co did not change the (002) growth direction. Furthermore, with the loading of Co, the average crystal size of these samples decreased from 28.78 nm (sample-S_0) to 14.39 nm (sample-S0.2) and 5.08 nm (sample-S0.8). Co led to a smaller crystal size of the produced zinc oxide particles.

Table 2. Texture coefficients of the obtained ZnO samples with different amounts of Co.

Samples	$M_{Co}^{2+}:M_{Zn}^{2+}$	Texture Coefficient		
		$T_{c(100)}$	$T_{c(002)}$	$T_{c(101)}$
S_0	0:100	0.91	1.29	0.80
$S_{0.2}$	0.2:100	0.87	1.24	0.89
$S_{0.6}$	0.6:100	0.88	1.24	0.88

The energy-dispersive spectra (EDS) further revealed that Co, Zn, and O exist in Co-based samples (Figure 2). Quantitative analysis of the atomic concentration (atom%) is listed in Supplementary Table S1. It can be found that the Co content in the samples increased from 0.12 to 0.38 atom%, based on the Co-doping levels. However, less Co was located in Co-doped ZnO crystallites than that provided in the precursor solution, indicating that some Co ions remain in the parent solution and do not become incorporated into the crystals. Furthermore, the band edge of the Co-doped ZnO sample-S0.8 shifted to the lower energy side, compared to the pure ZnO sample-S_0 (Figure 2). The redshift phenomenon is mainly due to the sp–d exchange interactions between the band electrons and the localized d electrons of the Co^{2+} cations. These results confirmed that the addition of Co did not change the ZnO structure.

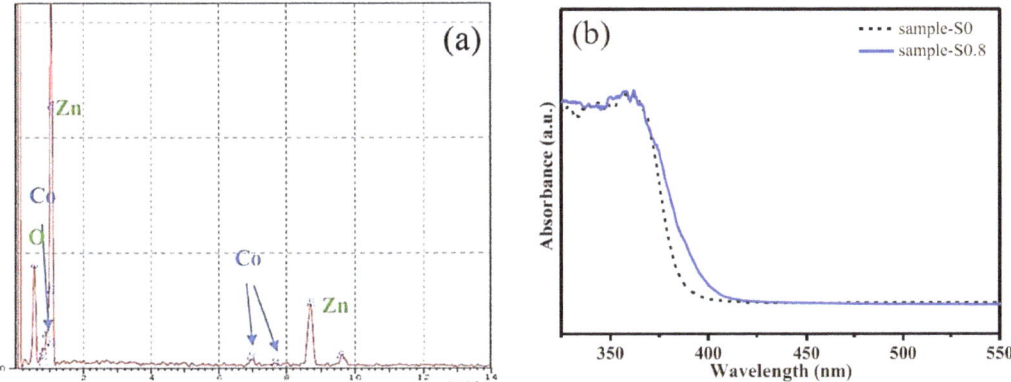

Figure 2. EDS (**a**) and UV (**b**) spectra of sample-S0.8.

3.2. Morphological Characteristics

From Figure 3a, it is apparent that pure ZnO exhibits an olive-like morphology with an average size of about 400 nm. With the addition of Co, the obtained sample-S0.2 also showed an olive-like morphology with a length of about 500 nm (Figure 3b,c). On the other hand, the morphology of sample-S0.8 produced after Co-doping changed significantly, and was mostly flower-like nanoclusters with an average size of 500 nm, which are comprised of much smaller nanorods (Figure 3d). Flower-like particles with enormous interface areas demonstrated good light-harvesting ability, and quickly transformed the light-generated charge in the photocatalytic processes, and had better catalytic ability than nanorods and nanoparticles [30]. These differences can be explained in terms of a thermodynamic barrier arising from the dopant (Co^{2+}) that slows the nucleation and inhibits further growth of ZnO around a doped nanocrystal.

3.3. Mechanism

3.3.1. Effects of Reaction Temperature on the Morphology of Co-Doped ZnO Particles

FESEM images of pure ZnO (sample-S_0*), sample-S0.2*, and sample-S0.8* are shown in Figure 4. When the reaction temperature was increased to 70 °C, the obtained ZnO particles of sample-S_0* with a size of about 90 nm were obtained, which is smaller than sample-S_0 obtained at 50 °C (400 nm, Figure 4a). Higher temperature accelerates the growth process and allows the synthesis of small ZnO particles [31]. With the addition of Co, the obtained sample-S0.2* and sample-S0.8* had a larger size of about 150 nm, but a similar 1D rod-like morphology to sample-S_0* (Figure 4b). This is quite different from particles obtained at the reaction temperature of 50 °C when 3D particles are formed. Higher temperature accelerates the growth process but does not change the growth characteristics of Co-doped ZnO samples. The morphology of Co-doped ZnO particles seems more dependent on reaction temperature than the doped amount of Co.

3.3.2. Effects of Cobalt Counter-Ions on the Morphology of Co-Doped ZnO Particles

Cobalt acetate tetrahydrate ($Co(CH_3COO)_2·4H_2O$, 0.2 mol% Co:ZnO) was also used as the Co source, and sample-C50 and sample-C70 were obtained when the reaction temperature was 50 and 70 °C, respectively. Figure 5a shows that when cobalt acetate tetrahydrate was used as the Co source instead of cobaltous chloride hexahydrate, flower-like particles composed of olives were produced (sample-C50, 50 °C). Interestingly, the flower-like particles of about 800 nm were larger than in sample-S0.8, which was obtained under nearly the same conditions, except for the amount of Co (0.8 mol% Co:ZnO). When the reaction was increased to 70 °C, rod-like particles similar to sample-S0.8* were obtained (sample-C70, Figure 5b). These results show that the cobalt counter-ion mainly affects the size of Co-doped ZnO particles.

Figure 3. FESEM and TEM images of pure ZnO sample-S0 (**a**), and Co-doped sample-S0.2 (**b**,**c**) and sample-S0.8 (**d**).

Figure 4. FESEM image of pure ZnO ((**a**) sample-S0*), and (**b**) the size distributions of sample-S0.2* and sample-S0.8*.

Figure 5. FESEM images of sample-C0.8#50 (**a**) and sample-C0.8#70 (**b**).

3.3.3. Photocatalytic Activity

The photochemical reactivity of ZnO-based composites could be used for oxidative degradation of a variety of emerging contaminants, such as pharmaceuticals associated with toxicological impacts to aquatic environments [32]. The photochemical reactivity of Co-doped ZnO catalysts synthesized with different molar ratios of Co^{2+} to Zn^{2+} for degradation of MB has been performed under UV light irradiation at room temperature. Figure 6a shows that blank ZnO enables a near 52% degradation rate of 1 mg of MB with 150 min of UV light irradiation. With the addition of Co (0.2 mol% Co:ZnO), the photocatalytic activity of the obtained sample-S0.2 was lower than the blank, and about 34% of the 1 mg of MB was degraded (Figure 6b). Further increasing the amount of Co, the photocatalytic degradation rate of the obtained samples gradually decreased from 20% (sample-S0.4, Figure 6c) to 18% (sample-S0.6, Figure 6d), and eventually to 16% (sample-S0.8, Figure 6e). In our previous work, samples with different morphologies displayed a large difference in activities, and flower-like ZnO showed the best photodegradation rates [11]. However, flower-like Co-doped ZnO composites had the worst photodegradation rates in this work, which is probably because of the addition of Co. The kinetic model is discussed here to better understand the photocatalytic activity of synthesized photocatalysts. In general, the kinetics of photocatalytic degradation of organic pollutants on the semiconducting oxide has been established and can be well-described by the apparent first-order reaction, $\ln(C_0/C_t) = k_{app}t$, where k_{app} is the apparent rate constant, C_0 is the concentration of dyes after darkness adsorption for 30 min, and C_t is the concentration of dyes at time t. Supplementary Figure S1 shows the relationship between illumination time and the degradation rate of dyes for UV illumination. The linear correlation of the plots of $\ln(C_0/C_t)$ versus time suggests a pseudo-first-order reaction for MB dye.

The photochemical reactivity of sample-S_0, sample-S0.2, sample-S_0*, and sample-S0.2* was performed under UV light irradiation at room temperature. From Supplementary Figure S2, it is obvious that blank ZnO (sample-S_0) obtained at 50 °C had better photocatalytic activity than that produced at 70 °C (sample-S_0*). With the addition of Co (0.2 mol% Co:ZnO), the photocatalytic activity of the obtained sample-S0.2 and sample-S0.2* was lower than their corresponding blanks. Furthermore, sample-S0.2* showed a worse photocatalytic activity, and about 25% of 1 mg of MB was degraded under UV light irradiation for 150 min.

Sample-S0.2 was taken as a typical sample to study the degradation process. Supplementary Figure S3 shows the variances in the absorbance of MB with sample-S0.2 as the photocatalyst. The prominent absorption peak centered at 663 nm, and no blue- or red-shifts were observed during the photocatalytic procedure, implying that the decolorating process for MB is not ascribed to the N-demethylation [33]. Furthermore, after irradiation for 150 min, the discoloration of MB aqueous solution was apparent (see the inset in Supplementary Figure S2).

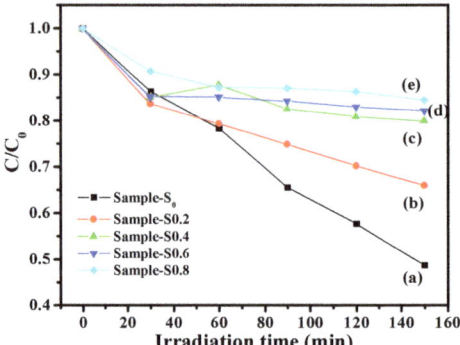

Figure 6. Extent of decomposition of the MB dye with respect to time intervals over ZnO (sample-S_0 (**a**), and Co-doped ZnO photocatalysts: sample-S0.2 (**b**), sample-S0.4 (**c**), sample-S0.6 (**d**), and sample-S0.8 (**e**), under UV irradiation.

We also evaluated these samples' photocatalytic hydrogen evolution activities (Figure 7). Pure ZnO exhibited the best H_2 production ability of about 50 μmol g^{-1} for 4 h (sample-S_0). Furthermore, the H_2 production rate also depends on the amount of Co. After doping with different mole fractions of Co, the photocatalytic H_2 production rate significantly decreased. The lowest H_2 production rate reached around 5.0 μmol h^{-1} g^{-1} over sample-S0.8, much lower than that over sample-S_0. The results suggest that doping Co into ZnO is not practical for enhancing the H_2 production rate.

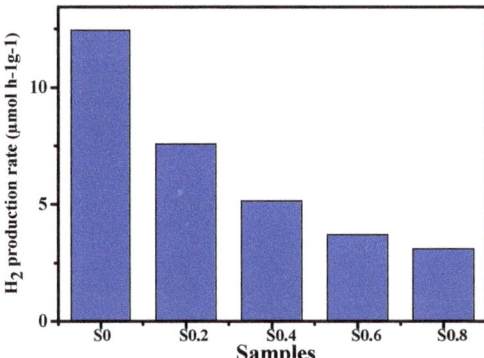

Figure 7. H_2 evolution over ZnO (sample-S_0), and Co-doped ZnO photocatalysts: sample-S0.2, sample-S0.4, sample-S0.6, and sample-S0.8.

It is known that illumination of the semiconductor material with a photon flux of energy greater than the bandgap width causes absorption of photons and generation of additional electron–hole pairs. As a result of the generation, an increase in the concentration of the excess electrons in the conduction band and the extra holes in the valence band occurs. The photocatalytic activity depends not only on intrinsic crystallographic structures of photocatalysts but also on the separation of the electron–hole pairs [34]. The transient photocurrent response of the prepared samples was measured to examine the charge separation and migration properties. In this way, the separation of electron–hole pairs generated by light absorption gives rise to a photocurrent. The photo-response curves were measured with light on/off cycles at 0 V versus SEC (saturated calomel electrode). The pure ZnO exhibited a fast and uniform photocurrent response for each light-on and light-off process, and the photocurrent density of the ZnO electrode was detected as 3.3 μA/cm^2 (Figure 8a). However, the photocurrent density dramatically decreased by about 50% after loading Co to ZnO (Figure 8b). The observed decrease in photocurrent intensity could be ascribed to combining the photo-generated electron–hole pairs with defects' formation by loading Co to ZnO. The ionic radii of Co and Zn are close to each other. The mechanism of reducing photocatalytic activity by Co-doping is therefore not considered to be by creating physical defects. The efficient recombination centers for photo-generated excitons would be the deep bandgap energy levels between the valence and conduction bands [35]. The results well-parallel the photocatalytic performances for dye degradation and H_2 production.

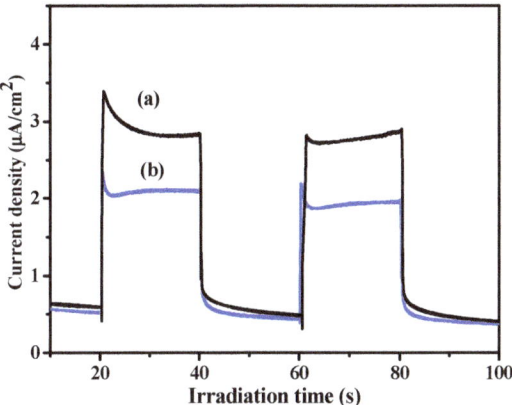

Figure 8. Photocurrent responses of ZnO (**a**) and Co-doped sample-S0.8 (**b**).

4. Conclusions

Co-doped ZnO particles directly from aqueous solutions at a low temperature (50 °C) have been successfully synthesized via a simple one-step solution route. The effects of Co-doping, reaction temperature, and Co counter-ions on the structural, morphological, and photocatalytic properties of doped ZnO particles were investigated. XRD spectra indicated that cobalt ions, in the oxidation state of Co2p, substitute Zn2p ions in the ZnO lattice without changing its wurtzite structure. According to the EDS spectra, the dopant content varied from 0.12% to 0.38%, depending on Co-doping levels. The reaction temperature and doped amount of Co significantly affected the morphology of doped ZnO particles. Pure ZnO exhibited an olive-like morphology, and the Co-ZnO particles took on the shape of olive-like and eventually flower-like nanoclusters depending on the amount of Co. The optical property of doped ZnO particles is determined by H_2 evolution and the degradation of MB. The photocurrent intensity test proved the formation of recombination centers, leading to the reduction of photocatalytic activity of the Co-doping samples. This work showed that doping is a double-edged sword, and attention should be paid not only to the doping itself but also the doping species. Further work can focus on the study of serial doping metal ions and the regulations of the structural and catalytic properties.

Supplementary Materials: The following supporting information can be downloaded, Figure S1: Apparent rate constants of MB dye in the presence of ZnO (sample-S0, a) and Co-doped ZnO photocatalysts: sample-S0.2 (b), sample-S0.4 (c), sample-S0.6 (d), and sample-S0.8 (e) under UV irradiation.; Figure S2: Extent of decomposition of the MB dye with respect to time intervals over pure ZnO sample-S0 and sample-S0* (a,c), and Co-dopped ZnO photocatalysts: sample-S0.2 (b), and sample-S0.2* (d) under UV irradiation.; Figure S3: The absorbance of MB with sample-S0.8 as a function of irradiation time. Table S1: The composition of ZnO nanocrystals dependence on Co-doping levels.

Author Contributions: Validation, Y.J., Z.Z. (Zhishang Zhu), Y.H. and J.W.; investigation, Y.J., Z.Z. (Zhishang Zhu) and Y.H.; writing—review and editing, L.T.; funding acquisition, L.T., Y.Z. and Z.Z. (Zhigang Zou). All authors have read and agreed to the published version of the manuscript.

Funding: This work was supported by the Major Research Plan of the National Natural Science Foundation of China (2018YFE0208500), the NSF of China (21773114, 21603183, 21473091), the NSF of Jiangsu Province (BK2012015 and BK20130425), the Key Laboratory for Advanced Technology in Environmental Protection of Jiangsu Province (JBGS004).

Institutional Review Board Statement: Not applicable.

Informed Consent Statement: Not applicable.

Data Availability Statement: The data presented in this study are available in Supplementary Materials.

Conflicts of Interest: The authors declare no conflict of interest.

Sample Availability: Samples of the compounds are not available from the authors.

References

1. Perera, F. Pollution from Fossil-Fuel Combustion is the Leading Environmental Threat to Global Pediatric Health and Equity: Solutions Exist. *Int. J. Environ. Res. Public Health* **2017**, *15*, 16. [CrossRef] [PubMed]
2. Majumder, S.; Chatterjee, S.; Basnet, P.; Mukherjee, J. ZnO based nanomaterials for photocatalytic degradation of aqueous pharmaceutical waste solutions—A contemporary review. *Environ. Nanotechnol. Monit. Manag.* **2020**, *14*, 100386–100401. [CrossRef]
3. Meldrum, F.; Cölfen, H. Controlling Mineral Morphologies and Structures in Biological and Synthetic Systems. *Chem. Rev.* **2008**, *108*, 4332–4432. [CrossRef] [PubMed]
4. Li, H.; Gu, S.; Sun, Z.; Guo, F.; Xie, Y.; Tao, B.; He, X.; Zhang, W.; Chang, H. In-built bionic "MoFe-cofactor" in Fe-doped two-dimensional $MoTe_2$ nanosheets for boosting the photocatalytic nitrogen reduction performance. *J. Mater. Chem. A* **2020**, *8*, 13038–13048. [CrossRef]
5. Li, H.; Deng, H.; Gu, S.; Li, C.; Tao, B.; Chen, S.; He, X.; Wang, G.; Zhang, W.; Chang, H. Engineering of bionic Fe/Mo bimetallene for boosting the photocatalytic nitrogen reduction performance. *J. Colloid Interface Sci.* **2022**, *607*, 1625–1632. [CrossRef] [PubMed]
6. Pietruszka, R.; Witkowski, B.S.; Gieraltowska, S.; Caban, P.; Wachnicki, L.; Zielony, E.; Gwozdz, K.; Bieganski, P.; Placzek-Popko, E.; Godlewski, M. New efficient solar cell structures based on zinc oxide nanorods. *Sol. Energy Mater. Sol. Cells* **2015**, *143*, 99–104. [CrossRef]
7. Momeni, K.; Attariani, H. Electromechanical properties of 1D ZnO nanostructures: Nanopiezotronics building blocks, surface and size-scale effects. *Phys. Chem. Chem. Phys.* **2014**, *16*, 4522–4527. [CrossRef]
8. Nandi, S.; Kumar, S.; Misra, A. Zinc oxide heterostructures: Advances in devices from self-powered photodetectors to self-charging supercapacitors. *Adv. Mater.* **2021**, *2*, 6768–6799. [CrossRef]
9. Nanto, H.; Minami, T.; Takata, S. Zinc-oxide thin-film ammonia gas sensors with high sensitivity and excellent selectivity. *J. Appl. Phys.* **1986**, *60*, 482–484. [CrossRef]
10. Rahman, F. Zinc oxide light-emitting diodes: A review. *Opt. Eng.* **2019**, *58*, 010901. [CrossRef]
11. Sun, L.; Shao, R.; Chen, Z.; Tang, L.; Dai, Y.; Ding, J. Alkali-dependent synthesis of flowerlike ZnO structures with enhanced photocatalytic activity via a facile hydrothermal method. *Appl. Surf. Sci.* **2012**, *258*, 5455–5461. [CrossRef]
12. Rahmane, S.; Aida, M.S.; Djouadi, M.A.; Barreau, N. Effects of thickness variation on properties of ZnO: Al thin films grown by RF magnetron sputtering deposition. *Superlattices Microstruct.* **2015**, *79*, 148–155. [CrossRef]
13. Wu, J.-J.; Liu, S.-C. Low-Temperature Growth of Well-Aligned ZnO Nanorods by Chemical Vapor Deposition. *Adv. Mater.* **2002**, *14*, 215–218. [CrossRef]
14. Jin, B.J.; Bae, S.H.; Lee, S.Y.; Im, S. Effects of native defects on optical and electrical properties of ZnO prepared by pulsed laser deposition. *Mater. Sci. Eng.* **2000**, *71*, 301–305. [CrossRef]
15. Kripal, R.; Gupta, A.K.; Srivastava, R.K.; Mishra, S.K. Photoconductivity and photoluminescence of ZnO nanoparticles synthesized via co-precipitation method. *Spectrochim. Acta Part A Mol. Biomol. Spectrosc.* **2011**, *79*, 1605–1612. [CrossRef] [PubMed]
16. Wadaa, S.; Tsurumia, T.; Chikamorib, H.; Nomab, T.; Suzukib, T. Preparation of nm-sized $BaTiO_3$ crystallites by a LTDS method using a highly concentrated aqueous solution. *J. Cryst. Growth* **2001**, *229*, 433–439. [CrossRef]
17. Yazdanbakhsh, M.; Khosravi, I.; Goharshadi, E.K.; Youssefi, A. Fabrication of nanospinel $ZnCr_2O_4$ using sol-gel method and its application on removal of azo dye from aqueous solution. *J. Hazard. Mater.* **2010**, *184*, 684–689. [CrossRef]
18. Sun, X.; Li, Z.; Wang, X.; Zhang, G.; Cui, P.; Shen, H. Single-Crystal Regular Hexagonal Microplates of Two-Dimensional α-Calcium Sulfate Hemihydrate Preparation from Phosphogypsum in Na_2SO_4 Aqueous Solution. *Ind. Eng. Chem. Res.* **2020**, *59*, 13979–13987. [CrossRef]
19. Boppella, R.; Anjaneyulu, K.; Basak, P.; Manorama, S.V. Facile Synthesis of Face Oriented ZnO Crystals: Tunable Polar Facets and Shape Induced Enhanced Photocatalytic Performance. *J. Phys. Chem. C* **2013**, *117*, 4597–4605. [CrossRef]
20. Tang, L.; Zhao, Z.; Zhou, Y.; Lv, B.; Li, P.; Ye, J.; Wang, X.; Xiao, M.; Zou, Z. Series of $ZnSn(OH)_6$ Polyhedra: Enhanced CO_2 Dissociation Activation and Crystal Facet-Based Homojunction Boosting Solar Fuel Synthesis. *Inorg. Chem.* **2017**, *56*, 5704–5709. [CrossRef]

21. Li, P.; Zhou, Y.; Zhao, Z.; Xu, Q.; Wang, X.; Xiao, M.; Zou, Z. Hexahedron Prism-Anchored Octahedronal CeO_2: Crystal Facet-Based Homojunction Promoting Efficient Solar Fuel Synthesis. *J. Am. Chem. Soc.* **2015**, *137*, 9547–9550. [CrossRef] [PubMed]
22. Tang, L.; Vo, T.; Fan, X.; Vecchio, D.; Ma, T.; Lu, J.; Hou, H.; Glotzer, S.C.; Kotov, N.A. Self-Assembly Mechanism of Complex Corrugated Particles. *J. Am. Chem. Soc.* **2021**, *143*, 19655–19667. [CrossRef] [PubMed]
23. Xu, C.; Cao, L.; Su, G.; Liu, W.; Qu, X.; Yu, Y. Preparation, characterization and photocatalytic activity of Co-doped ZnO powders. *J. Alloys Compd.* **2010**, *497*, 373–376. [CrossRef]
24. Šutka, A.; Käämbre, T.; Pärna, R.; Juhnevica, I.; Maiorov, M.; Joost, U.; Kisand, V. Co doped ZnO nanowires as visible light photocatalysts. *Solid State Sci.* **2016**, *56*, 54–62. [CrossRef]
25. Xie, F.; Guo, J.; Wang, H.; Chang, N. Enhancing visible light photocatalytic activity by transformation of Co^{3+}/Co^{2+} and formation of oxygen vacancies over rationally Co doped ZnO microspheres. *Colloids Surf. A Physicochem. Eng. Asp.* **2022**, *636*, 128157–128167. [CrossRef]
26. Chen, Z.; Fang, Y.; Wang, L.; Chen, X.; Lin, W.; Wang, X. Remarkable oxygen evolution by Co-doped ZnO nanorods and visible light. *Appl. Catal. B-Environ.* **2021**, *296*, 120369. [CrossRef]
27. Javed, A.; Shahzad, N.; Butt, F.; Khan, M.; Naeem, N.; Liaquat, R.; Khoja, A. Synthesis of bimetallic Co-Ni/ZnO nanoprisms (ZnO-NPr) for hydrogen-rich syngas production via partial oxidation of methane. *J. Environ. Chem. Eng.* **2021**, *9*, 106887–106899. [CrossRef]
28. Barret, C.S.; Massalski, T.B. *Structure of Metals*; Pergamon Press: Oxford, UK, 1980.
29. Romero, R.; Leinen, D.; Dalchiele, E.A.; Ramos-Barrado, J.R.; Martín, F. The effects of zinc acetate and zinc chloride precursors on the preferred crystalline orientation of ZnO and Al-doped ZnO thin films obtained by spray pyrolysis. *Thin Solid Film.* **2006**, *515*, 1942–1949. [CrossRef]
30. Kubacka, A.; Fernandez-Garcia, M.; Colon, G. Advanced nanoarchitectures for solar photocatalytic applications. *Chem. Rev.* **2012**, *112*, 1555–1614. [CrossRef]
31. Pourrahimi, A.M.; Liu, D.; Ström, V.; Hedenqvist, M.S.; Olsson, R.T.; Gedde, U.W. Heat treatment of ZnO nanoparticles: New methods to achieve high-purity nanoparticles for high-voltage applications. *J. Mater. Chem. A* **2015**, *3*, 17190–17200. [CrossRef]
32. Olatunde, O.C.; Kuvarega, A.T.; Onwudiwe, D.C. Photo enhanced degradation of contaminants of emerging concern in waste water. *Emerg. Contam.* **2020**, *6*, 283–302. [CrossRef]
33. Zhang, T.; Oyama, T.; Aoshima, A.; Hidaka, H.; Zhao, J.; Serpone, N. Photooxidative N-demethylation of methylene blue in aqueous TiO_2 dispersions under UV irradiation. *J. Photochem. Photobiol. A Chem.* **2001**, *140*, 163–172. [CrossRef]
34. Li, H.; Zhou, Y.; Tu, W.; Ye, J.; Zou, Z. State-of-the-Art Progress in Diverse Heterostructured Photocatalysts toward Promoting Photocatalytic Performance. *Adv. Funct. Mater.* **2015**, *25*, 998–1013. [CrossRef]
35. Mueller, T.; Malic, E. Exciton physics and device application of two-dimensional transition metal dichalcogenide semiconductors. *NPJ 2D Mater. Appl.* **2018**, *2*, 1–12. [CrossRef]

Article

Enhancement of Visible-Light Photocatalytic Degradation of Tetracycline by Co-Doped TiO$_2$ Templated by Waste Tobacco Stem Silk

Quanhui Li [1], Liang Jiang [1,*], Yuan Li [1], Xiangrong Wang [2], Lixia Zhao [1], Pizhen Huang [1], Daomei Chen [1] and Jiaqiang Wang [1,*]

[1] School of Materials and Energy, Yunnan Province Engineering Research Center of Photocatalytic Treatment of Industrial Wastewater, School of Chemical Sciences & Technology, School of Engineering, National Center for International Research on Photoelectric and Energy Materials, Yunnan University, Kunming 650091, China
[2] Kunming Academy of Eco-Environmental Sciences, Kunming 650032, China
* Correspondence: jiangliang@ynu.edu.cn (L.J.); jqwang@ynu.edu.cn (J.W.)

Abstract: In this study, Co-doped TiO$_2$ was synthesized using waste tobacco stem silk (TSS) as a template via a one-pot impregnation method. These samples were characterized using various physicochemical techniques such as N$_2$ adsorption/desorption analysis, diffuse reflectance UV–visible spectroscopy, X-ray diffraction, field-emission scanning electron microscopy, high-resolution transmission electron microscopy, X-ray photoelectron spectroscopy, photoluminescence spectroscopy, and electron paramagnetic resonance spectroscopy. The synthesized material was used for the photodegradation of tetracycline hydrochloride (TCH) under visible light (420–800 nm). No strong photodegradation activity was observed for mesoporous TiO$_2$ synthesized using waste TSS as a template, mesoporous Co-doped TiO$_2$, or TiO$_2$. In contrast, Co-doped mesoporous TiO$_2$ synthesized using waste TSS as a template exhibited significant photocatalytic degradation, with 86% removal of TCH. Moreover, owing to the unique chemical structure of Ti-O-Co, the energy gap of TiO$_2$ decreased. The edge of the absorption band was redshifted, such that the photoexcitation energy for generating electron–hole pairs decreased. The electron–hole separation efficiency improved, rendering the microstructured biotemplated TiO$_2$ a much more efficient catalyst for the visible-light degradation of TCH.

Keywords: visible-light photocatalysis; Co-doped TiO$_2$; one-pot impregnation; tetracycline hydrochloride degradation; biotemplate

Citation: Li, Q.; Jiang, L.; Li, Y.; Wang, X.; Zhao, L.; Huang, P.; Chen, D.; Wang, J. Enhancement of Visible-Light Photocatalytic Degradation of Tetracycline by Co-Doped TiO$_2$ Templated by Waste Tobacco Stem Silk. *Molecules* **2023**, *28*, 386. https://doi.org/10.3390/molecules28010386

Academic Editors: Hongda Li, Mohammed Baalousha and Victor A. Nadtochenko

Received: 1 December 2022
Revised: 26 December 2022
Accepted: 26 December 2022
Published: 2 January 2023

Copyright: © 2023 by the authors. Licensee MDPI, Basel, Switzerland. This article is an open access article distributed under the terms and conditions of the Creative Commons Attribution (CC BY) license (https://creativecommons.org/licenses/by/4.0/).

1. Introduction

In recent years, tetracycline (TC) has been misused, and its residues have significantly impacted the ecosystem and the physical and mental health of human beings owing to its improper degradation treatment [1–4]. Therefore, an effective and environmentally friendly method is required for TC degradation. Photocatalysis is one of the most effective and economical methods for the removal of organic pollutants. Recently, novel photocatalysts Ag$_3$PO$_4$@MWCNTs@PPy and Ag$_3$PO$_4$@NC with excellent photocatalytic activity and photostability were successfully synthesized [5,6]. TiO$_2$ is a widely used photocatalyst in this regard, owing to its low cost, low toxicity, and good stability [7,8]. However, TiO$_2$ has a wide forbidden bandwidth and reacts only to UV light from sunlight. Moreover, the photogenerated electrons and holes easily recombine, limiting its widespread application. Therefore, it is necessary to develop an efficient photocatalyst that can utilize most of the light from the solar spectrum to effectively degrade TC. Ion doping or the use of biotemplates is a common approach to enhance the photocatalytic performance of TiO$_2$. However, the effect of the simultaneous use of biotemplates and transition metal ion doping on the photocatalytic degradation efficiency has not been explored sufficiently.

Transition metal doping is an effective strategy for overcoming the limitations of TiO$_2$, as it can improve light absorption and conductivity, reduce carrier complexation on

TiO$_2$ surface [9], and improve the photocatalytic performance of quantum-sized TiO$_2$ by changing the band gap position and Fermi energy level [10]. The introduction of transition metals to improve the properties of TiO$_2$ has been often reported; among them, Mn^{2+} [11], Co^{2+} [10], Fe^{2+} [12], and Ni^{2+} [13] plasma doping have been reported in detail. Among these ions, Co^{2+} is preferred because the ionic radius of Co^{2+} (0.74 Å) is similar to that of Ti^{4+} (0.61 Å) and the former can easily replace Ti^{4+} in TiO$_2$ to form a stable structure. Doping with transition metal ions to change the crystal structure of TiO$_2$ also alters the forbidden band width and cell parameters and improves the photocatalytic activity [14,15]. A series of Co-doped TiO$_2$ materials synthesized by the sol–gel method were found to be effective for the visible-light degradation of methyl orange—increasing Co doping concentration enhanced the redshift of the UV–vis absorption spectrum. The dopant inhibited the growth of TiO$_2$ grains, causing them to aggregate. This shifted the absorption maximum of TiO$_2$ from the UV to visible region [16]. However, as electron complex centers, Co ions will decrease in the lifetime of the photogenerated electron–hole pairs. In this regard, controlling the valence state of Co ions to limit the utilization of electron–hole pairs is a more reasonable solution [17]. The present study was conducted to improve the photocatalytic efficiency by adjusting the Co^{2+}/Co^{3+} ratio—a method that has rarely been reported.

The strategy of using natural biomass to modulate the morphology of prepared materials and produce nanoscale catalysts with specific functions has attracted considerable attention [18,19]. Many TiO$_2$ materials with specific morphologies have been synthesized from biotemplates such as leaves, flower petals [20,21], foliage [22,23], bamboo [24], wood [25,26], and cotton [27]. Biotemplates can effectively improve the photocatalytic activity of TiO$_2$. Moreover, materials synthesized using biotemplates exhibit better photocatalytic degradation efficiency toward pollutants. Tobacco stem silk (TSS) is coarse and tobacco leaves have stratified veins. Consequently, these plant parts, whose microstructures are in the form of lamellar folds, are mostly discarded nowadays.

In our previous study, we prepared an efficient TiO$_2$ photocatalyst, TTS-ST(HF) [28], in which the morphology was modulated with TSS as a biotemplate. In this study, using tetrabutyl titanate (TBOT) and Co(NO$_3$)$_2$·6H$_2$O as the Ti and Co sources, respectively, an efficient Co-doped TiO$_2$ photocatalyst was synthesized using the one-pot impregnation method. TCH was used as a simulated pollutant to determine the photocatalytic degradation performance. The mechanism of heteroenergetic (Ti-O-Co) photocatalytic degradation has also been proposed.

2. Results

2.1. Synthesis of Photocatalysts by One-Pot Impregnation and Their Structural Characterization

In this study, we adopted a simple one-pot impregnation method (Figure 1), which is more concise than the conventional sol–gel biotemplate method, to replicate the complex structure of hard biotemplates in the microstructure of TiO$_2$. Doped metal ions act as charge complex centers and decrease the lifetime of the electron–hole pairs. Hence, we designed a scheme to reduce the extent of conversion between Co^{2+} and Co^{3+} ions by controlling the amount of doped Co ions to determine the ratio of Co^{2+} ions to Co^{3+}, thus enhancing the photocatalytic efficiency [17].

The N$_2$ adsorption–desorption curves (Figure S1) of all prepared samples were characteristic type IV isotherms with a H3-type hysteresis loops, indicating the presence of mesopores with a stacked pore structure.

The X-ray diffraction (XRD) patterns (Figure 2) of the prepared materials were acquired to investigate the crystal structure of the biotemplated TiO$_2$. The peaks at 2θ values of 25.34°, 37.01°, 37.85°, 38.64°, 48.14°, 53.97°, 55.18°, 62.24°, and 62.81° correspond to the (101), (103), (004), (112), (200), (105), (211), (213), and (204) crystal planes, respectively, consistent with standard card JCPDS 73-1764. The position of the characteristic diffraction peak of TiO$_2$ did not change, and no dichotomous peak related to Co was observed, which could be attributed to the low Co content [29]. It is worth noting that TiO$_2$ and the TiO$_2$-TSS materials were purely white in color, and the color gradually changed to green with increasing Co doping. Moreover, the crystalline shape of TiO_2 did not change after the introduction of the TSS filament template.

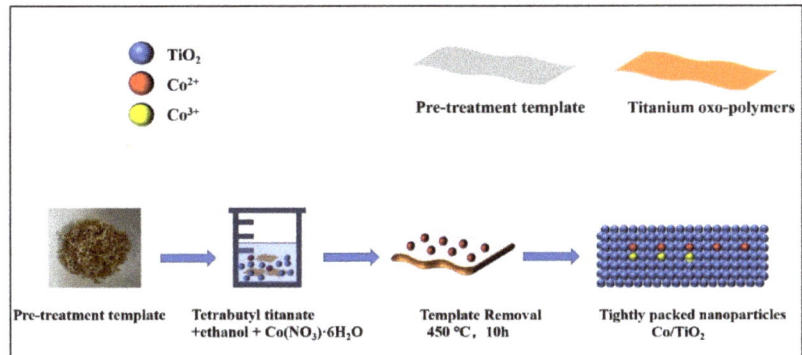

Figure 1. Schematic diagram of the synthesis through the one-pot impregnation method.

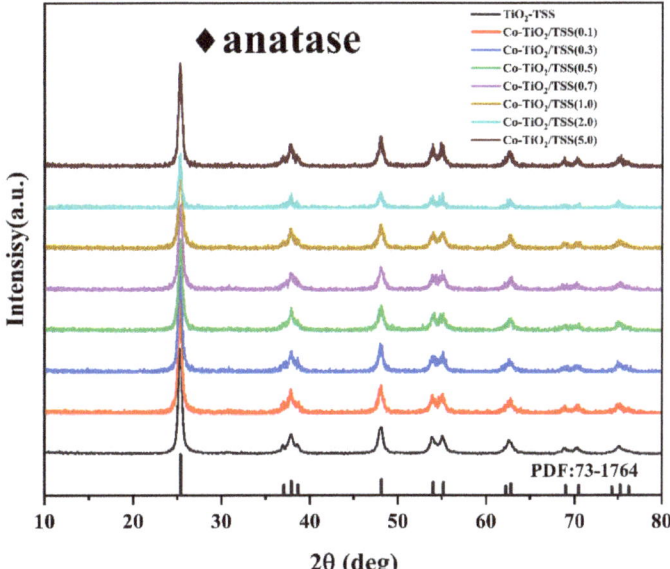

Figure 2. XRD patterns of Co-TiO_2/TSS(X).

We next studied the morphology of the prepared samples. Figure 3 shows the SEM images of the pure TSS templates. It is evident that the stalk filaments have lamellar structures with folds, and their sizes range from 200 to 300 μm. Figure 4 shows the SEM and transmission electron microscopy (TEM) images of Co-TiO_2/TSS(0.5) at different magnifications. Figure 4a–d indicates that Co is fully bound to the active sites on the surface of the TSS during the impregnation process and was successfully replicated after calcination at 450 °C, showing a lamellar folded morphology wherein the shrinkage and collapse of the pore structure resulted in size reduction after calcination at high temperatures [30]. Figure S2 shows the SEM images of Co-TiO_2/TSS(5.0). Figure S2a shows that with increasing Co doping, the structure becomes denser and the active sites may not be sufficiently bound, leading to decreased photocatalytic efficiencies. The TEM images also show the microstructure block of the material. Figure 4e shows the lattice stripes and lattice spacing of the Co-TiO$_2$/TSS(0.5). Analysis suggests that all the lattice spacings (d) are 0.351 nm, corresponding to the (101) crystal plane of the anatase crystal. Figure S3a,b shows the TEM image of Co-TiO$_2$/TSS(5.0); only TiO$_2$ lattice stripes are seen despite increased Co doping, indicating that the material has only an anatase crystalline phase. This is consistent with the XRD analysis [31]. The microstructural morphology (Figure 4f) of Co-TiO$_2$/TSS(0.5) shows closely spaced nanoparticles as microstructural building blocks that can exist in a lamellar state.

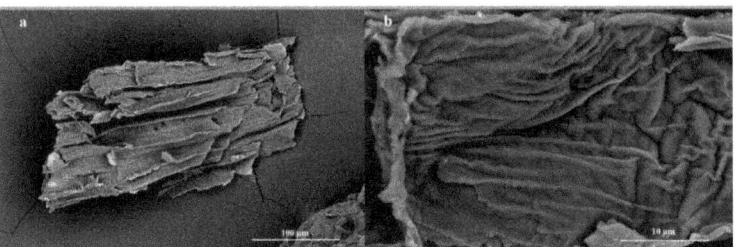

Figure 3. SEM images of original TSS at 1000× magnification (**a**) and 10,000× magnification (**b**).

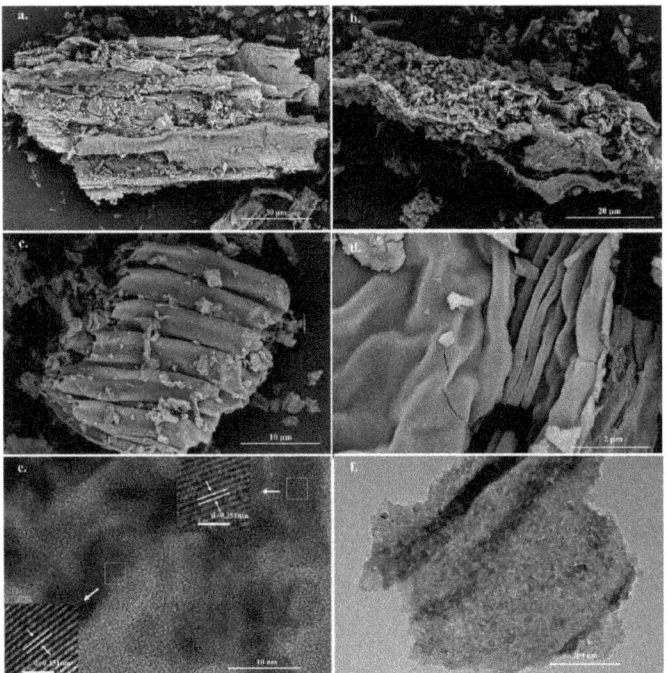

Figure 4. (**a**–**d**) SEM and (**e**,**f**) TEM images of Co-TiO$_2$/TSS(0.5).

2.2. Analysis of Photocatalytic Activity

The photodegradation of TCH (13 mg/L) was used to assess the visible-light photocatalytic activity of the prepared samples. For comparison, pure TiO$_2$ and Co-doped TiO$_2$, prepared using a method similar to that for Co-TiO$_2$/TSS, but without TSS, were used as the reference. As shown in Figure 5a, the removal rates with pure TiO$_2$, TiO$_2$-TSS, Co-TiO$_2$/TSS(0.5), Co-TiO$_2$/TSS(1.0), Co-TiO$_2$/TSS(2.0), Co-TiO$_2$/TSS(5.0), and Co-TiO$_2$, as calculated according to equations (1) and (2), are 12%, 65%, 86%, 76%, 54%, 62%, and 44%, respectively. The photocatalytic degradation rates of TCH over pure TiO$_2$, TiO$_2$-TSS,

Co-TiO$_2$/TSS(0.5), Co-TiO$_2$/TSS(1.0), Co-TiO$_2$/TSS(2.0), Co-TiO$_2$/TSS(5.0), and Co-TiO$_2$ are 10%, 62%, 84%, 74%, 50%, 57%, and 52%, respectively. During the photocatalysis, no significant degradation of TCH was observed in the absence of catalysts under visible-light irradiation. The performance of the biotemplate-modified TiO$_2$ was significantly improved, and the best results were obtained for the materials after modification with both Co ions and biotemplate. The best photocatalyst was found to be Co-TiO$_2$/TSS(0.5), with 86% TCH removal after 90 min of visible-light irradiation.

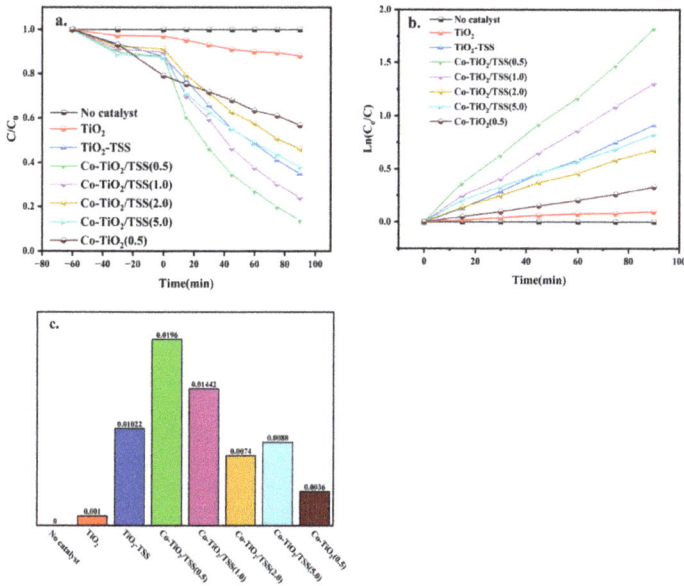

Figure 5. (**a**) TCH removal kinetics, (**b**) plot of ln (C$_e$/C) versus irradiation time, (**c**) pseudo-first-order rate constants, k (min^{-1}), for photocatalysis with pure TiO$_2$, TiO$_2$-TSS, Co-TiO$_2$, and Co-TiO$_2$/TSS(X).

Table 1 summarizes the comparison of the photodegradation efficiency of TCH over different photocatalysts. Obviously, UV light or simulated solar (500 W Xeon lamb) were used in some investigations. Our biotemplated TiO$_2$ was one of the efficient visible light photocatalysts.

Table 1. Comparison of photocatalytic activity for removal TCH over different photo catalysts.

Sample	Light Source and Irradiation Time	Initial Concentration and Catalyst Dosage	Degradation Rate (%)	Ref.
Ag$_3$PO$_4$@MWCNTs@PPy	300 W Xeon lamp, 2 min	20 mg/L 0.5 g/L	94.48	[5]
Cu$_2$O-TiO$_2$-Pal	500 W Xeon lamp, 4 h	30 mg/L 1.0 g/L	81.85	[32]
TiO$_{2-x}$/ultrathin g-C$_3$N$_4$/TiO$_{2-x}$	300 W Xeon lamp, 60 min	10 mg/L 1.0 g/L	87.70	[33]
Co-TiO$_2$/TSS	5 W LED lamp, 90 min	13 mg/L 0.5 h/L	86.00	This work

The photocatalytic degradation of TCH follows first-order kinetics, which can be derived from Equation (1). Figure 5b,c suggests that the first-order kinetic constants of pure TiO$_2$, TiO$_2$-TSS, Co-TiO$_2$, and Co-TiO$_2$/TSS(X) are 0.001, 0.0102, 0.0196, 0.01442, 0.0074 0.0036, and 0.0088, respectively. Doping with an appropriate amount of Co ions can improve the photocatalytic efficiency of materials. Consistent with this, the photocatalytic efficiency of Co-TiO$_2$/TSS(0.5) was 19 times higher than that of pure TiO$_2$.

Usually, the efficient degradation of TCH by biotemplated TiO$_2$ is achieved using UV irradiation [34]. Notably, the biotemplated photocatalyst prepared using the one-pot impregnation method can efficiently utilize the maximum percentage of the solar spectrum to degrade TCH.

2.3. Cyclic Stability Test

The stability experiments of Co-TiO$_2$/TSS(0.5) were carried out (Figure S5). As shown in Figure S5a, the removal rate decreased from 86% to 66% after 5 cycling runs. It can be due to the reduced adsorption rate in the dark adsorption process (Figure S5b). After the fifth cycle, the adsorption rate is reduced from 12.2% to less than 5%. The reason could be explained that the adsorbed intermediate products may block the pores and occupy adsorption sites of catalyst. However, the total amount of TCH removed by Co-TiO$_2$/TSS(0.5) for five cycling experiments was similar, which were 18.2, 19.2, 18.1, 16.9, and 16.7 mg/g, respectively. This result indicated that Co-TiO$_2$/TSS(0.5) was a stable photocatalyst for TCH degradation. After 5 cycles of experiments, the efficiency dropped by 8.2%.

2.4. Identification of the Active Species and Elucidation of Mechanism

We designed an experiment to capture the active species in order to determine the main active species for the photocatalytic degradation of TCH. EDTA-2Na, BQ, and AgNO$_3$ were used as trapping agents for h$^+$, ·O$_2^-$, and e$^-$, respectively. The results for the photocatalytic degradation of TCH by Co-TiO$_2$/TSS(0.5) are shown in Figure 6a. When no trapping agent was added, the removal rate was 86%. When EDTA-2Na and BQ were added, the photocatalytic degradation rate decreased to 44.9% and 59.6%, respectively, indicating that the main active species were h$^+$ and ·O$_2^-$. When AgNO$_3$ was added, the catalytic efficiency increased to 100%. It is likely that the photogenerated electrons were trapped, promoting the effective separation of photogenerated electrons and holes and generating more h$^+$, further indicating that h$^+$ was the main active species [28].

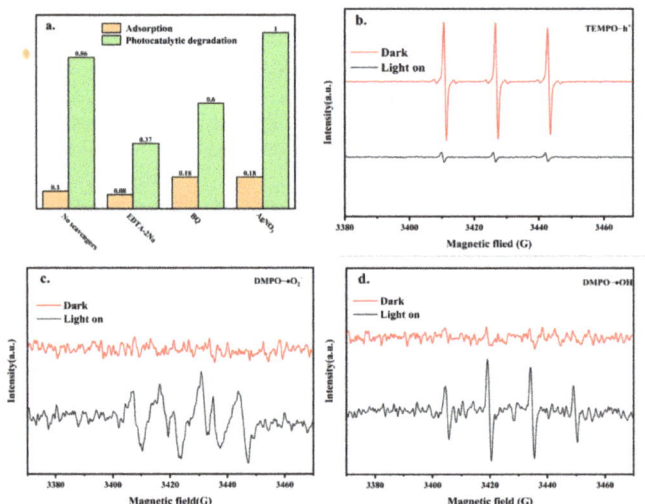

Figure 6. (a) Active species trapping experiments with Co-TiO$_2$/TSS(0.5). EPR spectra of Co-TiO$_2$/TSS(0.5) after the addition of (b) TEMPO to capture h$^+$, (c) DMPO to capture OH, and (d) DMPO to capture O$_2^-$.

To further understand the active species in the photocatalytic degradation of TCH by Co-TiO$_2$/TSS(0.5), the presence of ·OH, h$^+$, and ·O$_2^-$ was detected by electron paramagnetic resonance (EPR) spectroscopy. TEMPO was used to capture h$^+$ (Figure 6b), and a clear TEMPO signal was detected under dark conditions. When visible light was irradiated for 5 min, the signal from TEMPO significantly weakened, indicating the probable depletion of TEMPO due to h$^+$. In addition, this indicated the generation of cavities under light irradiation. The presence of ·OH, h$^+$, and O$_2^-$ was verified using DMPO (Figure 6c,d). No signal was detected under dark conditions, and weak signals from DMPO-OH and DMPO-O$_2^-$ were detected after 5 min of light irradiation. This indicated that the -OH and -O$_2^-$ active species were produced under light irradiation. Thus, the EPR experiments confirmed that these species played a role in the photocatalytic degradation. The dominant role in the photocatalytic degradation was played by h$^+$ and ·O$_2^-$, which is consistent with the results of the active species-capture experiments.

2.5. Photocatalytic Reaction Mechanism

2.5.1. X-ray Photoelectron Spectroscopy (XPS) Analysis

To analyze the chemical states of the sample surface, XPS analysis of the Co-TiO$_2$/TSS(X) materials was performed. As shown in Figure 7a, Co-TiO$_2$/TSS(X) consisted of four elements: C, Ti, O, and Co. The characteristic signal of Co was not obvious, probably because of its low content. However, the signal became stronger with increasing doping, and when Co: Ti $\geq$ 1, the peak intensity increased. Three chemical states of C can be observed in the high-resolution C 1 s spectra (Figure 7b). The peaks at 284.8 and 286.4 eV corresponded to carbon species present in the main chain and C-O bonds, attributable to C in indeterminate contaminants. Considering the presence of residual organic matter in the biotemplate [35], the species with a binding energy of 288.5 eV may be attributed to O-C=O, because the calcination of the remaining C species is incomplete [36]. Figure 7c shows a high-resolution O 1 s spectrum, with characteristic peaks of the Ti-O-Ti bond; that is, peak corresponding to lattice oxygen at 529.9 eV and that corresponding to the -OH group on the TiO$_2$ surface at 531.9 eV out [37]. The binding energy peaks of O 1 s of Co-TiO$_2$/TSS(0.5) appear at 529.32 and 530.86 eV, and the binding energy shifts slightly with increasing Co concentration, which may be due to the formation of the Ti-O-Co bonds [37,38]. Figure 7d shows the high-resolution Co 2p spectra. The Co 2p spectrum of CO-TiO$_2$/TSS(0.5) shows two main peaks at 781.8 and 796.83 eV, corresponding to Co 2p$_{3/2}$ and Co 2p$_{1/2}$, respectively. The small difference between the binding energies (Δ = 15.7 eV) of the Co 2p$_{1/2}$ and Co 2p$_{3/2}$ orbitals indicates that high-spin Co^{2+} is essentially in the oxidation state, and the two main peak difference (Δ = 15 eV) indicates that the low-spin Co^{3+} is essentially in the oxidation state [39]. When the Co:Ti ratio was $\geq$1, two different Co peaks were observed. With increasing Co doping, the peak area of Co^{3+} increased, and the catalytic activity decreased. This could be attributed to the hybridization of the appropriate energy levels of (Ti-O-Co^{3+}) and (Ti-O-Co^{2+}) [40]. Co^{3+} can capture the electrons excited under light irradiation and reduce to Co^{2+} [41]. The adsorbed oxygen molecules on the TiO$_2$ surface are reduced to ·O$_2^-$, following which Co^{2+} is oxidized to Co^{3+}. However, excess Co in the material is detrimental to the photocatalytic efficiency because the metal ions act as charge complex centers and reduce the lifetime of the electron–hole pairs [17]. Figure 7e shows the high-resolution Ti 2p spectrum, where two characteristic peaks of Co-TiO$_2$/TSS(0.5) Ti 2p$_{3/2}$ and Ti 2p$_{1/2}$, probably originating from spin-orbit splitting, can be observed [42]. The difference between the binding energies of Ti 2p$_{3/2}$ and Ti 2p$_{1/2}$ was 5.71 eV, consistent with previous reports [43,44]. The shoulder at 457.67 eV corresponds to Ti^{3+} of Ti$_2$O$_3$, and the slight shift in the binding energy and the shift in the intensity of the shoulder further indicate that the bandgap of Ti in the TiO$_2$ matrix decreases with the substitution of Co [42]. Moreover, a decrease in the bandgap leads to a shift in the binding energy.

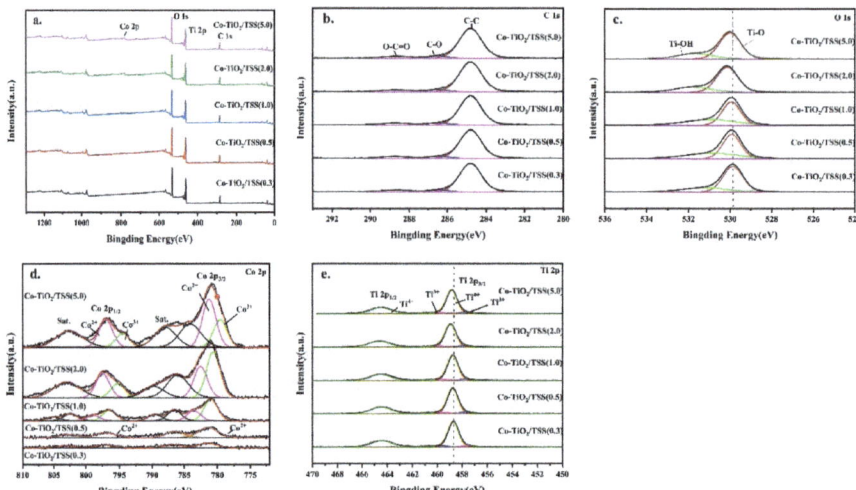

Figure 7. XPS spectra of (**a**) survey spectra, high-resolution XPS spectra of (**b**) C 1s, (**c**) O 1s, (**d**) Co 2p and (**e**) Ti 2p for as-prepared samples.

2.5.2. Ultraviolet–Visible (UV–vis) Diffuse Reflectance Spectral Analysis

Figure 8 shows the UV–vis diffuse reflectance spectra of TiO_2-TSS and Co-TiO_2/TSS (X). The Co-TiO_2/TSS(X) materials exhibited a higher light absorption ability than TiO_2-TSS in the visible region, with redshifted absorption band edges. The enhanced visible-light absorption and narrower band gap energy can be attributed to the sensitization of biomass carbon dopants in the samples induced by the incomplete removal of the biotemplate [21]. The light absorption gradually became more robust with increasing Co doping. Figure 8b shows that the valence band position of Co-TiO_2/TSS(0.5) is at 2.78 eV, indicating that Co doping has a negligible effect on the valence band position of TiO_2, while the forbidden band width of Co-TiO_2/TSS(0.5) was 3.01 eV. Figure 8a clearly shows that Co ions improve the photocatalytic activity by lowering the conduction band position, Ti-O-Co [45] chemical bond formation, which is consistent with the results of the XPS analysis.

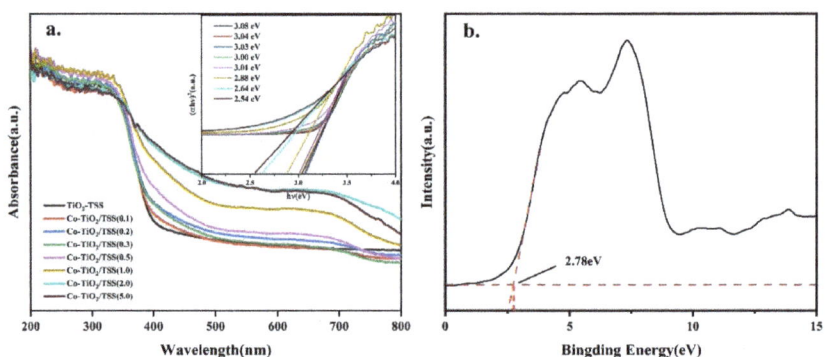

Figure 8. (**a**) UV–vis diffuse reflectance spectra of TiO_2-TSS and Co-TiO_2/TSS(X) and Tauc diagram (inset). (**b**) Valence band spectra of Co-TiO_2/TSS(0.5).

2.5.3. Photoluminescence (PL) Spectroscopy

To study the influence of the Ti-O-Co hybrid energy level formed upon photocatalysis and the characteristics of the photogenerated electron–hole pairs, we recorded the PL spectra and transient photocurrent response and performed electrochemical impedance spectroscopy (EIS) characterization of the materials. The separation efficiency of the photogenerated electrons, photogenerated electron–hole complexation, and migration efficiency of the modified TiO$_2$-TSS and Co-TiO$_2$/TSS(X) materials were determined from the PL spectra recorded at an excitation wavelength of 244 nm (Figure 9). As apparent from the figure, the higher the sample PL intensity, the higher the electron complexation efficiency [46]. The spectral intensity of the Co-TiO$_2$/TSS(X) materials is much lower than that of TiO$_2$-TSS, among which Co-TiO$_2$/TSS(1.0) has the lowest spectral intensity. The photogenerated electron and holes were not easily combined, which is in good agreement with the experimentally obtained results of the photocatalytic activity. The presence of Co^{3+} is not conducive to photocatalysis; thus, the lesser the Co^{3+} content, the better will be the photocatalysis, because Co, as an electron complex center, will reduce the lifetime of the photogenerated electron–hole pair [47].

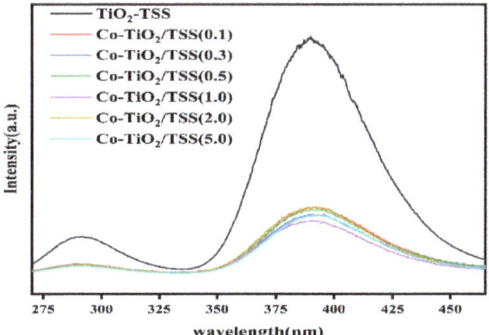

Figure 9. PL spectra of TiO$_2$-TSS and Co-TiO$_2$/TSS(X).

2.5.4. Electrochemical Analysis

To further investigate the photogeneration of electrons and the transport of the synthesized photocatalysts under visible light, the transient photocurrent response and EIS spectra of TiO$_2$, Co-TiO$_2$/(5.0), TiO$_2$-TSS, and Co-TiO$_2$/TSS(0.5) were recorded. The transient photocurrent responses of pure TiO$_2$, TiO$_2$-TSS, and Co-TiO$_2$/TSS(0.5) under visible-light irradiation are shown in Figure 10a. The transient photocurrent response of TiO$_2$, TiO$_2$-TSS, Co-TiO$_2$/TSS(5.0), and Co-TiO$_2$/TSS(0.5) increased sequentially, indicating an effective separation of the photogenerated electron and hole pairs of Co-TiO$_2$/TSS(0.5).

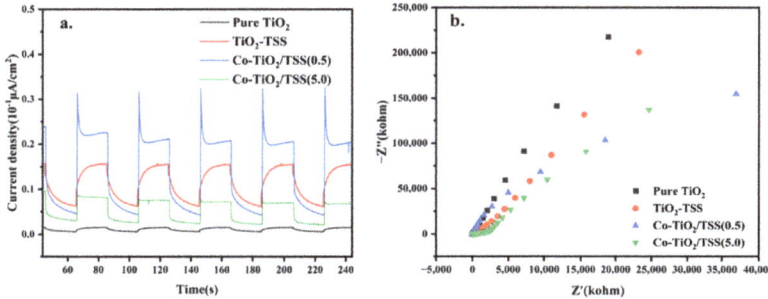

Figure 10. (a) Transient J-t photo response under visible-light illumination and (b) Nyquist plots of pure TiO$_2$, TiO$_2$-TSS, Co-TiO$_2$/TSS(0.5), and Co-TiO$_2$/TSS(5.0).

The interfacial charge transfer of pure TiO$_2$, TiO$_2$-TSS, Co-TiO$_2$/TSS(5.0), and Co-TiO$_2$/TSS(0.5) was also investigated using chemical impedance spectroscopy (Figure 10b). The impedances of pure TiO$_2$, TiO$_2$-TSS, Co-TiO$_2$/TSS(5.0), and Co-TiO$_2$/TSS(0.5) decreased sequentially, indicating an increased charge transfer efficiency of Co-TiO$_2$/TSS(0.5) in the photochemical system [48].

Our investigations reveal that the introduction of Co and regulation of the proportion of Co^{2+} to increase the lifetime of the photogenerated electron–hole pair can improve the photocatalytic activity.

2.5.5. Elucidation of Photocatalytic Mechanism

Based on the analysis of our experimental results, a possible photocatalytic degradation mechanism was proposed. The Ti-O-Co hybridization energy level formed under visible-light irradiation reduced the forbidden bandwidth of TiO$_2$, which was more favorable for electron excitation. During the catalytic process, a small amount of O$_2$ dissolved in water reacts with the photogenerated electrons to produce a small amount of ·O$_2^-$ [49]. Moreover, a small number of photogenerated holes left in the valence band react with H$_2$O to generate ·OH, which is responsible for the generation of -OH. The remaining holes, which are large in number, directly oxidize TCH, generating h$^+$, ·OH, and ·O$_2^-$ as the final active species. A schematic of the photocatalytic mechanism is shown in Figure 11.

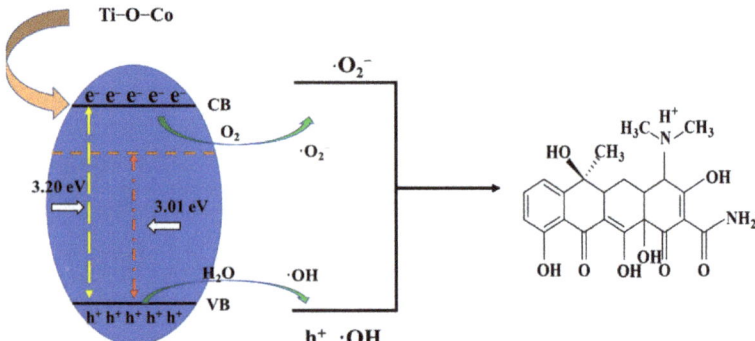

Figure 11. Reaction mechanism of visible-light-driven photocatalytic degradation of TCH on Co-TiO$_2$/TSS(0.5).

3. Materials and Methods

3.1. Materials

TSS (Kunming Cigarette Factory, Kunming, China), cobalt nitrate hexahydrate (Co(NO$_3$)$_2$·6H$_2$O, Shanghai Titan, Shanghai, China), TBOT (Adamas, Beijing, China), anhydrous ethanol (Xilong Chemical, Guangdong, China), hydrochloric acid (HCl, Chuandong Chemical, Chongqing, China), TCH (Adamas, Beijing, China), glutaraldehyde (Xilong Chemical, Beijing, China), p-benzoquinone (BQ, Adamas, Beijing, China), ethylenediamine disodium acetate (Acros Organics, Beijing, China), and silver nitrate (Ag(NO$_3$) Adamas, Beijing, China) were used for the experiments. All chemically synthesized photocatalysts were of analytical grade and used without further purification.

3.2. Preparation of Photocatalyst

To prepare the TSS biotemplate, TSS was pretreated by soaking it in 5% glutaraldehyde for 12 h and 5% HCl for 12 h, followed by gradient dehydration with ethanol. The dehydrated gradient material was dried overnight in an oven at 90 °C and then left to stand. Then, 2 g of the treated TSS was weighed in a 100 mL beaker, and 50 mL of ethanol was added to it, followed by the addition of 5 mL of TBOT and an appropriate amount of Co(NO$_3$)$_2$·6H$_2$O for 24 h. This process controlled the Co:Ti molar ratio to 0.1, 0.5, 1, 2, and 5. The solution was then poured and subjected to hydrolysis in petri dishes for 24 h. The hydrolyzed material was calcined in a muffle furnace at 450 °C for 10 h (2 °C/min), following which the temperature was reduced to room temperature to obtain the final

Article

Fabrication of Novel g-C$_3$N$_4$@Bi/Bi$_2$O$_2$CO$_3$ Z-Scheme Heterojunction with Meliorated Light Absorption and Efficient Charge Separation for Superior Photocatalytic Performance

Hongxia Fan, Xiaohui Ma, Xinyang Li, Li Yang, Yongzhong Bian * and Wenjun Li *

Beijing Key Laboratory for Science and Application of Functional Molecular and Crystalline Materials, University of Science and Technology Beijing, Beijing 100083, China
* Correspondence: yzbian@ustb.edu.cn (Y.B.); wjli_ustb@163.com (W.L.)

Abstract: Herein, a novel g-C$_3$N$_4$@Bi/Bi$_2$O$_2$CO$_3$ Z-scheme heterojunction was synthesized via simple methods. UV/Vis diffuse reflectance spectroscopy (DRS) revealed that the visible light absorption range of heterojunction composites was broadened from 400 nm to 500 nm compared to bare Bi$_2$O$_2$CO$_3$. The XRD, XPS and TEM results demonstrated that metal Bi was introduced into g-C$_3$N$_4$@Bi/Bi$_2$O$_2$CO$_3$ composites, and Bi may act as an electronic bridge in the heterojunction. Metal Bi elevated the separation efficiency of carriers, which was demonstrated by photocurrent and photoluminescence. The performance of samples was assessed via the degradation of Rhodamine B (RhB), and the results exhibited that g-C$_3$N$_4$@Bi/Bi$_2$O$_2$CO$_3$ possessed notably boosted photocatalytic activity compared with g-C$_3$N$_4$, Bi$_2$O$_2$CO$_3$ and other binary composites. The heterojunction photocatalysts possessed good photostability and recyclability in triplicate cycling tests. Radical trapping studies identified that h$^+$ and •O$_2^-$ were two primary active species during the degradation reaction. Based on the energy band position and trapping radical experiments, the possible reaction mechanism of the indirect Z-scheme heterojunction was also proposed. This work could provide an effective reference to design and establish a heterojunction for improving the photocatalytic activity of Bi$_2$O$_2$CO$_3$.

Keywords: photocatalytic; Bi$_2$O$_2$CO$_3$; g-C$_3$N$_4$; Z-scheme heterojunction

1. Introduction

With the increasingly serious problem of environmental pollution, people have been looking for available approaches to solve pollution challenges [1,2]. Advanced oxidation processes and photocatalysis in water treatment receive more and more attention because they could utilize solar energy to degrade organic pollutants [3,4]. Photocatalytic technology uses photocatalysts to produce active species (e.g., h$^+$, •O$_2^-$, •OH) to oxidize organic pollutants in the photocatalysis process. Therefore, the development of efficient visible light photocatalysts is decisive for photocatalytic technology [5,6]. Recently, bismuth-based photocatalysts have attracted extensive attention, such as BiVO$_4$ [7,8], Bi$_2$O$_2$CO$_3$ [9,10], BiOX [11,12], and Bi$_2$O$_3$ [13,14]. Among them, Bi$_2$O$_2$CO$_3$ (BOC) is deemed to be an attractive semiconductor for degrading organic dyes [15,16]. Deplorably, the poor activity of unmodified Bi$_2$O$_2$CO$_3$ limits its environmental application because of the wide band gap. At present, many efforts have been attempted to improve the photocatalytic activity of Bi$_2$O$_2$CO$_3$, such as by controlling morphology [17,18], constructing heterojunctions [19,20], and loading precious metals [21]. Constructing a heterojunction is considered as an ideal method because it can expand light absorption and improve carrier separation efficiency at the same time. Gao et al. [19] reported that the p-n BiOI/Bi$_2$O$_2$CO$_3$ heterojunction presented splendid photocatalytic performance for MB and RhB. Huang et al. [22] synthesized a Bi$_2$O$_2$CO$_3$/Bi$_2$WO$_6$ heterostructure, which possessed outstanding photocatalytic degradation activity for RhB. Even so, due to unsatisfactory activity, it is still necessary to explore other $Bi_2O_2CO_3$-based photocatalysts for organic degradation.

More recently, g-C_3N_4 (CN) was applied for photocatalytic degradation because of its stability, excellent absorption properties, easy fabrication, and suitable band gap [23–25]. Based on the matched energy band levels of BOC and CN, they could form a direct Z-scheme heterojunction for elevated photodegradation activity. The direct Z-scheme heterojunctions are not sufficiently efficacious to boost carrier separation, resulting from the unsatisfactory e^- transport ability of BOC. It has been reported that the use of metal (Au and Ag, etc.) as Z-scheme bridges to form indirect Z-scheme heterojunctions could further accelerate the separation efficiency of photocarriers and thus promote the photocatalytic activity of a direct Z-scheme heterojunction system [26,27]. Owing to its opportune work function and outstanding electrical conductivity, low-cost metal bismuth (Bi) is expected to become a substitute for precious metals as an electronic bridge and gradually come into people's view [28,29]. Additionally, a Bi bridge is acquired via the in situ reduction of BOC, which is awfully conducing to the constitution of Bi/BOC compact heterostructure. Thus, Bi as the bridge in the heterostructure g-C_3N_4@Bi/$Bi_2O_2CO_3$ heterojunction is anticipated to remarkably promote charge separation for high activity. The photocatalytic degradation of the g-C_3N_4@Bi/$Bi_2O_2CO_3$ heterojunction system has been not reported.

In this work, a novel g-C_3N_4@Bi/$Bi_2O_2CO_3$ Z-scheme heterojunction was successfully constructed. Benefiting from the formed heterojunction, g-C_3N_4@Bi/$Bi_2O_2CO_3$ composites possessed notably enhanced photocatalytic degradation activity of RhB due to promoted carriers separation efficiency and expanded photo-response range. We studied the g-C_3N_4@Bi/$Bi_2O_2CO_3$ heterojunction system with an electron-conduction bridge under a photodegradation experiment.

2. Results and Discussion

The microscopic structure and morphology of BOC, CN and BOC-CN-Bi photocatalysts are analyzed via SEM and EDS. Pure BOC is an irregular agglomerated grain structure (Figure 1A). As shown in Figure 1B, g-C_3N_4 possesses a sheet structure that is tens of nanometers thick. The microstructure of BOC-CN-Bi heterojunction composites is shown in Figure 1C. Figure 1E–H displays SEM images and elemental mapping images of BOC-CN-Bi composites, which manifests an even distribution of Bi, C, O and N elements. Table S1 (Supplementary Materials) shows that the Bi element is excessive, indicating the presence of metallic Bi in BOC-CN-Bi composites. The SEM results reveal that BOC-CN-Bi Z-scheme heterojunction composites are successfully prepared. Figure 2A–F shows TEM and HRTEM images of CN, BOC, and BOC-CN-Bi samples. The morphology shown by TEM images was in good agreement with the SEM results. In Figure 2D, CN possesses an amorphous structure as reported in the literature. Furthermore, pure BOC has good crystallinity and its (161) crystal plane spacing is 0.294 nm (Figure 2E). Interestingly, for BOC-CN-Bi (Figure 2F), both CN and BOC could be clearly observed. Meanwhile, metal Bi can also be found by reduction with EG. These results definitely illustrate the successful preparation of BOC-CN-Bi ternary heterostructure composites.

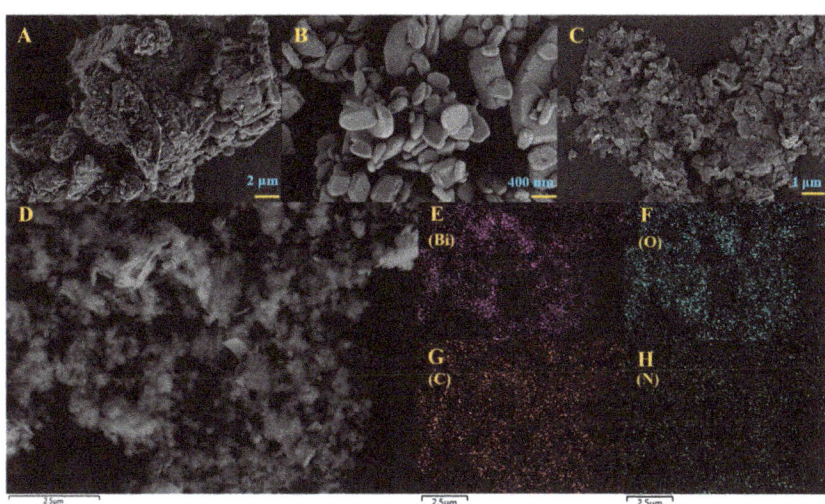

Figure 1. The SEM of BOC (**A**), CN (**B**) and BOC-CN-Bi (**C**,**D**), the mapping image of BOC-CN-Bi in (**E**) (Bi), (**F**) (O), (**G**) (C) and (**H**) (N).

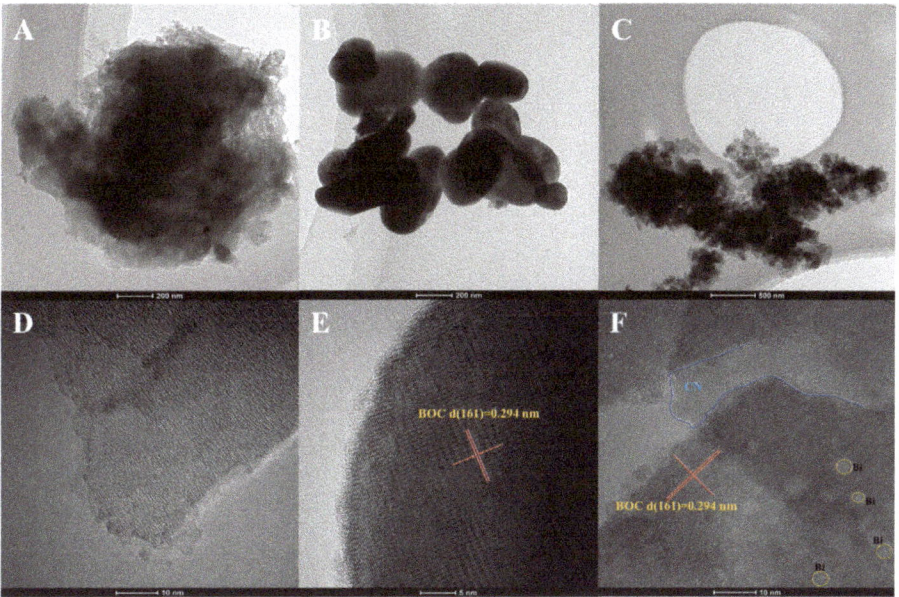

Figure 2. The TEM and HRTEM of CN (**A**,**D**), BOC (**B**,**E**) and BOC-CN-Bi (**C**,**F**).

Figure 3 displays the XRD patterns of as-prepared BOC, CN, BOC-Bi and BOC-CN-Bi photocatalysts. In the XRD patterns of BOC, BOC-Bi and BOC-CN-Bi composites, the peaks of BOC match with the orthorhombic phase of BOC (JCPDS card No. 84-1752). Eleven major peaks at 12.91°, 23.88°, 26.07°, 30.27°, 32.64°, 39.49°, 42.23°, 46.89°, 52.28°, 56.84° and 68.62° are indexed to the planes of (0 4 0), (1 2 1), (0 8 0), (1 6 1), (0 0 2), (0 12 2), (2 8 0), (2 0 2), (1 11 2), (1 6 3) and (4 0 0), respectively. Simultaneously, compared to BOC, there is a new weak peak at 27.16° in the patterns of BOC-Bi and BOC-CN-Bi composites, which can be indexed to metallic Bi [30]. The peaks ((0 0 2) and (1 0 0)) of CN are consistent with that reported in the literature [31]. The peaks of single BOC, CN and Bi could be observed in the patterns of BOC-CN-Bi composites. Additionally, no impurity peaks are detected in the XRD patterns. The aforementioned results indicate that BOC-CN-Bi Z-scheme heterojunction composites have been successfully prepared. The XRD results support SEM and EDS results and will be further determined by XPS analysis.

The surface chemical compositions and the oxidation states of BOC-CN-Bi composites are analyzed via XPS, further confirming the coexistence of BOC, CN and Bi in heterojunction composites. The peaks at 164.3 and 158.9 eV correspond to Bi $4f_{7/2}$ and Bi $4f_{5/2}$, respectively (Figure 4A). The bismuth is a Bi^{3+} cation in the BOC-CN-Bi composite [32]. The peak located at 157.7 eV is attributed to metallic Bi in the BOC-CN-Bi composites. Figure 4B shows two typical peaks of C 1s situate at 288.1 and 284.8 eV, which associates with N-C=N and C-C, respectively. The N-C=N is caused by the sp^2-bonded carbon in the nitrogenous aromatic ring. The C-C bond is caused by amorphous or graphitic carbons. In Figure 4C, the N 1s peak of composites is fitted as three peaks at 398.3, 399.6 and 400.9 eV, corresponding to sp2-hybridized nitrogen involved in triazine rings (C=N-C), tertiary N bonded in N-(C)$_3$ groups, and N-H groups, respectively [33,34]. As shown in Figure 4D, the peaks at 529.7, 530.7, and 532.8 eV are assigned to Bi-O binding, carbonate ions, and absorbed H_2O on the surface [35], respectively. Obviously, the aforementioned results confirm the formation of BOC-CN-Bi Z-scheme heterojunction.

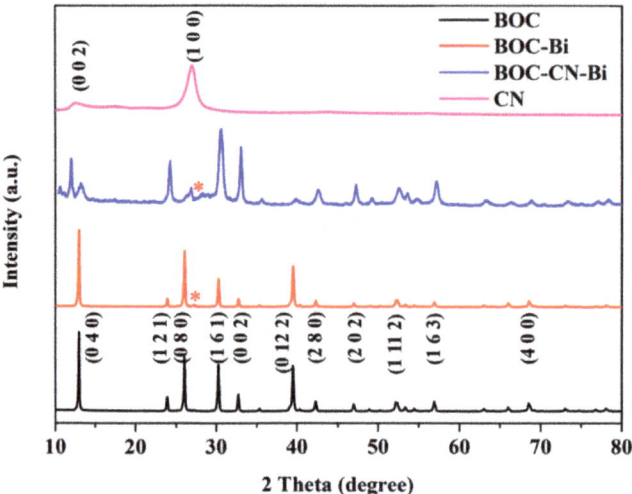

Figure 3. XRD patterns of BOC, CN, BOC-Bi and BOC-CN-Bi photocatalysts (* indicates peaks of metallic Bi).

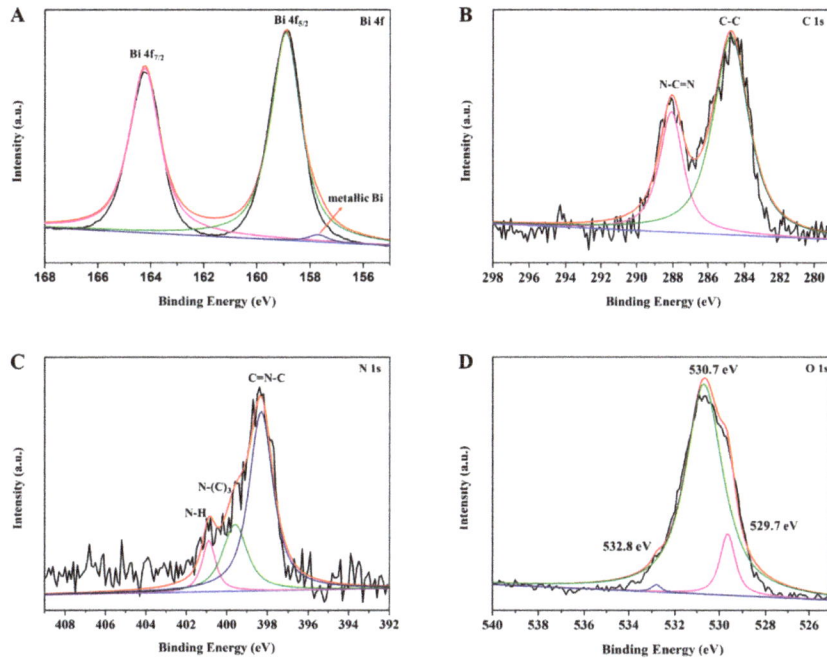

Figure 4. XPS spectra: (**A**) Bi 4f, (**B**) C 1s, (**C**) N 1s and (**D**) O 1s of BOC-CN-Bi composites.

The UV-vis diffuse reflectance spectra are acquired to assess the optical response range of all samples. Pristine BOC and CN reveal severally conspicuous absorption edges at approximately 400 and 500 nm (Figure 5A). The absorption edges for BOC-Bi and CN-Bi possess a slight redshift compared to pristine BOC and CN, which suggests that metal Bi could slightly ameliorate the visible light absorption capacity of photocatalysts [36,37]. Intriguingly, BOC-CN-Bi ternary photocatalysts have a wider visible light absorption range than BOC and BOC-Bi, insinuating the likelihood of better photocatalytic performance for BOC-CN-Bi ternary photocatalysts. The wider visible light absorption range assures that BOC-CN-Bi could provoke ample photoinduced e^--h^+ pairs. As presented in Figure 5B, the band gap of pristine BOC and CN are estimated as 3.17 and 2.38 eV via the Kubelka-Munk formula. Furthermore, the band gap of all composites is shown in Figure S1. The band gap for BOC-CN-Bi is 2.6 eV. Compared with BOC, the BOC-CN-Bi composite has a narrower band gap and higher carrier separation for efficient photocatalytic hydrogen production.

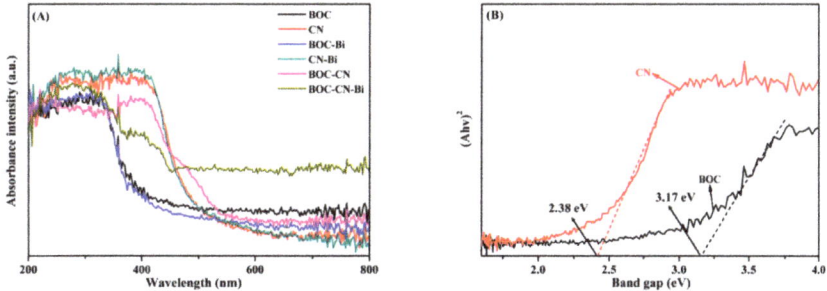

Figure 5. The DRS (**A**) of the as-obtained samples; the band gaps (**B**) of BOC and CN.

In Figure 6A, the photocatalytic performance of all photocatalysts is evaluated via the photodegradation of the rhodamine B (RhB) under visible light (≥420 nm). Before the photoreaction process, all samples were kept in the dark to realize the adsorption-desorption equilibrium. The value of RhB adsorption for all samples is shown in Table S2 (Supplementary Materials). Pristine BOC reveals an inconsequential photodegradation rate of 30% after 120 min because of its wide band gap. As for CN, only 50% of RhB is disintegrated after 120 min. However, after the introduction of metal Bi, the binary BOC-Bi and CN-Bi exhibit better degradation rates (40% and 60%, respectively). Simultaneously, the photodegradation rate of the RhB for the BOC-CN compound is 45%. Interestingly, the ternary composites (BOC-CN-Bi) reveal the best photocatalytic activity compared to other samples, which is approximately 93% after 120 min. Additionally, the ternary BOC-CN-Bi heterojunction composites still possess high photocatalytic performance for RhB degradation compared to the reported BOC-CN photocatalysts. Therefore, we can reach the conclusion that metal Bi has a remarkable effect on enhancing photocatalytic performance. To explore the kinetic reaction, the kinetics of RhB degradation for all samples are analyzed. The degradation data of all catalysts are linearly fitted according to the equation:

$$\text{Ln}\left(\frac{C}{C_0}\right) = kt + a$$

where C, C_0 and k are the concentration of RhB under the different times, the concentration of RhB after the adsorption-desorption equilibrium, and the apparent reaction rate constant, respectively. As shown in Figure 6B, the relationship between $\text{Ln}(C/C_0)$ and t of all samples conforms to the first-order kinetic model. The k of every sample is estimated and shown (Figure 6B). Obviously, BOC-CN-Bi possesses the fastest photodegradation rate, which is 6.2 and 3.5 times than that of BOC and CN. The aforementioned results forcefully verified that BOC-CN-Bi ternary composites possess a superior photocatalytic activity. We also explored the effects of pH, dye concentration and catalyst amount on photocatalytic degradation (Figure S2). Obviously, the more photocatalytic content, the better the degradation effect. However, the degradation efficiency and the catalyst increase disproportionately. Moreover, the concentration of organic pollutants increased, and the degradation efficiency decreased slightly. When pH increased, the degradation efficiency decreased significantly. When pH decreased, the catalyst completely adsorbed organic pollutants after darkness, and the degradation efficiency could not be measured (Figure S3, Supplementary Materials). It is well known that the stability of photocatalysts is crucial in practical applications. Therefore, the cycling experiments of BOC-CN-Bi composites were carried out. As shown in Figure 6C, the photocatalytic performance is not significantly reduced in three cycles, which confirms the outstanding stability of the composites.

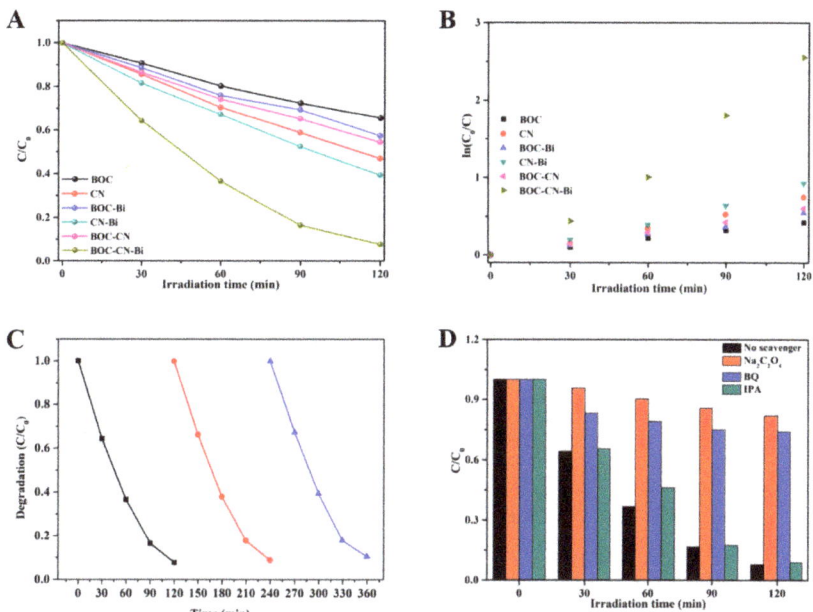

Figure 6. The photodegradation of different organic pollutants (**A**) and kinetics of the RhB decomposition (**B**) over as-prepared samples, the cycling experiments (**C**) and the trapping experiment (**D**) the result of BOC-CN-Bi ternary composites by degrading RhB.

Under photocatalytic degradation, organic pollutants are decomposed by active species, e.g., active holes (h^+), hydroxyl radicals ($\bullet OH$) and superoxide radicals ($\bullet O_2^-$). To verify the dominant active species of BOC-CN-Bi ternary composites under photocatalytic degradation, trapping experiments are performed by adding different scavengers (Figure 6D). In the experiment, benzoquinone (BQ), sodium oxalate ($Na_2C_2O_4$) and isopropanol (IPA) are used as a quencher of $\bullet O_2^-$, h^+ and $\bullet OH$. After adding IPA, the degradation efficiency decreases slightly, implying that $\bullet OH$ is not decisive active species in the photodegradation experiment. Evidently, when BQ and $Na_2C_2O_4$ are added, the degradation efficiency of pollutants is remarkably suppressed, which could declare that $\bullet O_2^-$ and h^+ species have a pivotal role in degrading pollutants. More importantly, the degradation rate of pollutants for $Na_2C_2O_4$ is obviously lower than that of BQ, suggesting that h^+ is a more vivacious species. The results evidently demonstrate that $\bullet O_2^-$ and h^+ are the main active species, and h^+ acts in a dominant role for RhB degradation.

Photoluminescence (PL) is used to study the electron-hole separation efficiency. The PL spectra of all samples are shown in Figure 7A,B. Commonly, the lower PL intensity signifies the lower recombination of electron-hole pairs [38,39]. CN-Bi has lower PL intensity compared with pure CN (Figure 7A). As shown in Figure 7B, the PL spectra of BCO-CN and BCO-CN-Bi obviously reduce, suggesting that CN and Bi can restrain the carrier recombination of BCO. Among all samples, BCO-CN-Bi has the lowest PL spectra, indicating that BCO-CN-Bi has the highest carrier separation efficiency. To further study the behaviors of charge transfer for as-prepared samples, the EIS and transient photocurrent responses are carried out. It is well known that the smaller the radius of the arc, the smaller the interfacial resistance [40,41]. As disclosed in Figure 7C, BOC-CN-Bi photocatalysts possess the smallest arc compared to pure BOC, CN and other binary samples, which implies the lowest charge transfer resistance. We also execute photocurrent measurement to further investigate the separation rate of e^--h^+ pairs for samples. The photocurrent intensity of BOC-CN-Bi photocatalysts was much stronger than BOC and CN alone (Figure 7D), explaining that the designed ternary Z-scheme heterojunction among BOC, Bi and CN boosts the separation efficiency of photoinduced carriers [42].

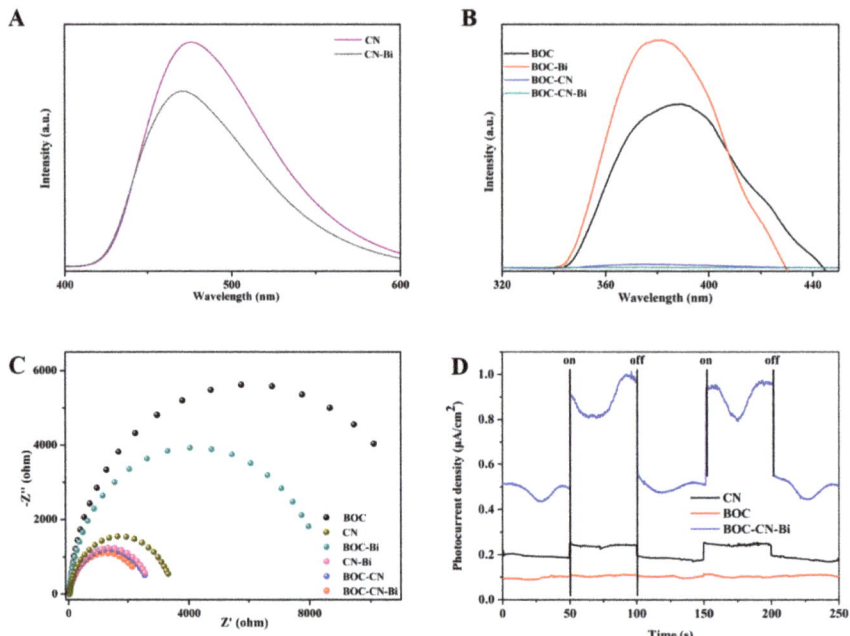

Figure 7. The PL (**A**,**B**), Nyquist (**C**) and photocurrent spectra (**D**) of over as-prepared samples.

Mott-Schottky was used to study the conduction band of CN. The conduction band of CN is −1.1 V. According to our previous studies, BOC has a conduction band of 0.33 V [43]. The valence band of BOC and CN is severally 3.5 and 1.28 V according to $E_g = E_{VB} - E_{CB}$. In consideration of the dominant active species, the elevated carrier separation efficiency and the relative band positions of BOC and CN, we designed and discussed a novel indirect heterojunction system to expatiate the feasible reaction mechanism of BOC-CN-Bi photocatalysts for degrading pollutants (Figure 8). When illuminated by visible light, both BOC and CN could produce photoinduced electrons in the conduction band. Simultaneously, the valence band of BOC and CN left a lot of holes. Metal Bi with excellent conductivity could readily capture and transmit electrons. Therefore, the electrons of BOC tended to transfer to metal Bi and consumed the holes of CN because of their different band level, greatly improving the separation efficiency of e^--h^+ pairs. The bi bridge was responsible for the elevated separation efficiency of carriers between BOC and CN, which was proved by the mentioned electrochemical and PL results. As a result, the electrons of BOC combined with the holes of CN resulted in the agglomeration of h^+ in the VB of BOC (3.5 V) and e^- in the CB of CN (−1.1 V) for efficient carrier separation. The reductive electrons in the CB of CN could reduce the absorbed O_2 to O_2^- due to $E(O_2/O_2^-) = -0.33$ V [44,45]. Ultimately, the holes of BOC together with the produced $·O_2^-$ were able to degrade the adsorbed organic pollutants. The remarkably better photocatalytic degradation performance for the ternary BOC-CN-Bi composites was beneficial from the intensive separation of carriers and the extended visible-light absorption range.

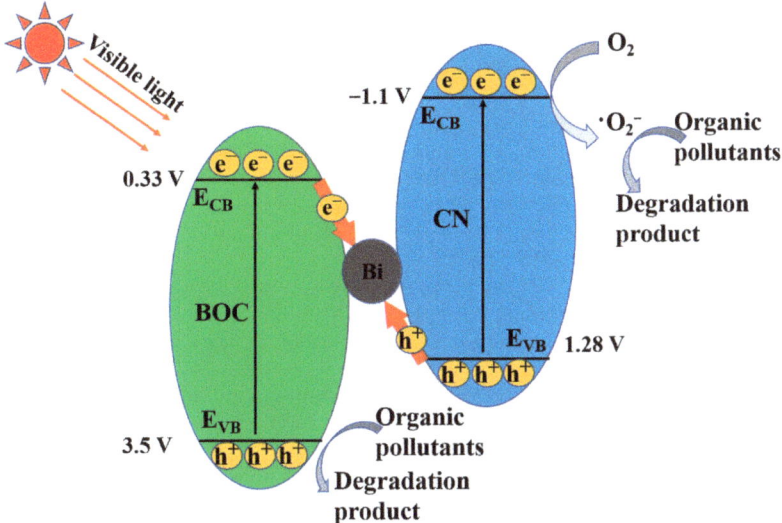

Figure 8. The schematic of BOC-CN-Bi Z-scheme heterojunction by degrading RhB.

3. Materials and Methods

3.1. Preparation of Photocatalyst

The preparation of g-C_3N_4: firstly, the melamine was thoroughly ground. The obtained powders were then calcined at 550 °C (heating rate: 5 °C min^{-1}) for 3 h in a muffle furnace. Until they reached room temperature, the products were ground for further use and labeled as CN.

The preparation of $Bi_2O_2CO_3$ and g-C_3N_4-$Bi_2O_2CO_3$: 0.238 g g-C_3N_4, 0.5 g cetyl trimethyl ammonium bromide and 2.99 g Na_2CO_3 were added into 40 mL deionized water under continuous stirring, and this was labeled as solution A. Meanwhile, 1.71 g $Bi(NO_3)_3 \cdot 5H_2O$ and 10 mL HNO_3 (1 mol L^{-1}) were added into 40 mL deionized water under continuous stirring to form solution B. We then mixed solution A and B and continued stirring for 3 hours. After that, the products were obtained via centrifugation, washed and dried, labeling them as BOC-CN. Without adding CN, $Bi_2O_2CO_3$ (BOC) was synthesized.

The preparation of g-C_3N_4-Bi, Bi/$Bi_2O_2CO_3$ and g-C_3N_4@Bi/$Bi_2O_2CO_3$: 1.71 g $Bi(NO_3)_3 \cdot 5H_2O$ was added into ethylene glycol (EG). After stirring for 2 h, 0.238 g g-C_3N_4 was added to the above solution and was stirred for 3 h. Finally, the mixed solution was heated at 180 °C for 10 h in a reaction kettle. The products were obtained via centrifugation, washed and dried, and labeled as CN-Bi. Similarly, Bi/$Bi_2O_2CO_3$ and g-C_3N_4@Bi/$Bi_2O_2CO_3$ were prepared by replacing 0.238 g g-C_3N_4 with 0.238 g $Bi_2O_2CO_3$ or g-C_3N_4-$Bi_2O_2CO_3$ during synthesis, and labeled as BCO-Bi and BOC-CN-Bi, respectively.

3.2. Materials Characterization

The elementary composition, phase structure and morphology of samples were measured by X-ray photoelectron spectroscopy (XPS), X-ray diffraction (XRD), scanning electron microscopy (SEM), and transmission electron microscopy (TEM), respectively. The UV-vis diffuse reflectance spectra (DRS) and photoluminescence (PL) of samples were obtained by a Lambda 950 Spectrophotometer and Fluorescence Spectrophotometer. The electrochemical impedance spectroscopy (EIS) and photocurrent response were recorded via an electrochemical workstation (CHI660E). The detailed condition of the electrochemical tests is the same as we reported before.

3.3. Photocatalytic Experiments

The degradation of RhB dyes was utilized to determine the photocatalytic degradation performance of all samples via a 400 W Xe lamp ($\geq$420 nm). Before each experiment, 30 mg samples were adequately dispersed in a 30 mL RhB solution. In the dark, the obtained suspension solution was consumingly stirred for 1 h on behalf of an adsorption-desorption equilibrium. During the photoreaction process, 3 mL solution was collected every 30 minutes and centrifuged. A UV-vis spectrophotometer was used to test the RhB concentration in the culture at 553 nm.

4. Conclusions

To sum up, we have successfully fabricated a novel BOC-CN-Bi Z-scheme heterojunction. As shown in UV-vis DRS, the BOC-CN-Bi composites have a wider visible-light absorption range than pure BOC. Bi metal as an electron bridge significantly improves the separation efficiency of photogenerated carriers in a heterojunction composite, which is demonstrated by the photocurrent and PL. The BOC-CN-Bi composites possess significantly superior photocatalytic degradation activity compared to single BOC and CN. Radical trapping studies identify that h^+ and $\bullet O_2^-$ are two primary active species during the degradation reaction. Cycling tests manifest the remarkable stability of the BOC-CN-Bi heterojunction composites. This work implies that BOC-CN-Bi photocatalysts might be efficient and promising novel materials for removing organic pollutants from wastewater.

Supplementary Materials: The following supporting information can be downloaded at: https://www.mdpi.com/article/10.3390/molecules27238336/s1, Figure S1: The band gaps of BOC-Bi, CN-Bi, BOC-CN, and BOC-CN-Bi; Figure S2: The photodegradation different organic pollutant of BOC-CN-Bi under different conditions; Figure S3: The color of the solution for BOC-CN-Bi after the dark at pH = 4; Figure S4: Mott-Schottky plots of CN; Table S1: Element content of BOC-CN-Bi composites by EDS.

Author Contributions: Conceptualization, H.F.; Methodology, H.F. and Y.B.; Formal analysis, X.L., L.Y. and Y.B.; Investigation, X.M.; Data curation, X.M.; Writing—original draft, H.F.; Writing—review & editing, H.F.; Project administration, W.L.; Funding acquisition, W.L. All authors have read and agreed to the published version of the manuscript.

Funding: We gratefully acknowledge the financial support provided by the National Natural Science Foundation of China (Grant No. 21271022).

Institutional Review Board Statement: Not applicable.

Informed Consent Statement: Not applicable.

Data Availability Statement: Not applicable.

Conflicts of Interest: The authors declare no conflict of interest.

References

1. Sun, Y.Y.; Feng, G.; Chen, C.; Liu, Y.H.; Zhang, X. Gram-Scale Synthesis of Polymeric Carbon Nitride-Supported Copper: A Practical Catalyst for Ullmann-Type C-N Coupling Modifying Secondary Pyrimidin-2-amines without Additional Ligand. *Chin. J. Org. Chem.* **2021**, *41*, 1216–1223. [CrossRef]
2. Zhang, S.; Zhang, Z.; Li, B.; Dai, W.; Si, Y.; Yang, L.; Luo, S. Hierarchical Ag_3PO_4@$ZnIn_2S_4$ nanoscoparium: An innovative Z-scheme photocatalyst for highly efficient and predictable tetracycline degradation. *J. Colloid Interface Sci.* **2021**, *586*, 708–718. [CrossRef] [PubMed]
3. Adhikari, S.; Mandal, S.; Kim, D.-H. Z-scheme 2D/1D MoS_2 nanosheet-decorated $Ag_2Mo_2O_7$ microrods for efficient catalytic oxidation of levofloxacin. *Chem. Eng. J.* **2019**, *373*, 31–43. [CrossRef]
4. Wang, Z.; Chen, Y.; Zhang, L.; Cheng, B.; Yu, J.; Fan, J. Step-scheme CdS/TiO_2 nanocomposite hollow microsphere with enhanced photocatalytic CO_2 reduction activity. *J. Mater. Sci. Technol.* **2020**, *56*, 143–150. [CrossRef]
5. Liu, G.; Wang, T.; Ouyang, S.; Liu, L.; Jiang, H.; Yu, Q.; Kako, T.; Ye, J. Band-structure-controlled BiO(ClBr)(1−x)/2Ix solid solutions for visible-light photocatalysis. *J. Mater. Chem. A* **2015**, *3*, 8123–8132. [CrossRef]
6. Xie, T.; Xu, L.; Liu, C.; Wang, Y. Magnetic composite $ZnFe_2O_4/SrFe_{12}O_{19}$: Preparation, characterization, and photocatalytic activity under visible light. *Appl. Surf. Sci.* **2013**, *273*, 684–691. [CrossRef]

7. Kim, T.W.; Choi, K.S. Nanoporous BiVO$_4$ Photoanodes with Dual-Layer Oxygen Evolution Catalysts for Solar Water Splitting. *Science* **2014**, *343*, 990–994. [CrossRef]
8. Zhang, Y.; Li, G.; Yang, X.; Yang, H.; Lu, Z.; Chen, R. Monoclinic BiVO4 micro-/nanostructures: Microwave and ultrasonic wave combined synthesis and their visible-light photocatalytic activities. *J. Alloys Compd.* **2013**, *551*, 544–550. [CrossRef]
9. Zheng, Y.; Duan, F.; Chen, M.; Xie, Y. Synthetic Bi$_2$O$_2$CO$_3$ nanostructures: Novel photocatalyst with controlled special surface exposed. *J. Mol. Catal. A Chem.* **2010**, *317*, 34–40. [CrossRef]
10. Chen, L.; Huang, R.; Yin, S.-F.; Luo, S.-L.; Au, C.-T. Flower-like Bi$_2$O$_2$CO$_3$: Facile synthesis and their photocatalytic application in treatment of dye-containing wastewater. *Chem. Eng. J.* **2012**, *193–194*, 123–130. [CrossRef]
11. Xiong, J.; Cheng, G.; Qin, F.; Wang, R.; Sun, H.; Chen, R. Tunable BiOCl hierarchical nanostructures for high-efficient photocatalysis under visible light irradiation. *Chem. Eng. J.* **2013**, *220*, 228–236. [CrossRef]
12. Hu, J.; Fan, W.; Ye, W.; Huang, C.; Qiu, X. Insights into the photosensitivity activity of BiOCl under visible light irradiation. *Appl. Catal. B Environ.* **2014**, *158–159*, 182–189. [CrossRef]
13. Yan, Y.; Zhou, Z.; Cheng, Y.; Qiu, L.; Gao, C.; Zhou, J. Template-free fabrication of α- and β-Bi$_2$O$_3$ hollow spheres and their visible light photocatalytic activity for water purification. *J. Alloys Compd.* **2014**, *605*, 102–108. [CrossRef]
14. Cheng, H.; Huang, B.; Lu, J.; Wang, Z.; Xu, B.; Qin, X.; Zhang, X.; Dai, Y. Synergistic effect of crystal and electronic structures on the visible-light-driven photocatalytic performances of Bi$_2$O$_3$ polymorphs. *Phys. Chem. Chem. Phys.* **2010**, *12*, 15468–15475. [CrossRef]
15. Dong, F.; Ho, W.-K.; Lee, S.C.; Wu, Z.; Fu, M.; Zou, S.; Huang, Y. Template-free fabrication and growth mechanism of uniform (BiO)$_2$CO$_3$ hierarchical hollow microspheres with outstanding photocatalytic activities under both UV and visible light irradiation. *J. Mater. Chem.* **2011**, *21*, 12428–12436. [CrossRef]
16. Feng, X.; Zhang, W.; Deng, H.; Ni, Z.; Dong, F.; Zhang, Y. Efficient visible light photocatalytic NOx removal with cationic Ag clusters-grafted (BiO)$_2$CO$_3$ hierarchical superstructures. *J. Hazard. Mater.* **2017**, *322*, 223–232. [CrossRef]
17. Dong, F.; Li, Q.; Sun, Y.; Ho, W.-K. Noble Metal-Like Behavior of Plasmonic Bi Particles as a Cocatalyst Deposited on (BiO)$_2$CO$_3$ Microspheres for Efficient Visible Light Photocatalysis. *ACS Catal.* **2014**, *4*, 4341–4350. [CrossRef]
18. Zhao, Z.; Zhou, Y.; Wang, F.; Zhang, K.; Yu, S.; Cao, K. Polyaniline-decorated {001} facets of Bi$_2$O$_2$CO$_3$ nanosheets: In situ oxygen vacancy formation and enhanced visible light photocatalytic activity. *ACS Appl. Mater. Interfaces* **2015**, *7*, 730–737. [CrossRef]
19. Cao, J.; Li, X.; Lin, H.; Chen, S.; Fu, X. In situ preparation of novel p-n junction photocatalyst BiOI/(BiO)$_2$CO$_3$ with enhanced visible light photocatalytic activity. *J. Hazard. Mater.* **2012**, *239–240*, 316–324. [CrossRef]
20. Xu, Y.-S.; Zhang, W.-D. Anion exchange strategy for construction of sesame-biscuit-like Bi$_2$O$_2$CO$_3$/Bi$_2$MoO$_6$ nanocomposites with enhanced photocatalytic activity. *Appl. Catal. B Environ.* **2013**, *140–141*, 306–316. [CrossRef]
21. Peng, S.; Li, L.; Tan, H.; Wu, Y.; Cai, R.; Yu, H.; Huang, X.; Zhu, P.; Ramakrishna, S.; Srinivasan, M.; et al. Monodispersed Ag nanoparticles loaded on the PVP-assisted synthetic Bi$_2$O$_2$CO$_3$ microspheres with enhanced photocatalytic and supercapacitive performances. *J. Mater. Chem. A* **2013**, *1*, 7630–7638. [CrossRef]
22. Huang, X.; Chen, H. One-pot hydrothermal synthesis of Bi$_2$O$_2$CO$_3$/Bi$_2$WO$_6$ visible light photocatalyst with enhanced photocatalytic activity. *Appl. Surf. Sci.* **2013**, *284*, 843–848. [CrossRef]
23. He, Y.Q.; Zhang, F.; Ma, B.; Xu, N.; Binnah Junior, L.; Yao, B.; Yang, Q.; Liu, D.; Ma, Z. Remarkably enhanced visible-light photocatalytic hydrogen evolution and antibiotic degradation over g-C$_3$N$_4$ nanosheets decorated by using nickel phosphide and gold nanoparticles as cocatalysts. *Appl. Surf. Sci.* **2020**, *517*, 146187. [CrossRef]
24. Han, C.; Su, P.; Tan, B.; Ma, X.; Lv, H.; Huang, C.; Wang, P.; Tong, Z.; Li, G.; Huang, Y.; et al. Defective ultra-thin two-dimensional g-C$_3$N$_4$ photocatalyst for enhanced photocatalytic H$_2$ evolution activity. *J. Colloid Interface Sci.* **2021**, *581*, 159–166. [CrossRef]
25. Wang, W.; Fang, J.; Huang, X. Different behaviors between interband and intraband transitions generated hot carriers on g-C$_3$N$_4$/Au for photocatalytic H$_2$ production. *Appl. Surf. Sci.* **2020**, *513*, 145830. [CrossRef]
26. Liu, X.; Jin, A.; Jia, Y.; Xia, T.; Deng, C.; Zhu, M.; Chen, C.; Chen, X. Synergy of adsorption and visible-light photocatalytic degradation of methylene blue by a bifunctional Z-scheme heterojunction of WO$_3$/g-C$_3$N$_4$. *Appl. Surf. Sci.* **2017**, *405*, 359–371. [CrossRef]
27. Wen, M.Q.; Xiong, T.; Zang, Z.G.; Wei, W.; Tang, X.S.; Dong, F. Synthesis of MoS$_2$/g-C$_3$N$_4$ nanocomposites with enhanced visible-light photocatalytic activity for the removal of nitric oxide (NO). *Opt. Express* **2016**, *24*, 10205–10212. [CrossRef]
28. Zhao, Z.; Zhang, W.; Sun, Y.; Yu, J.; Zhang, Y.; Wang, H.; Dong, F.; Wu, Z. Bi Cocatalyst/Bi$_2$MoO$_6$ Microspheres Nanohybrid with SPR-Promoted Visible-Light Photocatalysis. *J. Phys. Chem. C* **2016**, *120*, 11889–11898. [CrossRef]
29. Dong, X.a.; Zhang, W.; Sun, Y.; Li, J.; Cen, W.; Cui, Z.; Huang, H.; Dong, F. Visible-light-induced charge transfer pathway and photocatalysis mechanism on Bi semimetal@defective BiOBr hierarchical microspheres. *J. Catal.* **2018**, *357*, 41–50. [CrossRef]
30. Liu, H.; Zhou, H.; Liu, X.; Li, H.; Ren, C.; Li, X.; Li, W.; Lian, Z.; Zhang, M. Engineering design of hierarchical g-C$_3$N$_4$@Bi/BiOBr ternary heterojunction with Z-scheme system for efficient visible-light photocatalytic performance. *J. Alloys Compd.* **2019**, *798*, 741–749. [CrossRef]
31. Xia, J.; Di, J.; Yin, S.; Li, H.; Xu, H.; Xu, L.; Shu, H.; He, M. Solvothermal synthesis and enhanced visible-light photocatalytic decontamination of bisphenol A (BPA) by g-C$_3$N$_4$/BiOBr heterojunctions. *Mater. Sci. Semicond. Process.* **2014**, *24*, 96–103. [CrossRef]
32. Huo, Y.; Zhang, J.; Miao, M.; Jin, Y. Solvothermal synthesis of flower-like BiOBr microspheres with highly visible-light photocatalytic performances. *Appl. Catal. B Environ.* **2012**, *111–112*, 334–341. [CrossRef]

33. Wang, M.; Zhang, Y.; Jin, C.; Li, Z.; Chai, T.; Zhu, T. Fabrication of novel ternary heterojunctions of Pd/g-C$_3$N$_4$/Bi$_2$MoO$_6$ hollow microspheres for enhanced visible-light photocatalytic performance toward organic pollutant degradation. *Sep. Purif. Technol.* **2019**, *211*, 1–9. [CrossRef]
34. Wang, J.; Tang, L.; Zeng, G.; Liu, Y.; Zhou, Y.; Deng, Y.; Wang, J.; Peng, B. Plasmonic Bi Metal Deposition and g-C$_3$N$_4$ Coating on Bi$_2$WO$_6$ Microspheres for Efficient Visible-Light Photocatalysis. *ACS Sustain. Chem. Eng.* **2016**, *5*, 1062–1072. [CrossRef]
35. Xiong, T.; Wen, M.; Dong, F.; Yu, J.; Han, L.; Lei, B.; Zhang, Y.; Tang, X.; Zang, Z. Three dimensional Z-scheme (BiO)$_2$CO$_3$/MoS$_2$ with enhanced visible light photocatalytic NO removal. *Appl. Catal. B Environ.* **2016**, *199*, 87–95. [CrossRef]
36. Weng, S.; Chen, B.; Xie, L.; Zheng, Z.; Liu, P. Facile in situ synthesis of a Bi/BiOCl nanocomposite with high photocatalytic activity. *J. Mater. Chem. A* **2013**, *1*, 3068–3075. [CrossRef]
37. Zhang, X.; Ji, G.; Liu, Y.; Zhou, X.; Zhu, Y.; Shi, D.; Zhang, P.; Cao, X.; Wang, B. The role of Sn in enhancing the visible-light photocatalytic activity of hollow hierarchical microspheres of the Bi/BiOBr heterojunction. *Phys. Chem. Chem. Phys.* **2015**, *17*, 8078–8086. [CrossRef]
38. Ma, X.; Ren, C.; Li, H.; Liu, X.; Li, X.; Han, K.; Li, W.; Zhan, Y.; Khan, A.; Chang, Z.; et al. A novel noble-metal-free Mo$_2$C-In$_2$S$_3$ heterojunction photocatalyst with efficient charge separation for enhanced photocatalytic H$_2$ evolution under visible light. *J. Colloid Interface Sci.* **2021**, *582*, 488–495. [CrossRef]
39. Ma, X.; Li, W.; Ren, C.; Li, H.; Liu, X.; Li, X.; Wang, T.; Dong, M.; Liu, S.; Chen, S. A novel noble-metal-free binary and ternary In$_2$S$_3$ photocatalyst with WC and "W-Mo auxiliary pairs" for highly-efficient visible-light hydrogen evolution. *J. Alloys Compd.* **2021**, *875*, 160058. [CrossRef]
40. Xiaohui, M.; Wenjun, L.; Hongda, L.; Mei, D.; Xinyang, L.; Liang, G.; Hongxia, F.; Yanyan, L.; Hong, Q.; Tianyu, W. Fabrication of novel and noble-metal-free MoP/In$_2$S$_3$ Schottky heterojunction photocatalyst with efficient charge separation for enhanced photocatalytic H2 evolution under visible light. *J. Colloid Interface Sci.* **2022**, *617*, 284–292. [CrossRef]
41. Ma, X.; Li, W.; Ren, C.; Li, H.; Li, X.; Dong, M.; Gao, Y.; Wang, T.; Zhou, H.; Li, Y. Fabrication of novel noble-metal-free ZnIn$_2$S$_4$/WC Schottky junction heterojunction photocatalyst: Efficient charge separation, increased active sites and low hydrogen production overpotential for boosting visible-light H$_2$ evolution. *J. Alloys Compd.* **2022**, *901*, 163709. [CrossRef]
42. Ma, X.; Li, W.; Ren, C.; Dong, M.; Geng, L.; Fan, H.; Li, Y.; Qiu, H.; Wang, T. Construction of novel noble-metal-free MoP/CdIn$_2$S$_4$ heterojunction photocatalysts: Effective carrier separation, accelerating dynamically H$_2$ release and increased active sites for enhanced photocatalytic H$_2$ evolution. *J. Colloid Interface Sci.* **2022**, *628*, 368–377. [CrossRef] [PubMed]
43. Qiu, F.; Li, W.; Wang, F.; Li, H.; Liu, X.; Ren, C. Preparation of novel p-n heterojunction Bi$_2$O$_2$CO$_3$/BiOBr photocatalysts with enhanced visible light photocatalytic activity. *Colloids Surf. A Physicochem. Eng. Asp.* **2017**, *517*, 25–32. [CrossRef]
44. Qiu, F.; Li, W.; Wang, F.; Li, H.; Liu, X.; Sun, J. In-situ synthesis of novel Z-scheme SnS$_2$/BiOBr photocatalysts with superior photocatalytic efficiency under visible light. *J. Colloid Interface Sci.* **2017**, *493*, 1–9. [CrossRef] [PubMed]
45. Liu, H.; Zhou, H.; Li, H.; Liu, X.; Ren, C.; Liu, Y.; Li, W.; Zhang, M. Fabrication of Bi$_2$S$_3$@Bi$_2$WO$_6$/WO$_3$ ternary photocatalyst with enhanced photocatalytic performance: Synergistic effect of Z-scheme/traditional heterojunction and oxygen vacancy. *J. Taiwan Inst. Chem. Eng.* **2019**, *95*, 94–102. [CrossRef]

Article

Enhance ZnO Photocatalytic Performance via Radiation Modified g-C$_3$N$_4$

Yayang Wang [1,2,†], Xiaojie Yang [2,†], Jiahui Lou [2], Yaqiong Huang [2], Jian Peng [3,*], Yuesheng Li [2,*] and Yi Liu [1,4,*]

1. School of Chemistry and Chemical Engineering, Wuhan University of Science and Technology, Wuhan 430074, China
2. School of Nuclear Technology and Chemistry & Biology, Hubei Key Laboratory of Radiation Chemistry and Functional Materials, Hubei University of Science and Technology, Xianning 437000, China
3. Institute for Superconducting and Electronic Materials, Australian Institute for Innovative Materials, Innovation Campus, University of Wollongong, Squires Way, North Wollongong, NSW 2522, Australia
4. College of Chemistry and Chemical Engineering, Tiangong University, Tianjin 300387, China
* Correspondence: jianp@uow.edu.au (J.P.); frank78929@163.com (Y.L.); yiliuchem@whu.edu.cn (Y.L.)
† These authors contributed equally to this work.

Abstract: Environmental pollution, especially water pollution, is becoming increasingly serious. Organic dyes are one type of the harmful pollutants that pollute groundwater and destroy ecosystems. In this work, a series of graphitic carbon nitride (g-C$_3$N$_4$)/ZnO photocatalysts were facilely synthesized through a grinding method using ZnO nanoparticles and g-C$_3$N$_4$ as the starting materials. According to the results, the photocatalytic performance of 10 wt.% CN-200/Z-500 (CN-200, which g-C$_3$N$_4$ was 200 kGy, referred to the irradiation metering. Z-500, which ZnO was 500 °C, referred to the calcination temperature) with the CN-200 exposed to electron beam radiation was better than those of either Z-500 or CN-200 alone. This material displayed a 98.9% degradation rate of MB (20 mg/L) in 120 min. The improvement of the photocatalytic performance of the 10 wt.% CN-200/Z-500 composite material was caused by the improvement of the separation efficiency of photoinduced electron–hole pairs, which was, in turn, due to the formation of heterojunctions between CN-200 and Z-500 interfaces. Thus, this study proposes the application of electron-beam irradiation technology for the modification of photocatalytic materials and the improvement of photocatalytic performance.

Keywords: g-C$_3$N$_4$; g-C$_3$N$_4$/ZnO; electron beam irradiation; photocatalytic performance

1. Introduction

With increasing industrialization, environmental pollution is becoming more and more serious, especially water pollution [1–3]. Therefore, the treatment of industrial wastewater from various sources has attracted extensive attention from researchers. It is an important challenge to remove pollutants from bodies of water. Among the various types, organic dyes are some of the most harmful pollutants that pollute groundwater and destroy ecosystems. Photocatalysis can effectively degrade organic dyes and is widely used in the degradation of these types of pollutants [4,5].

Photocatalytic technology, due to the excellent stability of the catalyst, its low cost, and its environmental friendliness, has been widely studied by researchers [6–8]. ZnO, with good chemical stability, has low toxicity, low cost, and many other advantages, thus, it has been extensively studied in the field of photocatalytic degradation [9–12]. The narrow spectral absorption (Eg = 3.37 eV) and low degree of separation of photogenerated electron–hole pairs of ZnO, however, limit its applications in photocatalysis [13]. Improving its light absorption capacity and promoting the separation of photogenerated electron–hole pairs can effectively improve the photocatalytic performance of ZnO [14–17]. Graphitic carbon nitride (g-C$_3$N$_4$) possesses many good characteristics, such as a low band gap (Eg = 2.7 eV), stable physical and chemical properties, good absorption capacity of visible

light, simple synthetic methods, thermal stability, and low cost [18–22]. To further improve the performance of g-C_3N_4, electron beam radiation technology is used [23]. Enhancement of the photocatalytic activity of ZnO can be achieved by compounding g-C_3N_4 with it [24–28]. Ding et al. [29] reported a composite catalyst, AgCl/ZnO/g-C_3N_4, which was prepared via calcination, hydrothermal reaction, and in-situ deposition processes. The catalyst was assessed for its photocatalytic efficiency in eliminating tetracycline hydrochloride from pharmaceutical wastewater under visible light. Ganesh et al. [30] prepared ZnO, Sb-doped ZnO, and ZnO: Sb/g-C_3N_4 nanocomposite using a simple chemical route. ZnO: Sb and g-C_3N_4, which reduced luminous intensity due to increased charge transfer, can effectively improve photocatalytic performance. The degradation efficiency of the ZnO: Sb/g-C_3N_4 sample was 86% in 60 min. The efficiency of the degradation of methylene blue (MB) (1×10^{-5} M) by ZnO: Sb/g-C_3N_4 was 86%. Based on these studies, we modified the materials by electron beam radiation to study their degradation efficiency against dyes in high concentrations.

In this study, ZnO was prepared at different annealing temperatures and its photocatalytic properties were examined. The carbon nitride photocatalysts, g-C_3N_4 (CN-X (where X is the radiation absorption dose and X = 100, 200, 300, and 400 kGy)), were prepared by direct calcination of urea and CN-200 was prepared by the electron beam radiation technique. The presence of the CN-200/Z-500 heterostructures produced a large number of catalytic active sites, enhancing the catalytic performance. It was found that 10 wt.% CN-200/Z-500 particles degraded 98.9% of MB (20 mg/L) in 120 min. The improvement of the photocatalytic performance of 10 wt.% CN-200/Z-500 nanocomposite resulted from the improvement of the separation efficiency of photoinduced electron–hole pairs, which was due to the formation of heterojunctions between CN-200 and ZnO interfaces.

2. Results and Discussion

2.1. XRD and FT-IR Analysis

The crystal phases of the samples were examined by XRD. Figure 1a–d shows the XRD patterns of g-C_3N_4, ZnO, and CN-200/Z-500. g-C_3N_4 has its strongest peak at 2θ = 27.33° corresponding to the (002) planes (Figure 1a) due to the typical interplanar stacking structures of graphitic materials. The g-C_3N_4, which was irradiated by an electron beam, shows a red shift for the reflection of the (002) crystal planes. Electron beam irradiation was beneficial in increasing the spacing between crystal planes. To investigate the influence of different annealing temperatures on the crystalline properties of ZnO, the prepared materials were investigated by XRD, as presented in Figure 1b. Obviously, the peak position did not change with annealing temperature, suggesting that the crystal form did not change after calcination due to the material temperature. The crystal forms of CN-200, Z-500, and 10 wt.% CN-200/Z-500 are shown in Figure 1c. The diffraction peak of CN-200 at 27.65° corresponds to the (002) planes (JCPDS NO. 87-1526). The high-intensity diffraction peak of CN-200 at 27.65° corresponds to the packing of the conjugated aromatic system. Z-500 exhibits prominent characteristic peaks corresponding to the hexagonal wurtzite phase at 31.77°, 34.42°, 36.25°, 47.54°, 56.60°, 62.86°, 66.38°, 67.96°, 69.10°, 72.56°, and 76.95°, corresponding to the (100), (002), (101), (102), (110), (103), (200), (112), (201), (004), and (202) planes (JCPDS NO. 36-1451), respectively. The diffraction peaks at different positions are well-matched with the wurtzite diffraction peaks. This showed that the prepared hexagonal wurtzite zinc oxide samples have high crystallinity. Figure 1d presents the FT-IR spectra of CN, CN-200, Z-500, and 10 wt.% CN-200/Z-500. The peak at 815 cm^{-1} in the spectrum of CN corresponds to the s-triazine ring system. The peaks at 1242.35 cm^{-1} and 1627.17 cm^{-1} correspond to the C–N and C=N stretching vibrations, respectively. Comparing the infrared spectra of CN and CN-200, it can be seen that the electron beam irradiation did not change the structure of graphitic carbon nitride. In the case of Z-500, the band at 438.32 cm^{-1} due to the Zn–O stretching mode indicates the formation of ZnO crystal. The FT-IR spectrum of 10 wt.% CN-200/Z-500 was similar to the characteristic spectra of both CN-200 and

Z-500, suggesting the retention of the typical graphitic structure of CN-200 after it was compounded with Z-500.

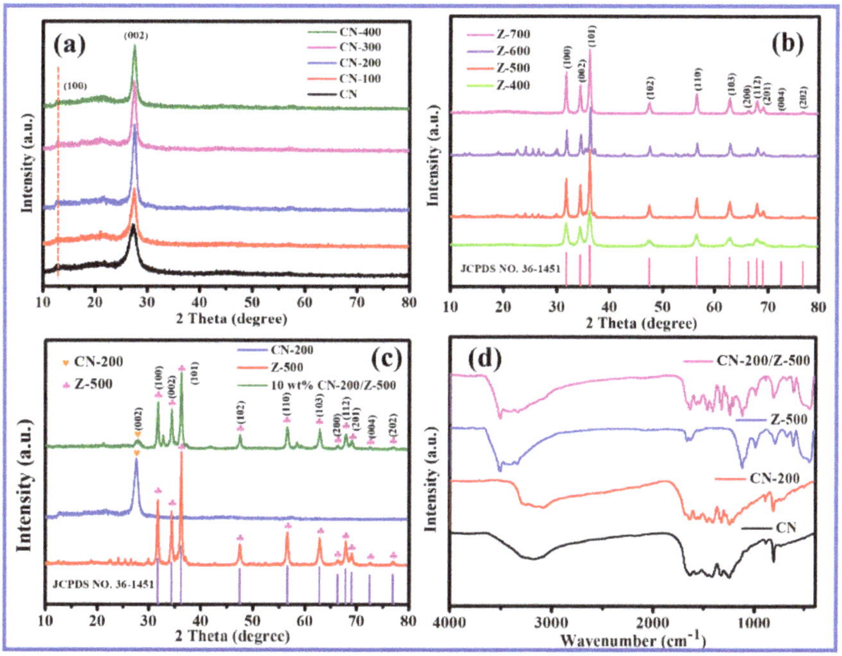

Figure 1. (**a**) XRD patterns of CN prepared under different absorbed radiation doses; (**b**) XRD patterns of ZnO prepared at different calcination temperatures; (**c**) XRD patterns of CN-200, Z-500, and 10 wt.% CN-200/Z-500; (**d**) infrared characterization of CN, CN-200, Z-500, and 10 wt.% CN-200/Z-500.

2.2. XPS Analysis

X-ray photoelectron spectroscopy (XPS) was used to analyze the chemical elements on the surface of the composite material. The XPS spectra of C 1s, N 1s, O 1s, and Zn 2p are shown in Figure 2. Three peaks were observed, which were at 288.54 eV, 287.13 eV, and 285.15 eV, respectively, corresponding to the C=N, C–N, and C–C bonds in Figure 2a. The peak at 287.13 eV belongs to the N–C=N bonds with sp^2 hybridization. The peak at 288.54 eV was assigned to sp2-bonded C–NH$_2$. As shown in Figure 2b, the N 1s spectrum of 10 wt.% CN-200/Z-500 has three characteristic peaks with binding energies of 401.19, 399.09, and 398.87 eV. The peak at 401.19 eV was attributed to C–N–H bonds, while the other peaks located at 399.09 and 398.87 eV were ascribed to N–(C)3 and C=N–H, respectively. The XPS spectrum of Zn 2p exhibits two peaks at 1045.17 and 1022.05 eV (Figure 2c), which correspond to Zn 2p$_{1/2}$ and Zn 2p$_{3/2}$, respectively. The Zn 2p peaks of 10 wt.% CN-200/Z-500 showed a negative shift in comparison with the pristine ZnO, which was reported by the authors of [31,32]. As shown in Figure 2d, three peaks were resolved from the O 1s spectrum of 10 wt.% CN-200/Z-500. The peak with binding energy of 530.50 eV was determined to belong to the lattice oxygen of Zn–O, which further verified the successful combination of Z-500 and CN-200. The other two peaks, located at 532.55 and 532.45 eV, were assigned to water molecules and surface hydroxyl, respectively.

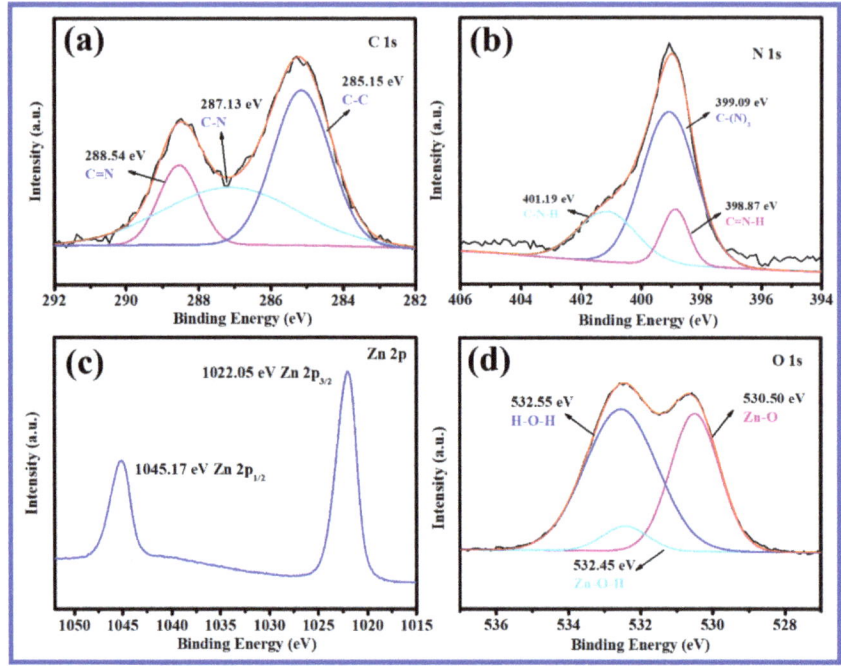

Figure 2. XPS spectra of 10 wt.% CN-200/Z-500: (**a**) C 1s; (**b**) N 1s; (**c**) Zn 2p; (**d**) O 1s.

2.3. FE-SEM and HR-TEM Analysis

The FE-SEM images of the CN, CN-200, Z-500, and CN-200/Z-500 samples are shown in Figure 3. A typical multilayer stacking structure of the CN is shown in Figure 3a,b. Compared with CN, CN-200 shows better dispersion, as shown in Figure 3c,d. Figure 3e,f indicates that Z-500 was mainly composed of non-uniform spherical particles. It is clear from Figure 3g,h that the composite material CN-200/Z-500 was formed from nanosheets of CN-200 and nanoparticles of Z-500. Figure 3g,h presents an SEM image of 10 wt.% CN-200/Z-500 photocatalyst. As seen above, the Z-500 photocatalyst was composed of irregular nanoparticles, which contributed to its low photocatalytic activity under visible light. The SEM image of the 10 wt.% CN-200/Z-500 shows that the Z-500 was evenly distributed on the surface of the CN-200, which favored the formation of heterojunctions and resulted in enhanced photocatalytic activity.

The 10 wt.% CN-200/Z-500 with the best photocatalytic activity was further analyzed by TEM and high-resolution TEM (HR-TEM). Figure 4a,b revealed typical TEM images of the composite photocatalyst 10 wt.% CN-200/Z-500. Z-500 dispersed on the surface of CN-200 can be seen in Figures 3 and 4. This observation is consistent with the XRD and SEM results.

2.4. Photocatalytic Activity

The photoluminescence (PL) emission spectra was conducted to determine the recombination rate of photoinduced charge carriers. The properties of photogenerated charge carriers between CN-200, Z-500, and 10 wt.% CN-200/Z-500 were studied using 365 nm as the excitation wavelength. As shown in Figure 5a, the PL strength of 10 wt.% CN-200/Z-500 was much weaker than that of pure CN-200, indicating that the photoinduced charge carrier recombination rate of 10 wt.% CN-200/Z-500 composite is greatly reduced. Figure 5b showed UV–visible spectra of 10 wt.% CN-200/Z-500 composites, CN-200 and Z-500. The corresponding band gaps of ZnO and g-C_3N_4 were acquired from the indirect estimation of Tauc plot $(Ahv)^n$ vs. The estimated Eg values for Z-500, CN-200, and 10 wt.% CN-200/Z-500 composites were 3.24 eV, 2.89 eV, and 3.22 eV, respectively. Z-500 had a narrower band gap after doping with CN-200, which indicated that it can absorb more visible light and show a better light response under visible light irradiation.

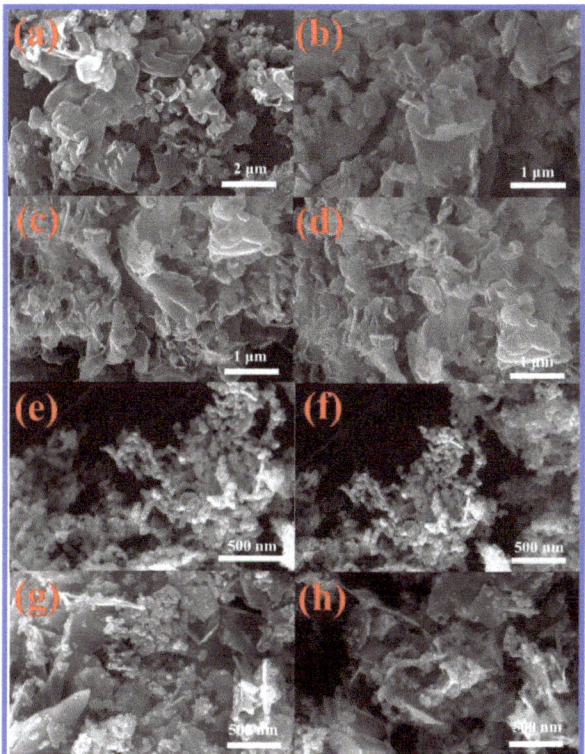

Figure 3. (**a**,**b**) SEM images of CN; (**c**,**d**) SEM images of CN-200; (**e**,**f**) SEM images of Z-500; (**g**,**h**) SEM images of CN-200/Z-500 photocatalyst.

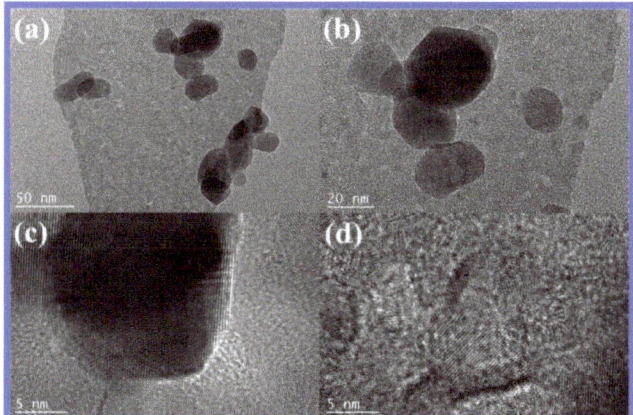

Figure 4. (**a**,**b**) TEM images of CN-200/Z-500 photocatalyst; (**c**,**d**) HR-TEM images of CN-200/Z-500 photocatalyst.

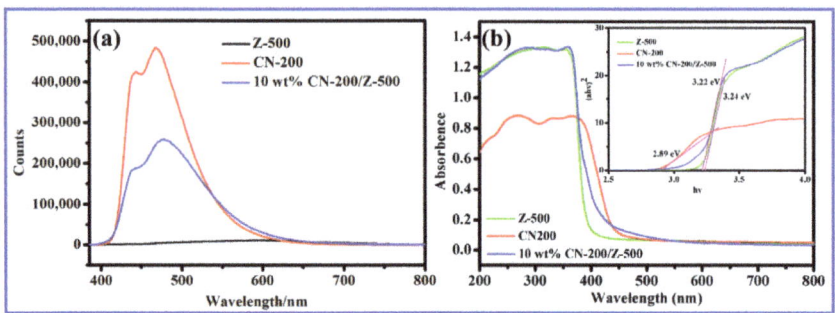

Figure 5. (a) Photoluminescence (PL) emission spectra of 10 wt.% CN-200/Z-500 composites, CN-200 and Z-500; (b) UV-Visible spectra of 10 wt.% CN-200/Z-500 composites, CN-200 and Z-500.

2.5. Photocatalytic Activity

Figure 6a,b shows the photocatalytic activities of different catalysts for MB (20 mg/L) degradation. The catalytic performance of ZnO prepared at different calcination temperatures while keeping the amount of CN-200 doping constant is shown in Figure 6a. The results showed that 10 wt.% CN-200/Z-500 exhibited the best catalytic activity. A very rapid degradation rate of MB was observed initially and then it slowed down until no more degradation occurred. As can be seen from Figure 6a, Z-500 showed better MB degradation than the other Z-T samples (T = 400, 600, and 700 °C). It is noteworthy that Z-500 and CN-200 played a crucial role in the construction of effective heterojunctions between the two components. As expected, the photocatalytic activity of CN-200/Z-500 composite was enhanced as compared to Z-500. Under visible light irradiation, the 10 wt.% CN-200/Z-500 decomposed MB solution (20 mg/L) with a high degradation rate of 98.9% within 120 min. This showed that CN-200 was well incorporated into Z-500, effectively enhancing the photocatalytic performance of the composite under visible light irradiation. Figure 6b shows the influence of different composite proportions on photocatalytic performance. It can be seen that, when the doping ratio was 10 wt.%, the photocatalytic performance was the best.

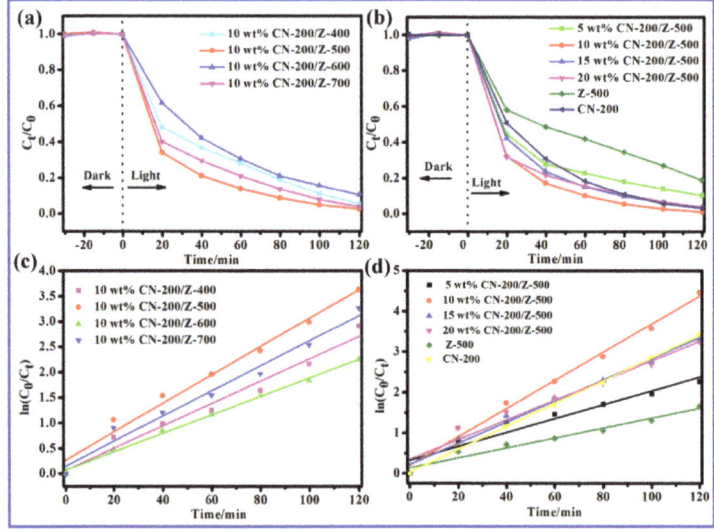

Figure 6. (a,b) Photocatalytic activity of MB degradation photocatalyst under visible light irradiation; (c,d) first-order kinetic curve of MB degradation photocatalyst under visible light irradiation.

MB dye was used to simulate pollutants and the photocatalytic activities of the catalysts were assessed by studying the MB degradation efficiency under visible light. The degradation of MB was monitored by observing the decrease in the intensity of the characteristic absorption peak at 662 nm with time. The degradation reaction kinetics of MB can be expressed quantitatively with the use of the pseudo-first-order model: $\ln(C_0/C_t) = kK_t = K't$, where C_0 refers to the original dye concentration; C_t refers to the MB concentration; K' represents the pseudo-first-order rate constant; and t represents the irradiation time. In accordance with this equation, K' can be acquired from a linear plot of $\ln(C_0/C_t)$ against t, which represents the degradation rate and is proportional to the photocatalytic degradation rate.

The kinetics of photo-oxidation/photoreduction of MB over the different samples are presented in Figure 6c,d. The MB photo-oxidation/photoreduction kinetics of different samples are shown in Figure 6c,d and K' is obtained after data fitting. Specific data are shown in Tables 1 and 2. Combined with the data analysis, the photocatalytic degradation of MB conforms to the first-order kinetic model, $R^2 > 0.92$, indicating a good degree of fit. Based on Figure 6d, the photocatalytic degradation efficiency is 98.86% for the 10 wt.% CN-200/Z-500, 80.97% for the Z-500, and 89.64% for the 5 wt.% CN-200/Z-500 after 120 min.

Table 1. Degradation rate of MB after compound of zinc oxide and CN prepared at different annealing temperatures.

	$y = \ln(C_0/C_t)$	R^2	Degradation Rate (%)
10% CN-200/Z-400	Y = 0.02194X + 0.07045	0.96537	94.6%
10% CN-200/Z-500	Y = 0.02796X + 0.27089	0.97759	97.376%
10% CN-200/Z-600	Y = 0.01819X + 0.07533	0.99565	89.529%
10% CN-200/Z-700	Y = 0.0247X + 0.15485	0.97284	96.195%

Table 2. Degradation rates of MB with different catalysts.

	$y = \ln(C_0/C_t)$	R^2	Degradation Rate (%)
5% CN-200/Z-500	Y = 0.01705X + 0.33072	0.92233	89.637%
10% CN-200/Z-500	Y = 0.03473X + 0.21363	0.98704	98.857%
15% CN-200/Z-500	Y = 0.02599X + 0.23163	0.98333	96.249%
20% CN-200/Z-500	Y = 0.02417X + 0.35889	0.95197	96.133%
Z-500	Y = 0.01223X + 0.14589	0.95744	80.971%
CN-200	Y = 0.02809X + 0.03753	0.99826	96.776%

The cyclic stability of 10 wt.% CN-200/Z-500 composite catalyst was evaluated by cyclic experiments of photocatalytic degradation of MB. The used catalyst would be centrifuged at high speed after each cycle test, washed, and vacuum-dried overnight at 60 °C for collection before use. Figure 7a showed the photocatalytic efficiency curves of 10 wt.% CN-200/Z-500 in three cycles. As shown in Figure 7a, after three cycles, the degradation rate of MB was 91.7%, while the degradation rate of the first cycle was 98.4%, indicating that the degradation efficiency of the catalyst did not decrease significantly with the progress of the photocatalytic process. The results indicated that the prepared 10 wt.% CN-200/Z-500 catalyst was stable and reusable in the photocatalytic degradation process. The trapping experiments were carried out to explore the photocatalytic mechanism. To test the role of these reactive species, Ethylenediaminetetraacetic acid disodium salt (EDTA-2Na), isopropyl alcohol (IPA), and β-benzoquinone (BQ) were employed as scavengers for h^+, •OH, and •O_2^-, respectively. As can be seen in Figure 7b, the degradation ratios for MB were all reduced after the addition of the three trapping reagents, indicating that the h^+, •OH, and •O_2^- are all responsible for the degradation process. When the EDTA-2Na was added into the reaction system, the degradation efficiency is remarkably decreased, suggesting that h^+ takes a crucial part in the degradation of organic pollutants.

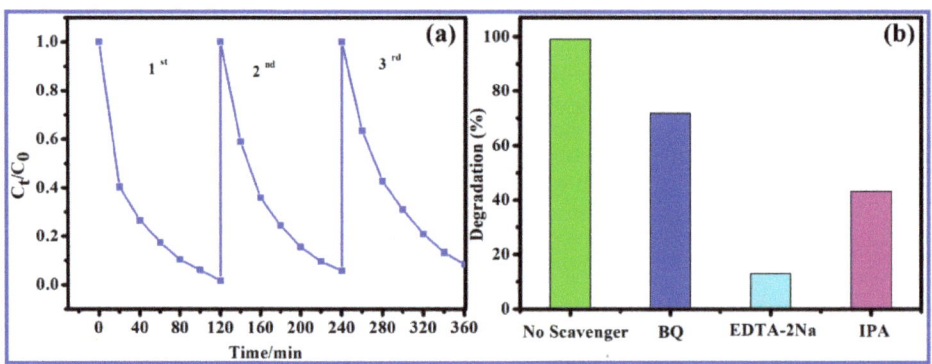

Figure 7. (a) Cycle experiments for degradation of MB over 10 wt.% CN-200/Z-500; (b) Influence of different scavengers on the degradation of MB over the 10 wt.% CN-200/Z-500.

On the basis of the results of the photodegradation and photogenerated carrier trapping test, a possible photodegradation mechanism for 10 wt.% CN-200/Z-500 was explained to clarify the enhanced photocatalytic activity of it for the degradation of MB. The degradation mechanism was shown in Figure 8. CN-200 absorbs visible light to induce Π–Π* transition, transporting the excited-state electrons from the HOMO to the lowest unoccupied molecular orbital (LUMO). The LUMO potential of CN-200 (−1.12 eV) is more negative than the conduction band (CB) edge of Z-500 (−0.5 eV), so the excited electron on CN-200 could directly inject into the CB of Z-500. Meanwhile, the holes on the VB of Z-500 transfer to the CN-200 because the VB edge potential of Z-500 was more positive than the CN-200. This can effectively restrain the recombination of photoinduced electron–hole pairs for Z-500 and CN-200. The O_2 molecules trapped photoelectron on the surface of 10 wt.% CN-200/Z-500 and generate the superoxide anion radical ($\bullet O_2^-$). Due to its high activity and disability of penetrating in bulk of the sample, $\bullet O_2^-$ mostly degrades the adsorbed MB molecules on the surface of the catalyst. Simultaneously, h^+ reacted with water molecules to produce hydroxyl radicals. The resulting $\bullet OH$ and h^+ oxidize the organic pollutants into small molecules, such as CO_2 and H_2O [33]. The photocatalytic reaction process is described below:

$$CN\text{-}200 + h\nu \rightarrow h^+ + e^-$$

$$Z\text{-}500 + h\nu \rightarrow h^+ + e^-$$

$$CN\text{-}200\ (e^-_{CB}) \rightarrow Z\text{-}500\ (e^-_{CB})$$

$$Z\text{-}500\ (h^+_{VB}) \rightarrow CN\text{-}200\ (h^+_{VB})$$

$$Z\text{-}500\ (e^-) + O_2 \rightarrow \bullet O_2^-$$

$$\bullet O_2^- + 2H^+ + e^- \rightarrow H_2O_2$$

$$H_2O_2 + e^- \rightarrow \bullet OH + OH^-$$

$$h^+ + \bullet OH + \bullet O_2^- + \text{pollutants} \rightarrow CO_2 + H_2O + \text{degradation intermediate}$$

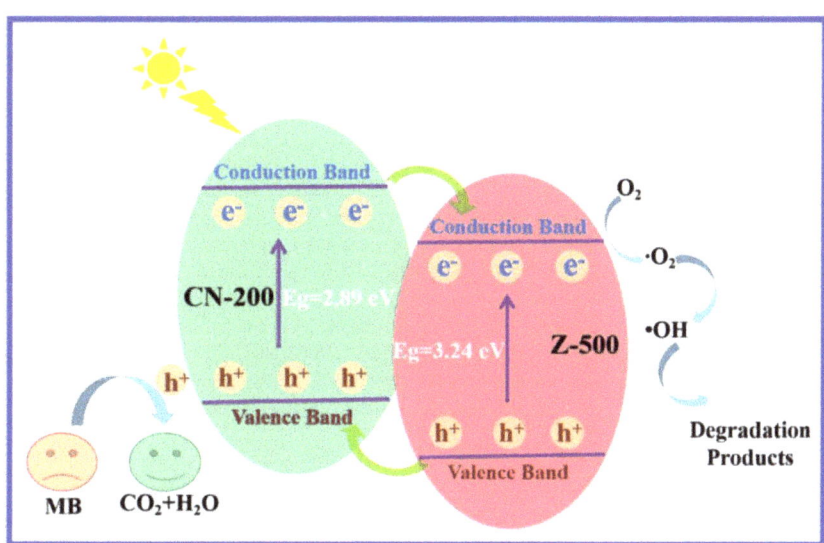

Figure 8. Schematic diagram for the proposed photocatalytic reaction mechanism of 10 wt.% CN-200/Z-500.

3. Experimental

3.1. Preparation of CN-X by Irradiation (X = 100, 200, 300, and 400 kGy)

g-C_3N_4 preparation: 20 g urea (CH_4N_2O) was calcined at 550 °C with a heating rate of 5 °C·min^{-1} for 180 min in an alumina crucible in a muffle furnace. The sample was crushed into powder after calcination.

Sample preparation for irradiation: 0.12 g g-C_3N_4, 12 mL isopropanol, 120 μL water, and 240 μL ammonia were mixed in a reagent bottle. The mixture was ultrasonically stirred for 10 min, which was repeated three times. The sample was transferred to a polyethylene bag and sealed under a vacuum. The sample was irradiated by a 1 MeV electron accelerator (1 MeV; Wasik Associates, Dracut, MA, USA) with a dose rate of 20 kGy/pass. The sample was repeatedly washed with ethanol and water, subsequently dried in an oven at 60 °C, and ground for later use. The samples with different irradiation doses of CN were denoted as CN-X, where X is the radiation absorption dose and X = 100, 200, 300, and 400 kGy, respectively.

3.2. Preparation of ZnO

ZnO was prepared using the co-precipitation method. Solution A was prepared by dissolving $ZnSO_4$ (100 mmol) and cetyltrimethylammonium bromide (CTAB, 4.3 mmol) in 125 mL of distilled water. Solution B consisted of ammonium bicarbonate (50 mmol) dissolved in 125 mL of distilled water. Solution B was then quickly poured into solution A, sonicated, and stirred. A gray precipitate resulted, which was then stirred for 60 min. The sample was washed several times with ethanol and water and then dried at 80 °C. Finally, it was calcined for 3 h at the rate of 2 °C·min^{-1} at 400, 500, 600, and 700 °C, respectively. The samples were labeled as Z-T (T = 400, 500, 600, and 700 °C).

3.3. Synthesis of CN-200/Z-500

During the earlier studies, it was found that the best photocatalytic efficiency for C_3N_4/ZnO was obtained when the annealing temperature was 500 °C and the radiation absorption dose was 200 kGy. Therefore, CN-200 and Z-500 were compounded together by the grinding method. CN-200 (50 mg) was added to the desired mass of Z-500. During the grinding process, 5 mL of absolute ethanol was added every 5 min, and after grinding for 30 min, the mixture was transferred to an oven at 60 °C and dried for 24 h.

3.4. Characterization

The crystalline samples were characterized by an X-ray diffractometer (XRD, DMAX-D8X; Rigaku, Tokyo, Japan). The dye degradation was studied by using an ultraviolet-visible (UV-Vis) (TU-1950; Persee, suzhou, China). The morphologies of the materials were analyzed by field-emission scanning electron microscopy (FE-SEM, SU8220; Tokyo, Japan) and field emission transmission electron microscopy (FE-TEM, FEI Tecnai G2 F30; Hillsboro, OR, USA). KBr pellets of the samples were made to analyze the samples using Fourier-transform infrared spectroscopy (FT-IR) (Thermo Ferret iS10 infrared spectrometer, Thermo Fisher Scientific, Waltham, MA, USA). Photoluminescence spectra (PL) of samples was recorded on an Edinburgh FLS1000 fluorescent spectrophotometer (UK). Ultraviolet–visible diffuse reflectance spectra (UV-Vis DRS; Japan) was performed on a Shimadzu, UV-3600i Plus UV-Vis spectrometer with a scan range of 200–800 nm.

3.5. Photocatalytic Experiments

The photocatalytic activities of the samples were evaluated by using the dye MB as the model. All experiments were performed in an open system at room temperature. The photocatalyst (50 mg) was dispersed in 50 mL of MB solution (20 mg/L) and stirred for 30 min in dark circumstances to achieve adsorption–desorption equilibrium between the photocatalyst and the MB dye in the photocatalysis study.

The catalysis reaction was performed under visible light by taking out 2 mL of solution from each sample suspension every 20 min. High-speed centrifugation was employed to remove photocatalysts. The concentrations of MB at various reaction times were measured with a UV-Vis spectrophotometer at λmax = 662 nm, and the degradation rate (%) was calculated using the following equation:

$$\text{Degradation Rate } (\%) = \frac{C_0 - C_t}{C_t} \times 100\% = \frac{A_0 - A_t}{A_0} \times 100\% \quad (1)$$

Here, C_t and C_0 represent pollutant concentrations at times t and t_0, respectively. A_0 and A_t are the absorbances of the MB solution before irradiation and after irradiation, respectively.

4. Conclusions

In this study, ZnO was prepared at different annealing temperatures and its photocatalytic properties were examined. The best photocatalytic degradation of 20 mg/L MB dye was achieved by ZnO prepared at 500 °C under visible light radiation. g-C_3N_4 photocatalysts were prepared by direct calcination of urea and CN-200 was prepared by irradiation with electron beams. The radiation dose of CN-200 was 200 kGy. A series of CN-200 doped ZnO hetero-structured photocatalysts were successfully prepared using the grinding method. The samples were characterized using XRD, FT-IR, FE-SEM, and HR-TEM. MB was chosen as the photodegradation target, and the photocatalytic activity of the composites was evaluated. Interestingly, 10 wt.% CN-200/Z-500 particles degraded 98.9% of the MB in 120 min, owing to the presence of a large number of catalytic active sites in CN-200/Z-500 heterostructures that enhanced its catalytic performance. Thus, the 10 wt.% CN-200/Z-500 has great potential to degrade pollutants. Thus, it can be concluded that electron beam irradiation technology can be effectively used for the modification of photocatalytic materials and the improvement of their photocatalytic performance.

Author Contributions: Conceptualization and methodology, Y.W. and X.Y.; validation, Y.W. and J.L.; formal analysis and investigation, Y.H.; writing—original draft preparation, X.Y.; writing—review and editing, J.P. and Y.L. (Yi Liu); provided technical and financial support, Y.L. (Yuesheng Li); proposed the guiding support and the final review of the paper, Y.L. (Yi Liu). All authors have read and agreed to the published version of the manuscript.

Funding: This work was supported by the Xianning City Program of Science & Technology (No. 2022ZRKX051), the Hubei University of Science and Technology Doctoral Research Initiation Project (No. BK202217), the Science Development Foundation of Hubei University of Science & Technology (No. 2021F005, No. 2021ZX14, Nos. 2020TD01, 2021ZX01, 2022FH09), the Hubei Provincial Colleges and Universities Outstanding Young and Middle-aged Technological Innovation Team Project (No. T2020022), and Xianning City Key Program of Science & Technology (No. 2021GXYF021).

Institutional Review Board Statement: Not applicable.

Informed Consent Statement: Not applicable.

Data Availability Statement: Not applicable.

Conflicts of Interest: The authors declare no conflict of interest.

Sample Availability: Samples of the compounds are available from the authors.

References

1. Zhang, Z.Y.; Sun, Y.S.; Wang, Y.L.; Yang, Y.; Wang, P.P.; Shi, L.F.; Feng, L.; Fang, S.Q.; Liu, Q.; Ma, L.Y.; et al. Synthesis and photocatalytic activity of g-C_3N_4/ZnO composite microspheres under visible light exposure. *Ceram. Int.* **2022**, *3*, 3293–3302. [CrossRef]
2. Liu, W.; Wang, M.; Xu, C.; Chen, S. Facile synthesis of g-C_3N_4/ZnO composite with enhanced visible light photooxidation and photoreduction properties. *Chem. Eng. J.* **2012**, *209*, 386–393. [CrossRef]
3. Wang, J.; Yang, Z.; Gao, X.X.; Yao, W.Q.; Wei, W.Q.; Chen, X.J.; Zong, R.L.; Zhu, Y.F. Core-shell g-C_3N_4@ZnO composites as photoanodes with double synergistic effects for enhanced visible-light photoelectrocatalytic activities. *Appl. Catal. B Environ.* **2017**, *217*, 169–180. [CrossRef]
4. Le, A.T.; Duy, H.L.T.; Cheong, K.Y.; Pung, S.Y. Immobilization of zinc oxide-based photocatalysts for organic pollutant degradation: A review. *J. Environ. Chem. Eng.* **2022**, *10*, 108505. [CrossRef]
5. Golli, A.E.; Fendrich, M.; Bazzanella, N.; Dridi, C.; Miotello, A.; Orlandi, M. Wastewater remediation with ZnO photocatalysts: Green synthesis and solar concentration as an economically and environmentally viable route to application. *J. Environ. Manag.* **2021**, *286*, 112226. [CrossRef]
6. Wang, H.J.; Li, X.; Zhao, X.X.; Li, C.Y.; Song, X.H.; Zhang, P.; Huo, P.W.; Li, X. A review on heterogeneous photocatalysis for environmental remediation: From semiconductors to modification strategies. *Chin. J. Catal.* **2022**, *43*, 178–214. [CrossRef]
7. Wang, H.P.; Yang, Y.L.; Zhou, Z.W.; Li, X.; Gao, J.F.; Yu, R.; Li, J.Q.; Wang, N.; Chang, H.Q. Photocatalysis-enhanced coagulation for removal of intracellular organic matter from Microcystis aeruginosa: Efficiency and mechanism. *Sep. Purif. Technol.* **2022**, *283*, 120192. [CrossRef]
8. Cheung, K.P.S.; Sarkar, S.; Gevorgyan, V. Visible light-induced transition metal catalysis. *Chem. Rev.* **2022**, *122*, 1543–1625. [CrossRef]
9. Lincho, J.; Zaleska-Medynska, A.; Martins, R.C.; Gomes, J. Nanostructured photocatalysts for the abatement of contaminants by photocatalysis and photocatalytic ozonation: An overview. *Sci. Total Environ.* **2022**, *837*, 155776. [CrossRef]
10. Qumar, U.; Hassan, J.Z.; Bhatti, R.A.; Raza, A.; Nazir, G.; Nabgan, W.; Ikram, M. Photocatalysis vs adsorption by metal oxide nanoparticles. *J. Mater. Sci. Technol.* **2022**, *131*, 122–166. [CrossRef]
11. Su, G.W.; Feng, T.Y.; Huang, Z.J.; Zheng, Y.N.; Zhang, W.X.; Liu, G.Z.; Wang, W.; Wei, H.Y.; Dang, L.P. MOF derived hollow CuO/ZnO nanocages for the efficient and rapid degradation of fluoroquinolones under natural sunlight. *Chem. Eng. J.* **2022**, *436*, 135119. [CrossRef]
12. Sanakousar, F.M.; Vidyasagar, C.C.; Jiménez-Pérez, V.M.; Prakash, K. Recent progress on visible-light-driven metal and non-metal doped ZnO nanostructures for photocatalytic degradation of organic pollutants. *Mater. Sci. Semicond. Process.* **2022**, *140*, 106390. [CrossRef]
13. Sirelkhatim, A.; Mahmud, S.; Seeni, A.; Kaus, N.H.M.; Ann, L.C.; Bakhori, S.K.M.; Hasan, H.; Mohamad, D. Review on zinc oxide nanoparticles: Antibacterial activity and toxicity mechanism. *Nano-Micro Lett.* **2015**, *7*, 219–242. [CrossRef] [PubMed]
14. Tanji, K.; Navio, J.A.; Chaqroune, A.; Naja, J.; Puga, F.; Hidalgo, M.C.; Kherbeche, A. Fast photodegradation of rhodamine B and caffeine using ZnO-hydroxyapatite composites under UV-light illumination. *Catal. Today* **2022**, *388–389*, 176–186. [CrossRef]
15. Meng, F.P.; Liu, Y.Z.; Wang, J.; Tan, X.Y.; Sun, H.Q.; Liu, S.M.; Wang, S.B. Temperature dependent photocatalysis of g-C_3N_4, TiO_2 and ZnO: Differences in photoactive mechanism. *J. Colloid Interface Sci.* **2018**, *532*, 321–330. [CrossRef]
16. Naseri, A.; Samadi, M.; Pourjavadi, A.; Ramakrishna, S.; Moshfegh, A.Z. Enhanced photocatalytic activity of ZnO/g-C_3N_4 nanofibers constituting carbonaceous species under simulated sunlight for organic dye removal. *Ceram. Int.* **2021**, *47*, 26185–26196. [CrossRef]
17. Li, N.; Tian, Y.; Zhao, J.H.; Zhang, J.; Zuo, W.; Kong, L.C.; Cui, H. Z-scheme 2D/3D g-C_3N_4@ZnO with enhanced photocatalytic activity for cephalexin oxidation under solar light. *Chem. Eng. J.* **2018**, *352*, 412–422. [CrossRef]
18. Zhang, Q.; Chen, J.; Che, H.N.; Wang, P.F.; Liu, B.; Ao, Y.H. Recent advances in g-C_3N_4-based donor-acceptor photocatalysts for photocatalytic hydrogen evolution: An exquisite molecular structure engineering. *ACS Mater. Lett.* **2022**, *4*, 2166–2186. [CrossRef]

19. Wang, L.Y.; Wang, K.H.; He, T.T.; Zhao, Y.; Song, H.; Wang, H. Graphitic carbon nitride-based photocatalytic materials: Preparation strategy and application. *ACS Sustain. Chem. Eng.* **2020**, *8*, 16048–16085. [CrossRef]
20. Tang, C.S.; Cheng, M.; Lai, C.; Li, L.; Yang, X.F.; Du, L.; Zhang, G.X.; Wang, G.F.; Yang, L. Recent progress in the applications of non-metal modified graphitic carbon nitride in photocatalysis. *Coordin. Chem. Rev.* **2023**, *474*, 214846. [CrossRef]
21. Bai, L.Q.; Huang, H.W.; Yu, S.X.; Zhang, D.Y.; Huang, H.T.; Zhang, Y.H. Role of transition metal oxides in g-C_3N_4-based heterojunctions for photocatalysis and supercapacitors. *J. Energy Chem.* **2022**, *64*, 214–235. [CrossRef]
22. Lin, B.; Xia, M.Y.; Xu, B.R.; Chong, B.; Chen, Z.H.; Yang, G.D. Bio-inspired nanostructured g-C_3N_4-based photocatalysts: A comprehensive review. *Chin. J. Catal.* **2022**, *43*, 2141–2172. [CrossRef]
23. Zhang, Y.L.; Zhang, H.J.; Cheng, L.L.; Wang, Y.J.; Miao, Y.; Ding, G.J.; Jiao, Z. Two physical strategies to reinforce a nonmetallic photocatalyst, g-C_3N_4: Vacuum heating and electron beam irradiation. *RSC Adv.* **2016**, *6*, 14002–14008. [CrossRef]
24. Jin, C.; Li, W.; Chen, Y.; Li, R.; Huo, J.B.; He, Q.Y.; Wang, Y.Z. Efficient photocatalytic degradation and adsorption of tetracycline over type-II heterojunctions consisting of ZnO nanorods and K-doped exfoliated g-C_3N_4 nanosheets. *Ind. Eng. Chem. Res.* **2020**, *59*, 2860–2873. [CrossRef]
25. Liu, Y.J.; Jin, Y.L.; Cheng, X.X.; Ma, J.Y.; Li, L.L.; Fan, X.X.; Ding, Y.; Han, Y.; Tao, R. K+-Doped ZnO/g-C_3N_4 heterojunction: Controllable preparation, efficient charge separation, and excellent photocatalytic VOC degradation performance. *Ind. Eng. Chem. Res.* **2022**, *61*, 187–197. [CrossRef]
26. Jiang, X.L.; Wang, W.T.; Wang, H.; He, Z.H.; Yang, Y.; Wang, K.; Liu, Z.T.; Han, B.X. Solvent-free aerobic photocatalytic oxidation of alcohols to aldehydes over ZnO/C_3N_4. *Green Chem.* **2022**, *24*, 7652–7660. [CrossRef]
27. Thuan, D.V.; Nguyen, T.B.H.; Pham, T.H.; Kim, J.; Chu, T.T.H.; Nguyen, M.V.; Nguyen, K.D.; Al-Onazi, W.A.; Elshikh, M.S. Photodegradation of ciprofloxacin antibiotic in water by using ZnO-doped g-C3N4 photocatalyst. *Chemosphere* **2022**, *308*, 136408. [CrossRef]
28. Sun, Q.; Sun, Y.; Zhou, M.Y.; Cheng, H.M.; Chen, H.Y.; Dorus, B.; Lu, M.; Le, T. A 2D/3D g-C_3N_4/ZnO heterojunction enhanced visible-light driven photocatalytic activity for sulfonamides degradation. *Ceram. Int.* **2022**, *48*, 7283–7290. [CrossRef]
29. Ding, C.M.; Zhu, Q.R.; Yang, B.; Petropoulos, E.; Xue, L.H.; Feng, Y.F.; He, S.Y.; Yang, L.Z. Efficient photocatalysis of tetracycline hydrochloride (TC-HCl) from pharmaceutical wastewater using AgCl/ZnO/g-C_3N_4 composite under visible light: Process and mechanisms. *J. Environ. Sci.* **2023**, *126*, 249–262. [CrossRef]
30. Ganesh, V.; Yahia, I.S.; Chidhambaram, N. Facile synthesis of ZnO:Sb/g-C_3N_4 composite materials for photocatalysis applications. *J. Clust. Sci.* **2022**, 1–10. [CrossRef]
31. Li, X.F.; Li, M.; Yang, J.H.; Li, X.Y.; Hu, T.J.; Wang, J.S.; Sui, Y.R.; Wu, X.T.; Kong, L.N. Synergistic effect of efficient adsorption g-C_3N_4/ZnO composite for photocatalytic property. *J. Phys. Chem. Solids* **2014**, *75*, 441–446. [CrossRef]
32. Nie, N.; Zhang, L.Y.; Fu, J.W.; Cheng, B.; Yu, J.G. Self-assembled hierarchical direct Z-scheme g-C_3N_4/ZnO microspheres with enhanced photocatalytic CO_2 reduction performance. *Appl. Surf. Sci.* **2018**, *441*, 12–22. [CrossRef]
33. Wang, Y.J.; Shi, R.; Lin, J.; Zhu, Y.F. Enhancement of photocurrent and photocatalytic activity of ZnO hybridized with graphite-like C_3N_4. *Energy Environ. Sci.* **2011**, *4*, 2922–2929. [CrossRef]

Article

Z-Scheme CuO$_x$/Ag/TiO$_2$ Heterojunction as Promising Photoinduced Anticorrosion and Antifouling Integrated Coating in Seawater

Xiaomin Guo [1], Guotao Pan [1], Lining Fang [2], Yan Liu [1,*] and Zebao Rui [1,*]

[1] School of Chemical Engineering and Technology, Guangdong Engineering Technology Research Center for Platform Chemicals from Marine Biomass and Their Functionalization, Sun Yat-sen University, Zhuhai 519082, China
[2] Department of Environmental Science, Hebei University of Environmental Engineering, Qinhuangdao 066102, China
* Correspondence: liuyan96@mail.sysu.edu.cn (Y.L.); ruizebao@mail.sysu.edu.cn (Z.R.)

Abstract: In the marine environment, steel materials usually encounter serious problems with chemical or electrochemical corrosion and fouling by proteins, bacteria, and other marine organisms. In this work, a green bifunctional Z-scheme CuO$_x$/Ag/P25 heterostructure coating material was designed to achieve the coordination of corrosion prevention and antifouling by matching the redox potential of the reactive oxygen species and the corrosion potential of 304SS. When CuO$_x$/Ag/P25 heterostructure was coupled with the protected metal, the open circuit potential under illumination negatively shifted about 240 mV (vs. Ag/AgCl) and the photoinduced current density reached 16.6 μA cm^{-2}. At the same time, more reactive oxygen species were produced by the Z-shape structure, and then the photocatalytic sterilization effect was stronger. Combined with the chemical sterilization of Ag and the oxide of Cu, the bacterial survival rate of CuO$_x$/Ag/P25 was low (0.006%) compared with the blank sample. This design provides a strategy for developing green dual-functional coating materials with photoelectrochemical anticorrosion and antifouling properties.

Keywords: antibacterial effect; antifouling effect; CuO$_x$/Ag/P25; photoelectrochemical cathodic protection; bifunctional coatings

1. Introduction

Metal corrosion and fouling in the marine environment are very serious because of the presence of oxygen, Cl$^-$ ions, sunlight, and micro-organisms [1–3], causing huge economic losses and significant harm to the human living environment. Corrosion generally results from chemical or electrochemical reactions between the metal and oxygen, while the metal surface submerged in seawater or in the humid marine atmosphere is often fouled by marine organisms such as proteins, bacteria, algae, and mollusks, which usually increase the metal weight and accelerate material surface damage [4,5].

Various methods, such as coatings [6], corrosion inhibitors [7], and cathodic protection [8], have been proposed to help reduce metal corrosion in the marine environment. However, these traditional anticorrosion techniques usually suffer from the high cost of resources or additional energy, together with the problem of environmental pollution. Photoelectrochemical cathodic protection (PECCP) technology, which couples a semiconductor material with the protected metal, is a new and green anticorrosion technology [1,3,9]. During the PECCP process, under illumination the semiconductor material generates photoelectrons, which are transferred to the protected metal surface to make the metal potential more negative than its corrosion potential. The typical semiconductor materials, TiO$_2$ [2,10,11], ZnO [12], and SrTiO$_3$ [9,13], have been evaluated for the PECCP process. Among them, TiO$_2$ is one of the most widely used photoelectric anode materials, and it has the advantages of high photoelectrochemical activity, low cost, high stability, and nontoxicity [14,15]. However, TiO_2 has a low utilization rate of sunlight (<5%), with a relatively wide band gap (~3.2 eV) [16] and a fast recombination rate of photogenerated carriers, which seriously limits its application in the PECCP field. Currently, forming a heterojunction by coupling narrow-band-gap metal oxides with TiO_2 is an effective way to improve the absorption of visible light and promote the separation of electron/hole pairs [17,18]. Tian et al. [3] deposited Cu_2O nanoparticles on the surface of TiO_2 nanotubes to prepare the Cu_2O/TiO_2 p-n heterojunction composite photoelectrodes to accelerate the photogenerated carriers' separation and improve PECCP performance.

Regarding the protection against microbial corrosion, the addition of antibacterial components to the protective paints or coatings has been commonly used to achieve antifouling and antibacterial effects [19]. Although many heavy metals and rare earth elements, such as Cu, Ag, Ce, La, etc., have a strong bactericidal effect [20–22], their adverse effects on the environment need to be considered [23]. It is reported that some semiconductors can be also used for photocatalytic sterilization and antifouling, which is correlated with the generation of reactive oxygen species (ROS, such as superoxide radicals $\bullet O_2^-$ and hydroxyl radicals $\bullet OH$) under irradiation and their inhibition to the growth of bacteria [24]. Wang et al. [24] reported that $AgVO_3/BiO_{2-x}$ inactivated bacteria in full spectrum, which is due to a large amount of ROS caused by internal structural defects and the formation of heterostructures. Moreover, the integration of chemical sterilization with photocatalytic sterilization has become a key research goal in the antimicrobial field [25–27]. Yang et al. [25] designed Cu_2O/Ag composites with strong and long-term antibacterial activities.

Apparently, both the PECCP performance and photocatalytic antibacterial properties of semiconductors are related to the light absorption, photoinduced charges separation, and the conduction/valence-band positions. In this sense, we herein design a bifunctional $CuO_x/Ag/TiO_2$ Z-type heterojunction, which is anticipated as coating material for metal anticorrosion/antifouling in the marine environment on the basis of the following considerations. (i) Both CuO_x and TiO_2 hold higher conduction-band positions against the potential of stainless steel (e.g., 304SS) and make the transfer of photoinduced electrons to the metal for PECCP possible. (ii) The combination of Ag nanoparticles (NP) with a local surface plasmon resonance (LSPR) effect and small-band-gap CuO_x can improve the visible light absorption of TiO_2 [28,29]. (iii) The unique Z-type $CuO_x/Ag/TiO_2$ heterojunction provides an effective charge transfer path for realizing the efficient separation of electron-hole pairs without sacrificing redox ability [30]. (iv) The bacteria, viruses, and other micro-organisms can be inhibited or chemically killed by Ag and CuO_x and by active ROS groups produced by photoinduced electrons and holes from the heterojunction [25–27,31,32]. As expected, the as-designed $CuO_x/Ag/TiO_2$ not only well demonstrates photocathodic protection performance for 304SS in the simulated seawater but also possesses chemical and photocatalytic synergistic bactericidal activities. Such a green bifunctional coating material shows great promise in metal anticorrosion and antifouling applications in the marine environment.

2. Results and Discussion

2.1. Structures, Compositions, and Morphologies

XRD measurements were employed to study the crystalline structures of samples. As shown in Figure 1a, the diffraction peaks for P25 could be assigned to anatase and rutile TiO_2 (JCPDS 84-1285 and JCPDS 86-0148). The diffraction peak intensity at $2\theta = 25.3°$ was strong, and the peak shape was symmetrical, indicating that the crystallinity of the material is very good. The reference material CuO_x presented characteristic diffraction peaks at $2\theta = 36.4$ and $42.3°$, corresponding to the (111) and (200) crystal planes of the Cu_2O phase of cubic hematite (JCPDS 05-0667). Compared with P25 and CuO_x, the diffraction peaks for $CuO_x/Ag/P25$ remained unchanged, indicating that CuO_x and Ag did not affect the crystal structure of P25, which may have been due to the small load, the small particle size, and the low crystallinity of CuO_x and Ag [3,33]. The morphologies and nanostructures of CuO$_x$/Ag/P25 were analyzed by TEM and HRTEM. As shown in Figure 1b, the lattice spacings of 2.5 and 3.6 Å corresponded to the (101) crystal plane of anatase TiO$_2$ and the (101) crystal plane of rutile TiO$_2$, respectively. The lattice spacings of 2.1 and 2.4 Å were ascribed to the (100) and (111) crystal planes of Cu$_2$O [3,16,34]. The (200) crystal plane of Ag with a lattice spacing of about 2.08 Å was found at the interface between the CuO$_x$ phase and the P25 phase [35]. Figure 1c shows that CuO$_x$/Ag/P25 was mainly composed of irregular particles with a diameter of 20~25 nm. The EDS element distribution mapping in Figure 1d exhibited that the Ti, O, Ag, and Cu elements were evenly distributed in CuO$_x$/Ag/P25. The corresponding element content is listed in Figure 1i.

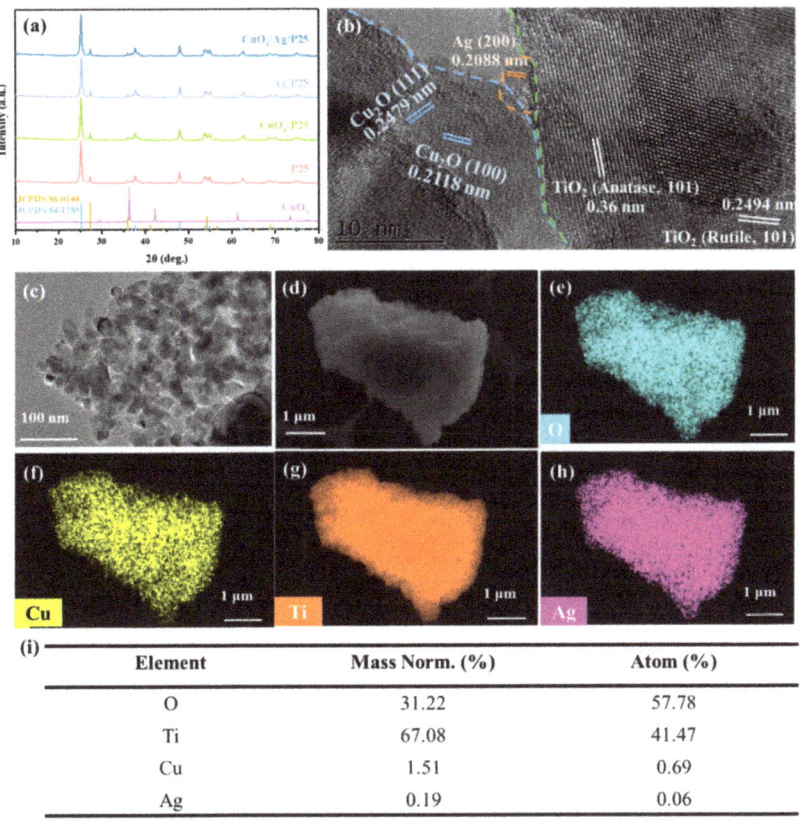

Element	Mass Norm. (%)	Atom (%)
O	31.22	57.78
Ti	67.08	41.47
Cu	1.51	0.69
Ag	0.19	0.06

Figure 1. (a) XRD patterns of P25, CuO$_x$, CuO$_x$/P25, Ag/P25, and CuO$_x$/Ag/P25; (b) HRTEM and (c) TEM images of CuO$_x$/Ag/P25; (d–h) EDS element mapping images of CuO$_x$/Ag/P25; (i) list of element content of CuO$_x$/Ag/P25.

The atomic valence states and energy-band structures of CuO$_x$/P25, Ag/P25, and CuO$_x$/Ag/P25 were further studied by XPS. The Ti 2p spectra for P25 in Figure 2a showed a pair of spin-orbital doublets at ~464.2 and ~458.5 eV, corresponding to Ti 2p$_{1/2}$ and Ti 2p$_{3/2}$ of Ti^{4+}, respectively [3,28,36]. Compared with P25, the peaks for CuO$_x$/Ag/P25 showed a shift to a higher binding energy, which was possibly due to the strong interaction between CuO$_x$, Ag, and P25 in the heterojunction structure of CuO$_x$/Ag/P25 [37]. The O1s spectra of the samples shown in Figure 2b can be divided into four types of peaks in total. Peak A, around 529.8 eV, belonged to the oxide peak, which was related to the lattice oxygen in TiO$_2$. The slightly shifted peak B, at ~530.9 eV, also belonged to the metal (Ag and Cu) oxide peak, which was attributed to the O atom near the oxygen vacancy [38].

Peak C, around ~532 eV, belonged to the -OH group, chemically adsorbed on the material surface [39], and peak D, around ~533 eV, belonged to adsorbed H_2O molecules on the surface [40]. Figure 2c shows that the Ag 3d spectra of Ag/P25 and CuO_x/Ag/P25 can be deconvoluted into two peaks, Ag $3d_{3/2}$ (373.5 eV) and Ag $3d_{5/2}$ (367.5 eV), indicating the existence of Ag metal in the material [30]. In the Cu 2p spectra for CuO_x (Figure 2d), the peaks with binding energies of 934 and 932.6 eV can belong to the Cu^{2+} [41] and Cu^+ [42] species, respectively. In addition, the satellite peaks corresponding to Cu^{2+} species can be seen between 940 and 945 eV. This indicates that Cu^+ and Cu^{2+} coexisted in CuO_x. For CuO_x/P25 and CuO_x/Ag/P25, only peaks at 932.6 eV corresponding to the binding state of Cu^+ species were observed, which demonstrates that the strong interaction between the carrier (P25) and the load (CuO_x and Ag) made the Cu species stable.

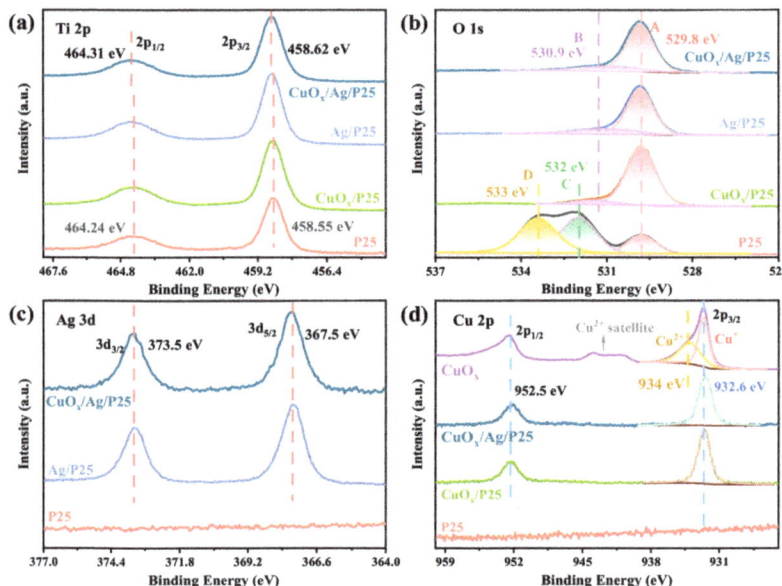

Figure 2. XPS spectra of the samples: (**a**) Ti 2p, (**b**) O 1s, (**c**) Ag 3d, and (**d**) Cu 2p.

2.2. Photoelectric Characterization

UV-visible diffuse reflectance spectra (or UV-vis DRS) were measured to investigate the light absorption of samples. Figure 3a showed a typical TiO_2 light-absorption range. The light-absorption range of as-prepared CuO_x/Ag/P25 was extended to the visible region with an edge of about 436 nm after loading CuO_x and/or Ag. At the same time, it can be observed that Ag/P25 and CuO_x/Ag/P25 had strong and wide absorption peaks near 550 nm, which were related to the LSPR effect of Ag NPs [43,44]. Figure 3a shows the band-gap results of the materials. CuO_x/Ag/P25 holds the narrowest band gap (2.17 eV) compared with P25 (3.01 eV), CuO_x/P25 (2.87 eV) and Ag/P25 (2.49 eV). The photoluminescence (PL) spectra of the catalysts at an excitation wavelength of 550 nm at room temperature were displayed in Figure 3b. The fluorescence peak intensity for CuO_x/Ag/P25 was lower than that of other compound materials, indicating the lowest recombination rate of e^-/h^+ pairs. The fluorescence intensity order was CuO_x/Ag/P25 < Ag/P25 < CuO_x/P25 < P25, which suggested that having Ag NP as the electron bridge promoted the separation and emigration of photogenerated electron-hole pairs [30,45]. By plotting the Kubelka–Munk function against the photon energy, the band0gap (E_g) values of P25 and CuO_x were estimated to be 3.01 and 2.05 eV, respectively (Figure 3c,d). Moreover, the E_g values of the composites, such as Ag/P25, CuO_x/P25, and CuO_x/Ag/P25, became narrower after loading CuO$_x$ and Ag, which confirmed the broadened light-response range. According to the XPS valence-band (or XPS-VB) spectra in Figure 3e,f, the corresponding $E_{VB,XPS}$ of P25 and CuO$_x$ were 2.55 and 1.5 eV, respectively. The VB positions of P25 and CuO$_x$ were calculated as 2.31 eV and 1.26 eV, respectively, by using the following equation: $E_{VB,NHE} = \varphi + E_{VB,XPS} - 4.44$, where φ is the work function of the instrument (4.2 eV [46]).

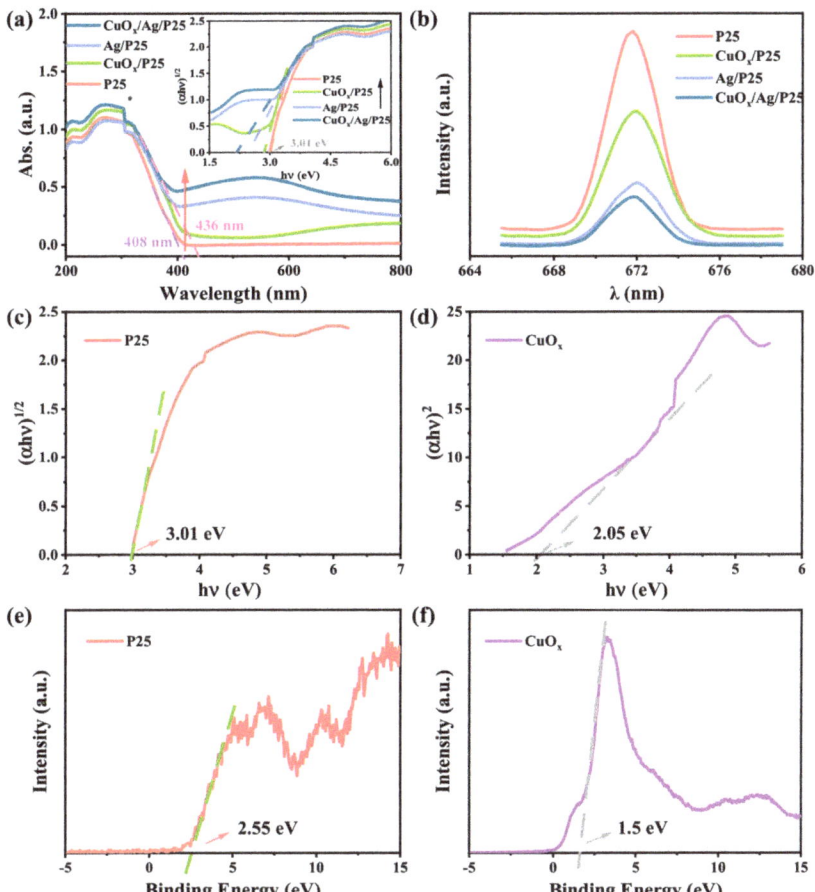

Figure 3. (**a**) UV−vis DRS, dependence of $(\alpha h\nu)^2$ on the photon energy (the inserts) and (**b**) photoluminescence spectra (λ_{ex} = 550 nm) of P25, CuO$_x$/P25, Ag/P25, and CuO$_x$/Ag/P25 at room temperature. Dependence of $(\alpha h\nu)^2$ on the photon energy for (**c**) P25 and (**d**) CuO$_x$. XPS valence-band spectra of (**e**) P25 and (**f**) CuO$_x$.

The electron paramagnetic resonance (EPR) spectra of free radical adducts trapped by DMPO in the dark and under irradiation at room temperature for the catalysts are shown in Figure 4. The EPR signals for DMPO−•OH and DMPO−•O$_2^-$ were almost negligible on P25, CuO$_x$/P25, Ag/P25, and CuO$_x$/Ag/P25 in darkness. Both DMPO−•OH and DMPO−•O$_2^-$ signals with four characteristic peaks with intensity ratios of 1:2:2:1 and 1:1:1:1 [47], respectively, can be observed in the suspension under simulated solar irradiation for 12 min. EPR signals of DMPO−•OH and DMPO−O$_2$•$^-$ were different in intensity. The intensity order of DMPO−•O$_2^-$ signals under irradiation was CuO$_x$/Ag/P25 > Ag/P25 > CuO$_x$/P25 > P25; the intensity order of DMPO−•OH signals under irradiation was CuO$_x$/Ag/P25 > Ag/P25 ≈ CuO$_x$/P25 > P25. CuO$_x$/Ag/P25 showed the highest intensity in the characteristic peaks of DMPO−•OH and DMPO−•O$_2^-$ when compared with those materials.

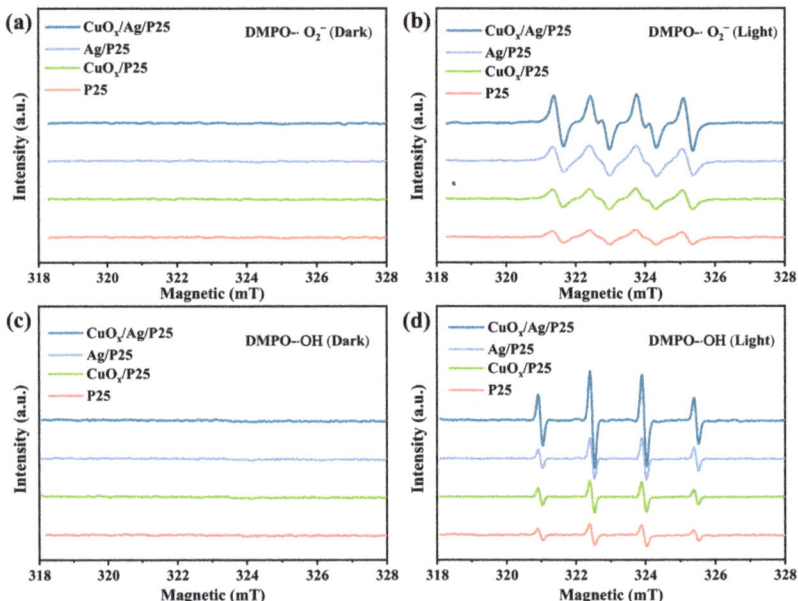

Figure 4. EPR spin-trapping spectra of DMPO−•O$_2^-$ and DMPO−•OH adducts (**a,c**) in the dark and (**b,d**) under illumination, respectively, for P25, CuO$_x$/P25, Ag/P25, and CuO$_x$/Ag/P25. The light used was full-spectrum light.

2.3. Photoelectrochemical Cathodic Protection Performance Evaluation

In general, photogenerated cathodic protection works by supplying electrons to the protected metal through an external photosensitive semiconductor, causing the metal potential to be more negative than the original corrosion potential under irradiation. Thus, the photoinduced OCP value is an important parameter to evaluate the PECCP performance of photoanode materials. Figure 5a shows the photogenerated OCP-time curves of samples. The photoelectrodes that were coupled with 304SS were used as working electrodes. P25 showed the minimum negative potential shift of 90 mV under light on/off cycles, which was a little lower than the self-corrosion potential of 304SS (−50 mV), indicating that P25 hardly achieved PECCP toward 304SS. The negative shift order of OCP values for samples was CuO$_x$/Ag/P25 > CuO$_x$/P25 > Ag/P25 > P25. Compared with CuO$_x$/P25 and Ag/P25, CuO$_x$/Ag/P25 displayed a significantly more negative potential shift of 240 mV under illumination, indicating that the introduction of CuO$_x$ and Ag improved the separation efficiency of electron-hole pairs, and more electrons were transferred to the coupled 304SS to provide photoelectrochemical cathodic protection. Figure 5b shows the photoelectric response of P25, CuO$_x$/P25, Ag/P25, and CuO$_x$/Ag/P25 coupled with 304SS in a 3.65 wt.% NaCl solution with an applied bias potential of 0 V (E_{ref}). P25, CuO$_x$/P25, Ag/P25, and CuO$_x$/Ag/P25 showed a positive current density change when the light was turned on, showing the characteristics of an n-type semiconductor [48]. It indicates that photoinduced electrons could be transferred from the semiconductor materials to the coupled 304SS. Compared with P25, the photocurrent densities of CuO$_x$/P25 and CuO$_x$/Ag/P25 greatly increased at the first light on/off switching. At the same time, using Ag particles as a conducting medium accelerated the separation of electron-hole pairs [3,30]. Thus, CuO$_x$/Ag/P25 had a photocurrent density of 16.6 μA cm^2, which was ~1.8 times greater than that of P25, indicating that it had the best PECCP performance. The migration ability of photogenerated carriers could be characterized by electrochemical impedance spectroscopy (EIS).

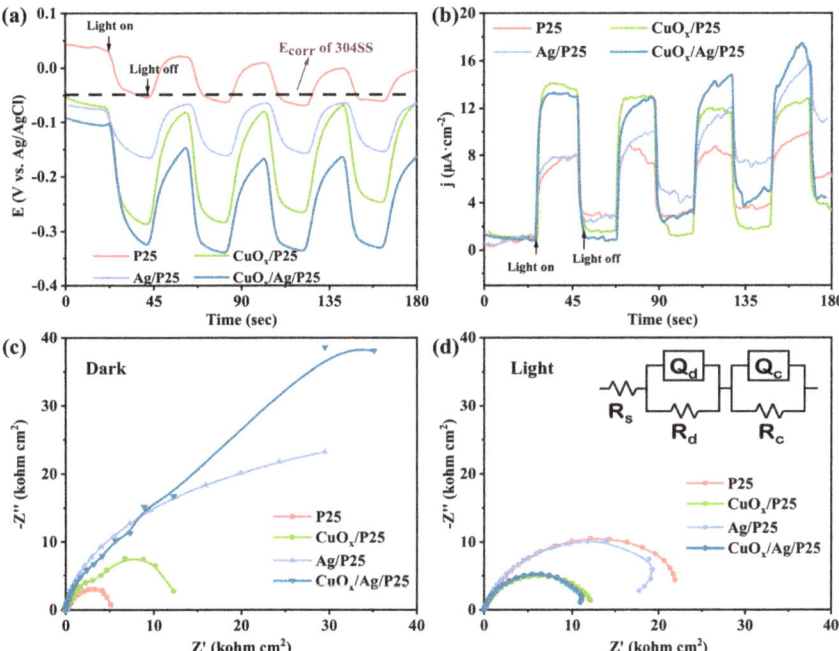

Figure 5. (**a**) The photoinduced OCP−time curves and (**b**) i−t curves of P25, CuO_x/P25, Ag/P25, and CuO_x/Ag/P25 coupled with 304SS. Nyquist plots measured (**c**) in the dark and (**d**) under illumination. Insert of (**d**) is the corresponding equivalent circuit.

Figure 5c,d show EIS results measured in the dark and under illumination, respectively. In darkness, the impedance arc radius of CuO_x/Ag/P25 was larger than that of the other materials, which indicates that the thin film electrode could be used as a coating material to protect 304SS from corrosion [18]. Under illumination, the smaller the impedance arc radius of the impedance spectrum is, the faster the photogenerated carriers migrate, resulting in better PECCP properties [49]. Figure 5d shows that CuO_x/Ag/P25 had the smallest arc radius, demonstrating the best carriers transfer efficiency and thus the greatest photochemical cathodic protection effect, which was consistent with the results of OCP response and the photoinduced current. In the corresponding equivalent circuit model, R_s, R_d, and R_c represented the solution resistance, the depletion layer resistance, and the charge transfer resistance [50], respectively. The fitting results in Table 1 showed that CuO_x/Ag/P25 had a higher R_d value than P25, which may have been due to the bending of the Fermi level caused by the formation of a Z-type heterostructure. Moreover, CuO_x/Ag/P25 had an R_c value of 1.23 Ω, much smaller than P25 (2.47×10^4), CuO_x/P25 (1.22×10^4), and Ag/P25 (300.5), respectively. Therefore, it could be understandable that CuO_x/Ag/P25 had the best PECCP performance for 304 SS in a 3.65 wt.% NaCl solution.

Table 1. EIS fitting results of photoanode films.

	R_s (Ω)	R_d (Ω)	R_c (Ω)
P25	9.24	31.68	2.47×10^4
CuO_x/P25	10.12	239.7	1.22×10^4
Ag/P25	11.50	2.10×10^4	300.5
CuO_x/Ag/P25	8.51	1.21×10^4	1.23

2.4. Antibacterial Performance Evaluation

The antibacterial activities of the materials were evaluated by using the CFU method, i.e., an *E. coli* colony–counting method. The optical pictures of *E. coli* colonies on nutrient agar plates in Figure 6a exhibited that CuO_x/Ag/P25 and Ag/P25 coatings had fewer *E. coli* colonies than the other samples did, indicating better antibacterial activities under 24 h of illumination. Taking the blank sample without active catalysts as a counterpart in Figure 6b, it can be calculated that the bacterial survival rate of P25 and CuO_x/P25 in a 1×10^6-times diluted bacterial solution were about 81.1% and 75.4%, respectively. Moreover, as shown in Figure 6c,d and Figure 7, the bacterial survival rates of Ag/P25 and CuO_x/Ag/P25 in a 100-times diluted solution were as low as 0.116% and 0.006%, respectively. It is concluded that CuO_x/Ag/P25 can effectively inhibit the growth of *E. coli*.

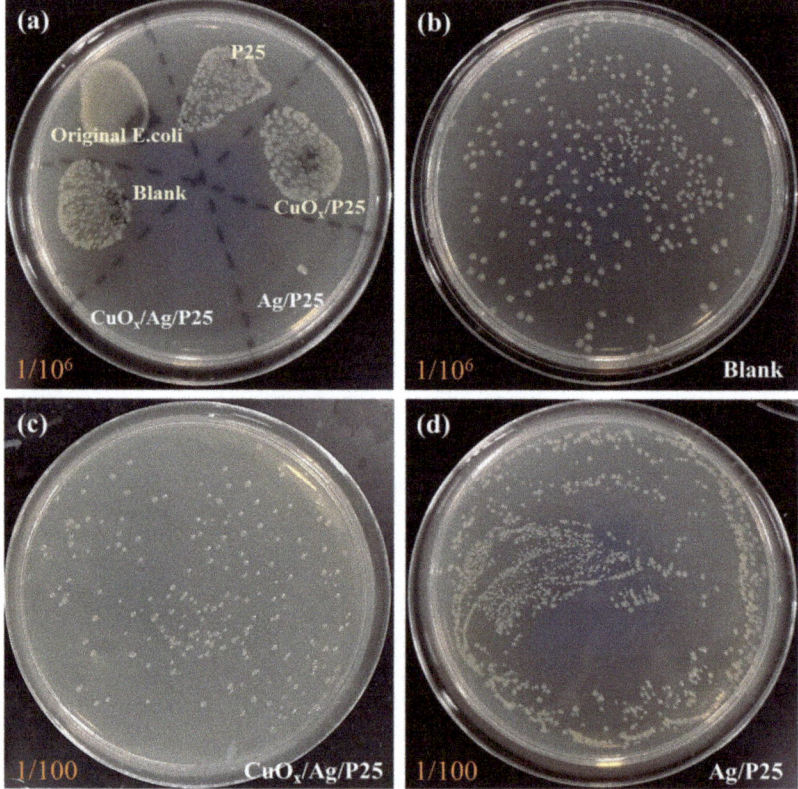

Figure 6. Optical photographs of agar plates coated with different catalysts and *E. coli* incubated solution with various dilutions: (**a**) comparison of the samples at 1×10^6-times dilution, (**b**) blank sample at 1×10^6-times dilution, (**c**) CuO_x/Ag/P25 at 100-times dilution, and (**d**) Ag/P25 at 100-times dilution.

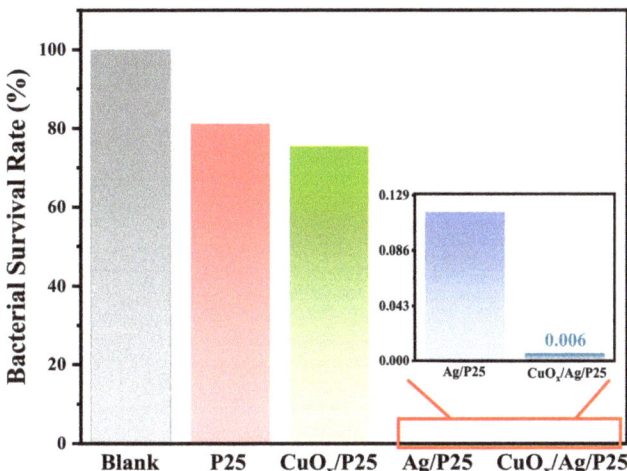

Figure 7. Comparison diagram of the bacterial survival rate in the media with different catalysts.

3. Discussion

The energy-band positions of CuO_x (−0.79 eV/1.26 eV) and P25 (−0.7 eV/2.31 eV) determined in this work (Figure 3c–f) and the reported Fermi level of Ag NP (~−0.22 eV) [44] are listed in Figure 8, indicating that a Z-scheme or type-II heterojunction structure was formed in CuO_x/Ag/P25 [28,30]. Moreover, because the VB potential of CuO_x is not positive enough for the production of •OH from H_2O oxidation, the stronger DMPO-•OH signal of CuO_x/Ag/P25, compared with that of Ag/P25 (Figure 4d), indicates that CuO_x/Ag/P25 is inclined to form a Z-scheme system in which the photoinduced electrons in Ag/P25 transfer and combine with the holes on the valance band (VB) of CuO_x, leaving the photoinduced holes at the VB of P25 for the production of •OH and the photoinduced electrons at the conduction band (CB) of CuO_x for the photocathode protection and production of superoxide ions (O_2•$^-$). Additionally, the incorporation of Ag NP with the LSPR effect and CuO_x with a narrow band gap greatly improved the visible light absorption (Figure 3a). All these together promoted the production of photoinduced charges and the ROS of •O_2^- and •OH radicals (Figure 4) under illumination. On one hand, the efficient production and transfer of the photoinduced charges in the unique Z-scheme heterojunction of CuO_x/Ag/P25 makes effective the transfer of photogenerated electrons from the CB of CuO_x to the coupled 304SS, achieving enough of a negative shift of the 304SS potential and photoelectrochemical cathodic polarization protection toward the metal (Figure 5a). On the other hand, their strong ROS (superoxide radicals •O_2^- and hydroxyl radicals •OH) production ability upon irradiation and their ability to inhibit the growth of bacteria, viruses, and other micro-organisms provide the antifouling ability of the CuO_x/Ag/P25 coating [26,27]. Moreover, both CuO_x and Ag NPs, especially the Cu_2O species stabilized by the photoinduced electrons through the unique Z-scheme transfer path (Figure 2d), effectively maintain chemical disinfection properties [25,31,32]. The superior antibacterial performance of CuO_x/Ag/P25 and their application potential as an antifouling coating are demonstrated in Figures 6 and 7.

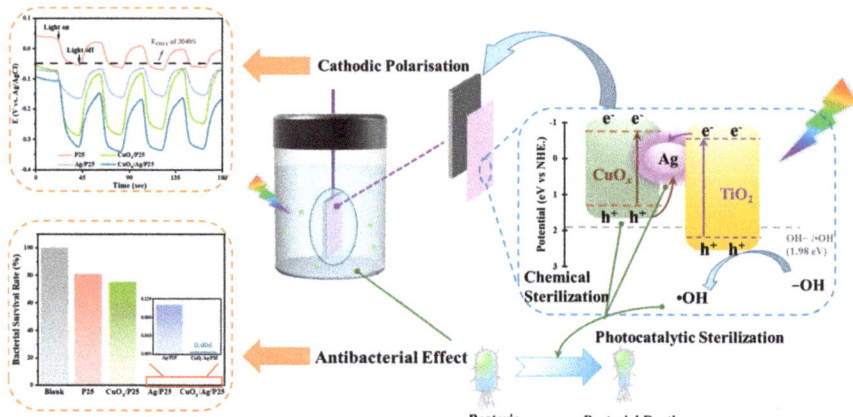

Figure 8. Schematic diagram of the antibacterial and anticorrosion dual mechanism for $Cu_xO/Ag/P25$.

4. Experimental Section

4.1. Synthesis of $CuO_x/Ag/TiO_2$

First, noble metal Ag was loaded on TiO_2 (P25) by using the wet chemical reduction method. Typically, 0.5 g of P25 was ultrasonically dispersed in 30 mL of deionized water. Then 5 mL of $AgNO_3$ aqueous solution with a concentration of 1 mg mL^{-1} was added to the above suspension under vigorous stirring. After fully mixing, an appropriate amount of $NaBH_4$ aqueous solution was added dropwise to reduce Ag^+. The molar ratio of $NaBH_4$ to $AgNO_3$ was 4:1. Under continuously stirring for 4 h, the resulting powder was obtained by centrifugation and washing with deionized water and absolute ethanol, separately, several times, followed by drying in an oven at 80 °C to obtain the Ag/P25 sample with a theoretical Ag loading of 1.0 wt.%.

Then, CuO_x was loaded on P25 or Ag/P25 by using the deposition-precipitation method. Next, 0.3 g of P25 or Ag/P25 and 0.0074 g of $CuSO_4·5H_2O$ were added into 60 mL of 0.2 mol mL^{-1} NaOH aqueous solution under magnetic stirring for 30 min, followed by the addition of 9.0 mL of 0.1 mol L^{-1} ascorbic acid aqueous solution, drop by drop. After reaction for 1.5 h, the product was centrifuged and washed with deionized water and absolute ethanol, separately, until the solution was neutral. At last, 1.0 wt.% $CuO_x/P25$ or 1.0 wt.% $CuO_x/Ag/P25$ was obtained after drying the sample overnight in the flowing N_2 at 80 °C.

4.2. Preparation of Coatings

The as-prepared sample was mixed with nafion/ethanol solution under ultrasonic treatment for the preparation of the coating material [51]. Typically, 1 mL of absolute ethanol was mixed with 100 μL of 5% nafion to obtain the solvent. Next, 0.5 mg of $CuO_x/Ag/P25$ (or other sample) was then ultrasonic dispersed in the above 200 μL solvent. The as-obtained ultrasonic slurry was coated on FTO glass substrates with a coating area of 1 cm × 1 cm. Finally, the photoanode coating was obtained after drying in an oven at 40 °C for 3 h.

4.3. Characterizations

The crystal structures of the prepared powders were analyzed by an X-ray powder diffractometer (Rigaku Ultima IV) equipped with a Cu Kα radiation source. The diffraction patterns in the range of 10 to 80° were recorded at a scan speed of 10° min^{-1}. The morphologies and element distribution of the samples were observed by high-resolution thermal field emission scanning electron microscope (FESEM, Gemini500). The crystal interface structures of the samples were analyzed by high-resolution transmission electron microscope (HRTEM, FEI Tecnai G2 F30) with an accelerating voltage of 300 kV. The element composition and valence state of the catalysts were analyzed by X-ray photoelectron spectroscopy/ESCA (XPS, Thermo Fisher Scientific, Nexsa). A monochromatic Al Kα X-ray source (hv = 1486.6 eV) under a vacuum degree of ~2×10^{-9} mbar was used. The surface pollution C1s (284.8 eV) was used as the standard for energy correction. The UV-vis diffuse reflectance spectroscopy (UV-vis DRS) response of catalysts in the wavelength range from 200 to 800 nm was analyzed by integrating a sphere UV-vis spectrophotometer (UV2600) with BaSO$_4$ as a reference. Photoluminescence (PL) spectra were characterized by using Edinburgh FS5 fluorescence spectrometer with an excitation wavelength of 550 nm. Electron paramagnetic resonance (EPR) spectrometer (JES X320) was used to obtain signals of photogenerated radicals spin trapped by 5,5-dimethyl-1-pyrroline-N-oxide (DMPO), e.g., DMPO-•OH and DMPO-O$_2$•$^-$, in the dark and under illumination (190–900 nm) at room temperature. The free radical signals were collected after 12 min.

4.4. Photoelectrochemical Measurements

The electrochemical measurements were performed at room temperature using Gamry electrochemical workstation (Gamry Interface1010E). All measurements were carried out after the open circuit potential (OCP) value was stable. A 300 W Xenon lamp (CEL-HXF300) was used as the light source with an optical power density of 200 mW cm^{-2}. In a three-electrode system, Pt plate was used as a counter electrode, Ag/AgCl electrode (saturated KCl) was used as a reference electrode, and the photoanode (catalyst coatings on FTO glass substrates, i.e., P25, Ag/P25, CuO$_x$/P25, or CuO$_x$/Ag/P25) was used as a working electrode. In the PECCP performance measurements, such as those from the OCP and photocurrent density-time (i-t) tests, the photoanode coupled with 304SS was used as a working electrode. Additionally, a 3.65 wt.% NaCl solution was used as an electrolyte solution to simulate the marine environment. Mott–Schottky (M–S) curves were measured in the potential scope of −1.5 to 0.2 V at a frequency of 1000 Hz. Electrochemical impedance spectroscopy (EIS) was performed in the AC voltage of 10 mV and the frequency range from 10^5 to 10^{-2} Hz with or without light illumination.

4.5. Antibacterial Performance Evaluation

The antibacterial activities of the materials were evaluated by using the CFU method (a plate-counting method) using *Escherichia coli* (*E. coli*) in a beef-extract-peptone (BEP) medium at 37 °C for 24 h. The concentration of the bacterial suspension was adjusted to $3 \times 10^7 \sim 3 \times 10^8$ CFU mL^{-1}. The catalysts were dispersed by deionized water to a concentration of 2 mg mL^{-1}. The mixture of 1 mL of catalyst, 1 mL of BEF, and 100 µL of *E. coli* suspensions was cultured on a shaking table (ZQLY-180GN, 150 rpm) under simulated solar irradiation at 37 °C for 24 h (16000LX). At the end of the incubation period, the culture medium was sampled to determine the viable counts of planktonic bacteria. The viable bacteria in the sampled suspension were counted by using a 10-times gradient dilution method. Specifically, 100 µL of a diluted sample was transferred to an LB agar plate, which was cultured in a bacterial incubator (MJX-150) at 37 °C for 24 h. Finally, the colonies were counted, and the antibacterial activities of the catalysts were compared according to the bacterial survival rate α, as calculated by

$$\alpha = \frac{\text{Number of viable bacteria in the control sample} \times \text{Dilution ratio}}{\text{Number of viable bacteria in the blank} \times \text{Dilution ratio}} \times 100\%$$

5. Conclusions

In summary, we have successfully prepared CuO$_x$/Ag/P25 coatings on FTO glass substrates, which displayed excellent photocathodic protection performance and antifouling activities under simulated solar illumination in a 3.65 wt.% NaCl solution. The photoelectrochemical measurements and characterizations revealed that a Z-type CuO$_x$/Ag/P25 heterostructure provided a photogenerated carrier transfer channel, which facilitated carrier separation and resulted in a more negative OCP shift of 240 mV. Meanwhile, the Z-type heterojunction of coatings had high-conduction-band and deep-valance-band potentials, which generated more reactive oxygen species of $\bullet O_2^-$ and $\bullet OH$ radicals, thus effectively killing *E. coli* bacteria, with a low survival rate of 0.006%.

Author Contributions: Conceptualization, Z.R.; methodology, Y.L. and X.G.; validation, X.G., Y.L. and Z.R.; formal analysis, X.G. and Z.R.; investigation, X.G. and G.P.; resources, Y.L. and Z.R.; data curation, X.G. and Z.R.; writing—original draft preparation, X.G.; writing—review and editing, L.F., Y.L. and Z.R.; supervision, Z.R.; funding acquisition, Y.L. and Z.R. All authors have read and agreed to the published version of the manuscript.

Funding: This study was financially supported by the Natural Science Funds of Guangdong for Distinguished Young Scholars (No. 2022B1515020098), National Natural Science Foundation of China (No. 22002192), and Science and Technology Program of Guangzhou (No. 202102020172).

Institutional Review Board Statement: Not applicable.

Informed Consent Statement: Not applicable.

Data Availability Statement: The data presented in this study are available on request from the corresponding author.

Conflicts of Interest: The authors declare no conflict of interest.

Sample Availability: Samples of the compounds are not available from the authors.

References

1. Wang, X.T.; Xu, H.; Nan, Y.B.; Sun, X.; Duan, J.Z.; Huang, Y.L.; Hou, B.R. Research progress of TiO_2 photocathodic protection to metals in marine environment. *J. Oceanol. Limnol.* **2020**, *38*, 1018–1044. [CrossRef]
2. Zhang, X.F.; Li, M.Y.; Kong, L.F.; Wang, M.; Yan, L.; Xiao, F.J. Research progress in metal photoelectrochemical cathodic protection materials and its anticorrosion function realization. *Surf. Technol.* **2021**, *50*, 128–140. [CrossRef]
3. Tian, J.; Chen, Z.Y.; Jing, J.P.; Feng, C.; Sun, M.M.; Li, W.B. Photoelectrochemical cathodic protection of Cu_2O/TiO_2 p-n heterojunction under visible light. *J. Oceanol. Limnol.* **2020**, *38*, 1517–1531. [CrossRef]
4. Vedaprakash, L.; Dineshram, R.; Ratnam, K.; Lakshmi, K.; Jayaraj, K.; Babu, S.M.; Venkatesan, R.; Shanmugam, A. Experimental studies on the effect of different metallic substrates on marine biofouling. *Colloids Surf. B* **2013**, *106*, 1–10. [CrossRef]
5. Lee, J.S.; Little, B.J. A mechanistic approach to understanding microbiologically influenced corrosion by metal-depositing bacteria. *Corrosion* **2019**, *75*, 6–11. [CrossRef]
6. Vazirinasab, E.; Jafari, R.; Momen, G. Application of superhydrophobic coatings as a corrosion barrier: A review. *Surf. Coat. Technol.* **2018**, *341*, 40–56. [CrossRef]
7. Raja, P.B.; Ismail, M.; Ghoreishiamiri, S.; Mirza, J.; Ismail, M.C.; Kakooei, S.; Rahim, A.A. Reviews on corrosion inhibitors: A short view. *Chem. Eng. Commun.* **2016**, *203*, 1145–1156. [CrossRef]
8. Hussain, A.K.; Seetharamaiah, N.; Pichumani, M.; Chakra, C.S. Research progress in organic zinc rich primer coatings for cathodic protection of metals-A comprehensive review. *Prog. Org. Coat.* **2021**, *153*, 106040. [CrossRef]
9. Bu, Y.Y.; Chen, Z.Y.; Ao, J.P.; Hou, J.; Sun, M.X. Study of the photoelectrochemical cathodic protection mechanism for steel based on the $SrTiO_3-TiO_2$ composite. *J. Alloys Compd.* **2018**, *731*, 1214–1224. [CrossRef]
10. Zheng, J.Y.; Lyu, Y.H.; Wang, R.L.; Xie, C.; Zhou, H.J.; Jiang, S.P.; Wang, S.Y. Crystalline TiO_2 protective layer with graded oxygen defects for efficient and stable silicon-based photocathode. *Nat. Commun.* **2018**, *9*, 3572. [CrossRef]
11. Lu, X.Y.; Liu, L.; Ge, J.W.; Cui, Y.; Wang, F.H. Morphology controlled synthesis of $Co(OH)_2/TiO_2$ p-n heterojunction photoelectrodes for efficient photocathodic protection of 304 stainless steel. *Appl. Surf. Sci.* **2021**, *537*, 148002. [CrossRef]
12. Yang, Y.; Cheng, Y.F. One-step facile preparation of ZnO nanorods as high-performance photoanodes for photoelectrochemical cathodic protection. *Electrochim. Acta* **2018**, *276*, 311–318. [CrossRef]
13. Jing, J.P.; Chen, Z.Y.; Bu, Y.Y.; Sun, M.M.; Zheng, W.Q.; Li, W.B. Significantly enhanced photoelectrochemical cathodic protection performance of hydrogen treated Cr-doped $SrTiO_3$ by Cr^{6+} reduction and oxygen vacancy modification. *Electrochim. Acta* **2019**, *304*, 386–395. [CrossRef]
14. Ge, M.Z.; Li, Q.S.; Cao, C.Y.; Huang, J.Y.; Li, S.H.; Zhang, S.N.; Chen, Z.; Zhang, K.Q.; Al-Deyab, S.S.; Lai, Y.K. One-dimensional TiO_2 nanotube photocatalysts for solar water splitting. *Adv. Sci.* **2017**, *4*, 1600152. [CrossRef]
15. Roy, P.; Berger, S.; Schmuki, P. TiO_2 nanotubes: Synthesis and applications. *Angew. Chem. Int. Ed.* **2011**, *50*, 2904–2939. [CrossRef]
16. An, X.Q.; Liu, H.J.; Qu, J.H.; Moniz, S.J.; Tang, J.W. Photocatalytic mineralisation of herbicide 2,4,5-trichlorophenoxyacetic acid: Enhanced performance by triple junction $Cu-TiO_2-Cu_2O$ and the underlying reaction mechanism. *New J. Chem.* **2015**, *39*, 314–320. [CrossRef]

17. Wang, M.Y.; Sun, L.; Lin, Z.Q.; Cai, J.H.; Xie, K.P.; Lin, C.J. p-n Heterojunction photoelectrodes composed of Cu_2O-loaded TiO_2 nanotube arrays with enhanced photoelectrochemical and photoelectrocatalytic activities. *Energy Environ. Sci.* **2013**, *6*, 1211–1220. [CrossRef]
18. Sun, W.X.; Wei, N.; Cui, H.Z.; Lin, Y.; Wang, X.Z.; Tian, J.; Li, J.; Wen, J. 3D $ZnIn_2S_4$ nanosheet/TiO_2 nanowire arrays and their efficient photocathodic protection for 304 stainless steel. *Appl. Surf. Sci.* **2018**, *434*, 1030–1039. [CrossRef]
19. Liu, D.; Yang, C.T.; Zhou, E.Z.; Yang, H.Y.; Li, Z.; Xu, D.K.; Wang, F.H. Progress in microbiologically influenced corrosion of metallic materials in marine environment. *Surf. Technol.* **2019**, *48*, 166–174. [CrossRef]
20. Mehtab, A.; Banerjee, S.; Mao, Y.; Ahmad, T. Type-II $CuFe_2O_4$/graphitic carbon nitride heterojunctions for high-efficiency photocatalytic and electrocatalytic hydrogen generation. *ACS Appl. Mater. Inter.* **2022**, *14*, 44317–44329. [CrossRef]
21. Jasrotia, R.; Verma, A.; Verma, R.; Ahmed, J.; Godara, S.K.; Kumar, G.; Mehtab, A.; Ahmad, T.; Kalia, S. Photocatalytic dye degradation efficiency and reusability of Cu-substituted Zn-Mg spinel nanoferrites for wastewater remediation. *J. Water Process. Eng.* **2022**, *48*, 102865. [CrossRef]
22. Jasrotia, R.; Verma, A.; Verma, R.; Godara, S.K.; Ahmed, J.; Mehtab, A.; Ahmad, T.; Puri, P.; Kalia, S. Photocatalytic degradation of malachite green pollutant using novel dysprosium modified Zn-Mg photocatalysts for wastewater remediation. *Ceram. Int.* **2022**, *48*, 29111–29120. [CrossRef]
23. Bhat, S.A.; Hassan, T.; Majid, S. Heavy metal toxicity and their harmful effects on living organisms-a review. *Int. J. Med. Sci. Diagn. Res.* **2019**, *3*, 106–122. [CrossRef]
24. Wang, R.; Liu, R.X.; Luo, S.J.; Wu, J.X.; Zhang, D.H.; Yue, T.L.; Sun, J.; Zhang, C.; Zhu, L.Y.; Wang, J.L. Band structure engineering enables to UV-Visible-NIR photocatalytic disinfection: Mechanism, pathways and DFT calculation. *Chem. Eng. J.* **2021**, *421*, 129596. [CrossRef]
25. Yang, Z.Q.; Ma, C.C.; Wang, W.; Zhang, M.T.; Hao, X.P.; Chen, S.G. Fabrication of Cu_2O-Ag nanocomposites with enhanced durability and bactericidal activity. *J. Colloid Interf. Sci.* **2019**, *557*, 156–167. [CrossRef]
26. You, J.H.; Guo, Y.Z.; Guo, R.; Liu, X.W. A review of visible light-active photocatalysts for water disinfection: Features and prospects. *Chem. Eng. J.* **2019**, *373*, 624–641. [CrossRef]
27. Wen, B.; Waterhouse, G.I.; Jia, M.Y.; Jiang, X.H.; Zhang, Z.M.; Yu, L.M. The feasibility of polyaniline-TiO_2 coatings for photocathodic antifouling: Antibacterial effect. *Synth. Met.* **2019**, *257*, 116175. [CrossRef]
28. Kong, J.J.; Rui, Z.B.; Liu, S.H.; Liu, H.W.; Ji, H.B. Homeostasis in Cu_xO/$SrTiO_3$ hybrid allows highly active and stable visible light photocatalytic performance. *Chem. Commun.* **2017**, *53*, 12329–12332. [CrossRef]
29. Yin, Z.; Wang, Y.; Song, C.Q.; Zheng, L.H.; Ma, N.; Liu, X.; Li, S.W.; Lin, L.L.; Li, M.Z.; Xu, Y.; et al. Hybrid Au-Ag nanostructures for enhanced plasmon-driven catalytic selective hydrogenation through visible light irradiation and surface-enhanced raman scattering. *J. Am. Chem. Soc.* **2018**, *140*, 864–867. [CrossRef]
30. Ji, W.K.; Rui, Z.B.; Ji, H.B. Z-scheme Ag_3PO_4/Ag/$SrTiO_3$ heterojunction for visible-light induced photothermal synergistic VOCs degradation with enhanced performance. *Ind. Eng. Chem. Res.* **2019**, *58*, 13950–13959. [CrossRef]
31. Ghasemi, N.; Jamali-Sheini, F.; Zekavati, R. CuO and Ag/CuO nanoparticles: Biosynthesis and antibacterial properties. *Mater. Lett.* **2017**, *196*, 78–82. [CrossRef]
32. Hans, M.; Erbe, A.; Mathews, S.; Chen, Y.; Solioz, M.; Mücklich, F. Role of copper oxides in contact killing of bacteria. *Langmuir* **2013**, *29*, 16160–16166. [CrossRef]
33. Chen, Y.F.; Huang, W.X.; He, D.L.; Situ, Y.; Huang, H. Construction of heterostructured g-C_3N_4/Ag/TiO_2 microspheres with enhanced photocatalysis performance under visible-light irradiation. *ACS Appl. Mater. Inter.* **2014**, *6*, 14405–14414. [CrossRef]
34. Sui, Y.M.; Fu, W.Y.; Yang, H.B.; Zeng, Y.; Zhang, Y.Y.; Zhao, Q.; Li, Y.G.; Zhou, X.M.; Leng, Y.; Li, M.H.; et al. Low temperature synthesis of Cu_2O crystals: Shape evolution and growth mechanism. *Cryst. Growth Des.* **2010**, *10*, 99–108. [CrossRef]
35. Zhao, X.Y.; Zhang, J.; Wang, B.S.; Zada, A.; Humayun, M. Biochemical synthesis of Ag/AgCl nanoparticles for visible-light-driven photocatalytic removal of colored dyes. *Materials* **2015**, *8*, 2043–2053. [CrossRef]
36. Li, Y.P.; Wang, B.W.; Liu, S.H.; Duan, X.F.; Hu, Z.Y. Synthesis and characterization of Cu_2O/TiO_2 photocatalysts for H_2 evolution from aqueous solution with different scavengers. *Appl. Surf. Sci.* **2015**, *324*, 736–744. [CrossRef]
37. Wang, F.Z.; Li, W.J.; Gu, S.N.; Li, H.D.; Wu, X.; Ren, C.J.; Liu, X.T. Facile fabrication of direct Z-scheme MoS_2/Bi_2WO_6 heterojunction photocatalyst with superior photocatalytic performance under visible light irradiation. *J. Photochem. Photobiol. A* **2017**, *335*, 140–148. [CrossRef]
38. Purvis, K.L.; Lu, G.; Schwartz, J.; Bernasek, S.L. Surface characterization and modification of indium tin oxide in ultrahigh vacuum. *J. Am. Chem. Soc.* **2000**, *122*, 1808–1809. [CrossRef]
39. Forget, A.; Limoges, B.; Balland, V. Efficient chemisorption of organophosphorous redox probes on indium tin oxide surfaces under mild conditions. *Langmuir* **2015**, *31*, 1931–1940. [CrossRef]
40. Tohsophon, T.; Dabirian, A.; De Wolf, S.; Morales-Masis, M.; Ballif, C. Environmental stability of high-mobility indium-oxide based transparent electrodes. *APL Mater.* **2015**, *3*, 116105. [CrossRef]
41. Avgouropoulos, G.; Ioannides, T.; Matralis, H. Influence of the preparation method on the performance of CuO-CeO_2 catalysts for the selective oxidation of CO. *Appl. Catal. B-Environ.* **2005**, *56*, 87–93. [CrossRef]
42. Lee, Y.H.; Leu, I.C.; Liao, C.L.; Chang, S.T.; Fung, K.Z. Fabrication and characterization of Cu_2O nanorod arrays and their electrochemical performance in Li-ion batteries. *Electrochem. Solid-State Lett.* **2006**, *9*, A207. [CrossRef]

43. Li, J.W.; Yang, X.Q.; Ma, C.R.; Lei, Y.; Cheng, Z.Y.; Rui, Z.B. Selectively recombining the photoinduced charges in bandgap-broken Ag_3PO_4/$GdCrO_3$ with a plasmonic Ag bridge for efficient photothermocatalytic VOCs degradation and CO_2 reduction. *Appl. Catal. B-Environ.* **2021**, *291*, 120053. [CrossRef]
44. Li, J.W.; Chen, J.Y.; Fang, H.L.; Guo, X.M.; Rui, Z.B. Plasmonic metal bridge leading type III heterojunctions to robust type B photothermocatalysts. *Ind. Eng. Chem. Res.* **2021**, *60*, 8420–8429. [CrossRef]
45. Zhang, Q.; Huang, Y.; Xu, L.F.; Cao, J.J.; Ho, W.K.; Lee, S.C. Visible-light-active plasmonic Ag-$SrTiO_3$ nanocomposites for the degradation of NO in air with high selectivity. *ACS Appl. Mater. Inter.* **2016**, *8*, 4165–4174. [CrossRef] [PubMed]
46. Fernández-Catalá, J.; Navlani-García, M.; Berenguer-Murcia, Á.; Cazorla-Amorós, D. Exploring Cu_xO-doped TiO_2 modified with carbon nanotubes for CO_2 photoreduction in a 2D-flow reactor. *J. CO2 Util.* **2021**, *54*, 101796. [CrossRef]
47. Yang, X.Q.; Liu, S.H.; Li, J.W.; Chen, J.Y.; Rui, Z.B. Promotion effect of strong metal-support interaction to thermocatalytic, photocatalytic, and photothermocatalytic oxidation of toluene on Pt/$SrTiO_3$. *Chemosphere* **2020**, *249*, 126096. [CrossRef]
48. Jing, J.P.; Chen, Z.Y.; Feng, C. Dramatically enhanced photoelectrochemical properties and transformed p/n type of g-C_3N_4 caused by K and I co-doping. *Electrochim. Acta* **2019**, *297*, 488–496. [CrossRef]
49. Zhang, J.; Hu, J.; Zhu, Y.F.; Liu, Q.; Zhang, H.; Du, R.G.; Lin, C.J. Fabrication of CdTe/ZnS core/shell quantum dots sensitized TiO_2 nanotube films for photocathodic protection of stainless steel. *Corros. Sci.* **2015**, *99*, 118–124. [CrossRef]
50. Li, W.F.; Wei, L.C.; Shen, T.; Wei, Y.N.; Li, K.J.; Liu, F.Q.; Li, W.H. Ingenious preparation of "layered-closed" TiO_2-$BiVO_4$-CdS film and its highly stable and sensitive photoelectrochemical cathodic protection performance. *Chem. Eng. J.* **2022**, *429*, 132511. [CrossRef]
51. Ding, J.; Liu, P.; Zhou, M.; Yu, H.B. Nafion-endowed graphene super-anticorrosion performance. *ACS Sustain. Chem. Eng.* **2020**, *8*, 15344–15353. [CrossRef]

Disclaimer/Publisher's Note: The statements, opinions and data contained in all publications are solely those of the individual author(s) and contributor(s) and not of MDPI and/or the editor(s). MDPI and/or the editor(s) disclaim responsibility for any injury to people or property resulting from any ideas, methods, instructions or products referred to in the content.

Review

Application of Photocatalysis and Sonocatalysis for Treatment of Organic Dye Wastewater and the Synergistic Effect of Ultrasound and Light

Guowei Wang and Hefa Cheng *

MOE Key Laboratory for Earth Surface Processes, College of Urban and Environmental Sciences, Peking University, Beijing 100871, China
* Correspondence: hefac@umich.edu; Tel.: +86-10-6276-1070; Fax: +86-10-6276-7921

Abstract: Organic dyes play vital roles in the textile industry, while the discharge of organic dye wastewater in the production and utilization of dyes has caused significant damage to the aquatic ecosystem. This review aims to summarize the mechanisms of photocatalysis, sonocatalysis, and sonophotocatalysis in the treatment of organic dye wastewater and the recent advances in catalyst development, with a focus on the synergistic effect of ultrasound and light in the catalytic degradation of organic dyes. The performance of TiO_2-based catalysts for organic dye degradation in photocatalytic, sonocatalytic, and sonophotocatalytic systems is compared. With significant synergistic effect of ultrasound and light, sonophotocatalysis generally performs much better than sonocatalysis or photocatalysis alone in pollutant degradation, yet it has a much higher energy requirement. Future research directions are proposed to expand the fundamental knowledge on the sonophotocatalysis process and to enhance its practical application in degrading organic dyes in wastewater.

Keywords: sonocatalysis; sonoluminescence; photocatalysis; organic dye; sonophotocatalysis; synergistic mechanism

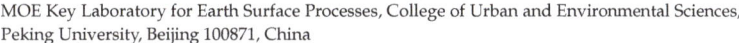

1. Introduction

In recent years, the textile industry has played a vital role in the global economy, but it is also a major contributor to environmental pollution, particularly in terms of organic dye wastewater [1–5]. The discharge of organic dye wastewater from textile production and utilization can cause significant damage to the aquatic ecosystem, and therefore, it is imperative to develop effective treatment methods to degrade these pollutants [6–8]. Among the various technologies geared toward the treatment of organic dye wastewater, photocatalysis, sonocatalysis, and sonophotocatalysis have received increasing attention owing to their high efficiency and potential for large-scale industrial applications [9–13].

Photocatalysis is a well-established technology for the treatment of organic pollutants in wastewater [14–17]. Qutub et al. investigated CdS/TiO_2 nanocomposites for photocatalytic degradation of organic pollutants in wastewater [18]. The results showed that $CdS-TiO_2$ nanocomposites exhibited the highest photocatalytic activity in the degradation of AB-29 dye, with a degradation efficiency of 84%, compared to 68% and 9% achieved by CdS and TiO_2 under comparable conditions, respectively. The enhanced photocatalytic performance of $CdS-TiO_2$ was attributed to reduced charge carrier recombination, improved charge separation, and expansion of the response of TiO_2 to visible light. In photocatalytic systems, semiconductor photocatalysts, such as TiO_2, are irradiated with light, generating electron-hole pairs that react with water or oxygen to form reactive species, e.g., hydroxyl radicals. These reactive species can later degrade the organic pollutants into harmless products [19–22]. Sonocatalysis, on the other hand, utilizes ultrasonic wave to generate cavitation bubbles in the solution, which collapse and produce high-energy conditions that can promote chemical reactions. Wang et al. developed a recyclable $WO_3/NiFe_2O_4/BiOBr$

(WNB) composite with dual Z-scheme heterojunction for the degradation of levofloxacin (LEV) in aqueous solution [23]. The WNB composite showed the highest removal efficiency (97.97%) for LEV within 75 min under ultrasonic irradiation. The ternary composite comprises three different semiconductors suitable for harvesting full-spectrum light. The combination of sonocatalysis and photocatalysis, known as sonophotocatalysis, can further improve the efficiency of both processes, as the cavitation bubbles can create local "hot spots" that increase the photocatalytic activity for the catalyst [24–28]. Despite the promising results of sonophotocatalysis in pollutant degradation, there are still significant challenges that need to be addressed in order to enhance its performance for the treatment of organic dye wastewater [29]. For example, the high energy requirement of sonophotocatalysis limits its practical application, as it has high electricity consumption for generating the ultrasonic wave and producing the light [30–36]. In addition, mechanistic understanding on the synergistic effect of ultrasound and light in sonophotocatalysis is still not lacking, and more studies are required to clarify the underlying mechanism for optimization of the process [37,38]. Wang et al. synthesized Fe_3O_4@SiO_2/PAEDTC@MIL-101(Fe), a mesoporous composite with a core-shell structure, and evaluated its sonophotocatalytic performance in degrading acid red 14 (AR14) [39]. The results showed that Fe_3O_4@SiO_2/PAEDTC@MIL-101 (Fe)/UV/US exhibited excellent activity in the removal of AR14 and total organic carbon.

In this review, we aim to present an overview on the mechanisms of photocatalysis, sonocatalysis, and sonophotocatalysis in the treatment of organic dye wastewater [40,41]. We compare the performance of TiO_2-based catalysts in photocatalytic, sonocatalytic, and sonophotocatalytic systems, with a focus on the synergistic effect of ultrasound and light in the catalytic degradation of organic dyes [42]. We further discuss the recent advances in catalyst development for sonophotocatalysis and point out the potential risks associated with the sonophotocatalytic process, such as the generation of toxic byproducts and the potential releases of nanoparticles into the environment [43–48]. It is essential to monitor the reaction products and assess their toxicity to ensure that the sonophotocatalytic process is safe for both the environment and human health. Finally, we propose future research directions to expand the fundamental knowledge on the sonophotocatalysis process and enhance its practical application in degrading organic pollutants [49–53]. The information presented in this review can provide valuable insights into the mechanisms and performance of photocatalysis, sonocatalysis, and sonophotocatalysis, and contribute to the development of more efficient and cost-effective treatment methods for organic dye wastewater. This review will be of great interest to researchers and practitioners in the field of environmental science and engineering, especially those involved in the development of sustainable wastewater treatment technologies.

2. Sonocatalytic and Photocatalytic Mechanisms

2.1. Sonocatalytic Mechanism

The sonocatalytic process is believed to be predominantly based on the "hot spots" and "sonoluminescence" that originate from the ultrasonic cavitation phenomenon [54]. Ultrasonic wave of a specific frequency and intensity can produce numerous small bubbles in liquids [55]. These minute bubbles trigger various physical and chemical transformations during their formation, oscillation, expansion, contraction, and ultimate collapse [56–58]. Figure 1 scehematically depicts the phenomenon of ultrasonic cavitation. It promotes the production of light with a range of wavelengths, called "sonoluminescence," and a large number of localized "hot spots" with very high temperatures (up to ~5000 K) and pressures (up to ~1000 atm) [59,60]. These localized "hot spots" can cause pyrolysis of H_2O molecules, producing hydroxyl radicals (•OH) [61], which can effectively oxidize organic pollutants and even mineralize them into CO_2 and H_2O [62–66].

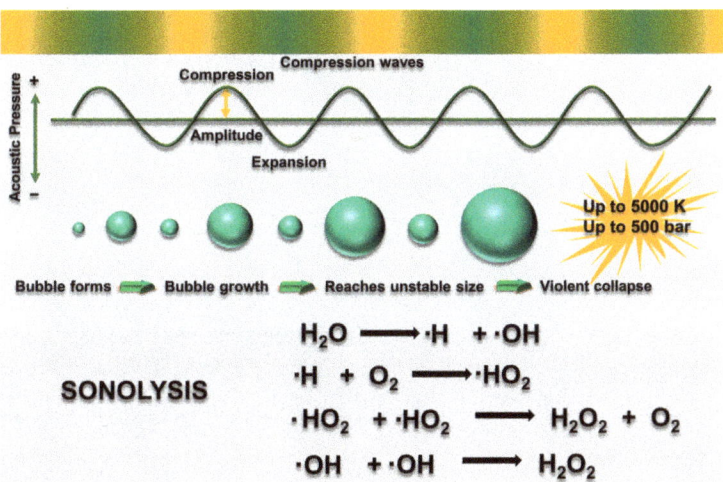

Figure 1. Schematic illustration of the acoustic generation of a cavitation bubble in water (after [66]).

In general, the sonolytic removal of organic pollutants involves oxidation through both pyrolysis and free radical attack [67]. However, due to the significant energy loss that occurs during thermal dissipation (exceeding 50%), radpid degradation often cannot occur when relying solely on ultrasound [68]. In recent years, the use of ultrasound in the presence of appropriate catalysts, known as sonocatalysis, has been increasingly used to degrade organic pollutants due to its numerous benefits, including convenient handling and low cost, as well as environmental friendliness [69]. Sonocatalytic degradation involves the use of a sonocatalyst to create additional active sites for the cavitation effect, leading to the formation of greater numbers with highly reactive radicals [70]. In general, these radicals could recombine to form H_2O, •OH, H_2O_2, and •O_2^- in water [71,72]:

$$H_2O +))) \rightarrow \bullet H + \bullet OH \quad (1)$$

$$\bullet H + \bullet OH \rightarrow H_2O_2 \quad (2)$$

$$O_2 + \bullet H \rightarrow \bullet HO_2 \quad (3)$$

$$2\bullet OH \rightarrow H_2O_2 \quad (4)$$

$$2\bullet HO_2 \rightarrow O_2 + H_2O_2 \quad (5)$$

$$H_2O + \bullet OH \rightarrow H_2O_2 + \bullet H \quad (6)$$

Sonocatalysis is a crucial technology in the degradation of pollutants, and radicals play an essential role in this process [73]. These radicals can initiate chain reactions that lead to the degradation of pollutants. To enhance the efficiency of the sonocatalytic process, it is essential to understand the underlying mechanisms of sonocatalysis [74]. Figure 2 depicts the major processes involved in sonocatalysis elucidated by extensive studies conducted in the field [75,76].

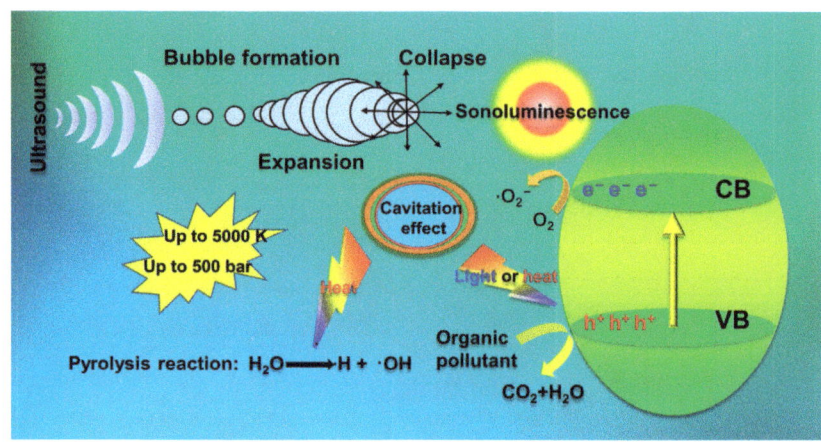

Figure 2. Schematic diagram of sonocatalytic mechanism (after [23]).

2.1.1. Heterogeneous Nucleation Mechanism

Semiconductor particles have been observed to induce preferential formation of nuclei at solid surfaces or phase boundaries, leading to increased formation of cavitation bubbles and free radicals, such as •OH [54]. The phenomenon of heterogeneous nucleation has been found to be more applicable than homogeneous cavitation in sonocatalysis [56]. This can be attributed to the fact that the thermodynamic nucleation barriers on interfaces are generally lower than their bulk counterparts, promoting surface nucleation [18]. The relationship between the maximum energy barriers of heterogeneous and homogeneous nucleation processes can be expressed as [57,60]:

$$\Delta G^{*}{}_{het} = \frac{16\pi\sigma^3}{3P^2} f(\theta) = \Delta G^{*}{}_{hom} f(\theta) \qquad (7)$$

where ΔG_{het} and ΔG_{hom} are the maximum energy barrier for heterogeneous and homogeneous reactions, respectively, σ is the surface tension of water (J/m^2), θ is the contact angle between the liquid and solid, and P is the sum of the partial pressure of the entrapped gas [47].

It is expected that preferential nucleation will transpire on hydrophobic surfaces, notably on solid surfaces [63]. Furthermore, the rate of bubble nucleation at the solid surface can be significantly influenced by many factors. Sonication parameters, such as ultrasonic power, frequency, as well as changes in surface energy, aqueous temperature, and type of absorbed gas, can greatly affect this process [59]. In addition, physicochemical properties of the solid particles, such as roughness, particle size, pore size, and wettability, can play a crucial role in influencing the nucleation rate.

2.1.2. Photo-Excitation Mechanism

Sonoluminescence (SL) is a light-emitting phenomenon caused by the collapse of cavitational bubbles. The light emitted during sonoluminescence has high intensity and covers a wide range of wavelengths, typically between 200 and 700 nm [77]. In the presence of a semiconductor catalyst during ultrasonication, the energy from the light generated can exceed the band gap of the semiconductor, leading to the excitation of electrons from the valence band (VB) to the conduction band (CB) [78]. This process generates holes in the valence band, which are caused by the excited electrons [79]. When photogenerated electron-hole pairs react with dissolved oxygen, they create highly reactive radicals. This process in sonocatalysis is akin to that in photocatalysis [80].

2.1.3. Thermal Excitation Mechanism

The "hot spots" hypothesis proposes that elevated temperatures in a specific area may result in thermal excitation of the semiconductor, causing the formation of electron-hole pairs [81]. This phenomenon has been observed in numerous studies [82], demonstrating that certain semiconductors can be stimulated by high temperatures to generate electron-hole pairs. At room temperature, TiO_2 displays low catalytic activity, but its performance improves significantly after being heated to temperatures ranging between 350 and 500 °C [83]. Such enhancement is attributed to the abundant highly oxidative holes that arise due to the thermal excitation of semiconductors [84].

2.2. Photocatalytic Mechanism

The photocatalytic process occurs when a semiconductor catalyst is exposed to light of greater energy than the semiconductor's bandgap [85], as depicted in Figure 3. When this happens, electrons in the VB may become excited and jump into the CB, forming a hole (h^+_{VB}) (Equation (8)). Subsequently, the electron-hole pairs that are generated by the absorption of light recombine together, leading to the emission of energy (Equation (9)) [86]. The poor quantum efficiency of the semiconductor is attributed to this recombination, which leads to low light-to-energy conversion rates [87,88]. If the photogenerated carriers do not recombine, light-generated electron-hole (e^-, h^+) pairs separate and move to the material's surface, reacting with the adsorbed molecules [89]. When photo-excited electrons come into contact with dissolved oxygen molecules (O_2) in an aqueous solution, they can react and form superoxide radical anions ($\bullet O_2^-$), as indicated by Equation (10) [90]. At the same time, the holes may directly oxidize pollutants or H_2O molecules to produce hydroxyl radicals ($\bullet OH$) (Equation (11)). The reactive radicals generated ($\bullet OH$, $\bullet O_2^-$) are highly reactive oxidizing agents [91], andthey may readily mineralize many organic molecules, producing water and carbon dioxide (Equations (12) and (13)).

$$\text{Semiconductor} + h\nu \to e^-_{CB} + h^+_{VB} \tag{8}$$

$$e^-_{CB} + h^+_{VB} \to \text{energy} \tag{9}$$

$$e^- + O_2 \to \bullet O_2^- \tag{10}$$

$$h^+ + H_2O \to \bullet H + \bullet OH \tag{11}$$

$$O_2^- + \text{Pollutant} \to H_2O + CO_2 \tag{12}$$

$$\bullet OH + \text{Pollutant} \to H_2O + CO_2 \tag{13}$$

2.3. Comparison of Sonocatalytic and Photocatalytic Mechanisms

Comparison of the mechanisms of sonocatalysis and photocatalysis can help better understand the unique features and advantages of sonocatalysis in promoting efficient and sustainable chemical transformations. The similarity and difference between these two types of mechanisms are detailed below.

2.3.1. Similarity

Semiconductor catalyst plays a vital role in lowering the energy barrier for the formation of cavitation bubbles, which is similar to the way that a traditional catalyst reduces the activation energy of a chemical reaction [92]. This is achieved by providing a surface for the accumulation and stabilization of gas or vapor pockets within the fluid medium, effectively reducing the threshold pressure required for bubble nucleation [93].

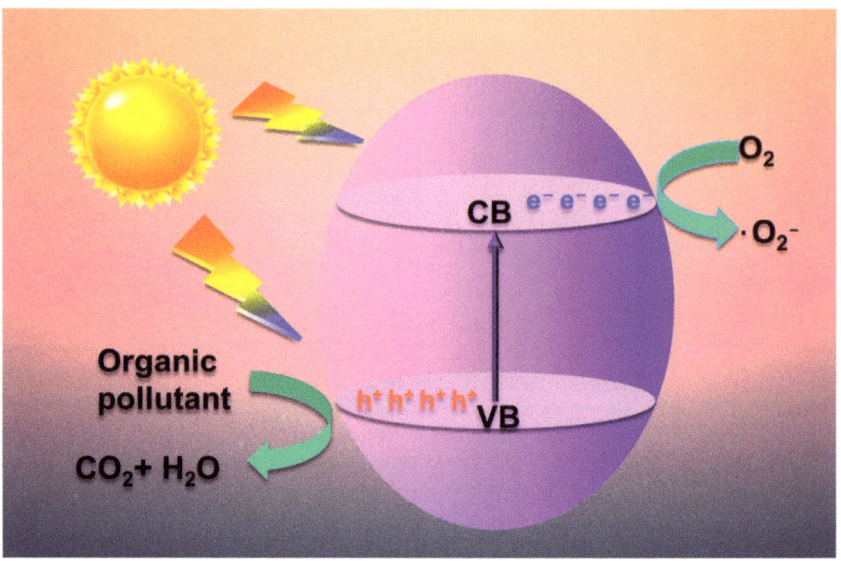

Figure 3. Schematic illustration of photocatalytic mechanism (after [23]).

Undoubtedly, photocatalysts have the potential to serve as effective sonocatalysts, leveraging the phenomenon of sonoluminescence generated by cavitation [94]. Given their inherent properties and unique chemical compositions, photocatalysts can harness the energy released by cavitation bubbles to enhance catalytic reactions and promote efficient chemical transformations [95].

2.3.2. Difference

The formation of cavitation bubbles is primarily driven by physical processes, involving the rapid formation and collapse of small pockets of gas or vapor within a fluid medium [96]. This can occur due to the changes in pressure and temperature that cause the fluid to reach its boiling point, resulting in the generation of these bubbles [97]. The effect of cavitation can be significant, leading to the erosion of solid surfaces and the generation of shockwave that can have profound impacts on the surrounding environment [98].

Acoustic cavitation is a key phenomenon in sonocatalysis, whereby high-intensity sound wave generates microscopic bubbles in a liquid medium [99]. During the cavitation process, these bubbles release energy in the form of heat, shockwave, and free radicals, which can induce chemical reactions in the solution [100]. As the bubbles collapse, they generate extremely high temperatures and pressures in localized regions of the solution [101]. The sudden and intense energy release can result in large increases in temperature, which can accelerate the rate of chemical reactions in the solution [102]. Moreover, the high temperatures generated by acoustic cavitation can lead to thermal excitation of the catalyst, thereby promoting the generation of reactive species, such as electron-hole pairs [103]. This, in turn, can lead to enhanced catalytic activity and selectivity in sonocatalysis.

3. Sonophotocatalytic Process

3.1. Sonophotocatalytic Mechanism

Sonophotocatalysis is essentially a combination of light, ultrasound, and catalyst that accelerates the degradation rates of organic pollutants via increasing the production of active radicals [104–106]. The highly efficient degradation of organic pollutants in sonophotocatalytic process is principally based on the synergistic effect of sonocatalysis and photocatalysis [107,108]. Figure 4 depicts the mechanism for the synergistic effect of photocatalysis and sonocatalysis. The key advantage of combining these two technologies is the greater number of cavitation bubbles generated via ultrasound, and more radicals generated via electron-hole pair separation in semiconductor photocatalysts. In addition, ultrasound continuously cleans the surface of the photocatalyst, which helps maintain the catalyst activity for extended periods. The combination of these two technologies can degrade hydrophobic and hydrophilic organic pollutants [109].

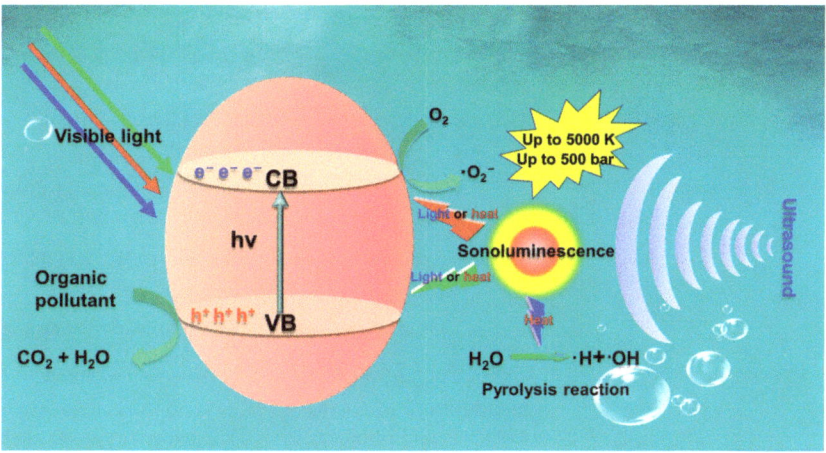

Figure 4. Schematic diagram of sonophotocatalytic mechanism (after [23]).

The sonophotocatalytic degradation of organic pollutants include the phenomena of both sonocatalysis and photocatalysis. Upon irradiation of ultrasonic wave, sonoluminescence and "hot spot" are generated due to cavitation in the aqueous solution. Moreover, the "hot spots" formed through ultrasonic cavitation may cause the pyrolysis of water molecules in contact with the surface of the sonocatalyst, generating hydroxyl radicals (•OH) and hydrogen radicals (•H) (Equations (14)–(19)) [110]. Subsequently, the light generated with a wide range of wavelength from sonoluminescence can excite the catalyst, facilitating charge carriers' formation and the generation of electron-hole pairs in the CB and VB (Equation (20)). Additionally, irradiation of the catalyst's surface with light increases the generation of electron-hole pairs and active radical species during sonophotocatalysis (Equation (21)) [111–113]. In the VB, holes react with water molecules adsorbed on the catalyst surface to generate •OH (Equation (22)). At the same time, electrons generated in the CB react with dissolved oxygen to generate $•O_2^-$, •OH, and H_2O_2 (Equations (23)–(25)). Subsequently, these active species react with organic pollutants to generate different degradation intermediates and even mineralization products (i.e., H_2O and CO_2) (Equation (26)) [114].

$$H_2O + \text{heat (hot spot)} \rightarrow •OH + •H \tag{14}$$

$$H_2O + H• \rightarrow H_2 + •OH \tag{15}$$

$$•OH + •OH \rightarrow H_2O_2 \tag{16}$$

$$O_2 + H• \rightarrow HO_2 \tag{17}$$

$$HO_2 + HO_2 \rightarrow O_2 + H_2O_2 \tag{18}$$

$$H_2O_2 + \text{heat (hot spot)} \rightarrow 2\bullet OH \tag{19}$$

$$\text{Semiconductor Catalyst} + US \rightarrow h^+/VB + e^-/CB \tag{20}$$

$$\text{Semiconductor Catalyst} + h\nu \rightarrow h^+/VB + e^-/CB \tag{21}$$

$$h^+/VB + H_2O \rightarrow \bullet OH \tag{22}$$

$$e^-/CB + O_2 \rightarrow \bullet O_2^- \tag{23}$$

$$2H_2O + 2\bullet O_2^- \rightarrow 2OH\text{-} + H_2O_2 + O_2 \tag{24}$$

$$2\bullet O_2^- + H_2O_2 \rightarrow OH\text{-} + \bullet OH + O_2 \tag{25}$$

$$\text{Organic pollutants} + \text{Reactive oxidative species} \rightarrow H_2O + CO_2 \tag{26}$$

3.2. Summary of the Synergistic Effect during Sonophotocatalytic Process

In order to compare the effects of sonophotocatalysis with those of separate processes (sonocatalysis and photocatalysis), it is necessary to assess the synergistic contribution to the elimination of organic pollutants during the degradation process by sonophotocatalysis. The synergistic effect of a sonophotocatalysis process can be assessed using the synergistic index. This index is calculated as the ratio of the rate constant of sonophotocatalysis to the sum of the rate constants of the individual processes, and is commonly employed to analyze the degree of synergistic enhancement in dye decolorization. The generic expression of the synergy index can be expressed as ([115,116]):

$$\text{Synergy Index} = \frac{k_{\text{sonophotocatalysis}}}{k_{\text{sonocatalysis}} + k_{\text{photocatalysis}}} \tag{27}$$

where k represents the pseudo-first-order rate constants of the photocatalytic, sonophotocatalytic, and sonocatalytic degradation processes, a synergistic index value of >1 means the efficiency of the sonophotocatalytic degradation is higher than the cumulative value of the individual processes (sonocatalytic or photocatalytic).

The synergistic effect of sonophotocatalysis in organic pollutant degradation has been demonstrated in many studies. Mosleh et al. reported that the pseudo-first-order rate constant for sonophotocatalytic degradation of trypan blue was 26.33×10^{-2} min^{-1}, while the sum of the rate constants of photocatalysis and sonocatalysis was only 9.88×10^{-2} min^{-1}, resulting in a synergistic index of 2.53 [117]. Babu et al. reported a synergistic index of 3.7 for the sonophotocatalytic degradation of Methyl orange using CuO-TiO$_2$/rGO nanocatalysts [118]. The authors concluded that the high synergy probably resulted from the combined action of hydroxyl radicals generated by the sonolytic and photocatalytic systems. Benomara et al. reported that the pseudo-first-order rate constants for the degradation of methyl violet 2B were 6.8×10^{-3} for sonocatalysis, 22.9×10^{-3} for photocatalysis, and 39.7×10^{-3} min^{-1} for sonophotocatalysis, demonstrating the significant synergistic effect of sonophotocatalysis [119]. Ahmad et al. investigated the degradation of Rhodamine B (RhB) in photocatalytic, sonocatalytic, and sonophotocatalytic systems, and found that the sonophotocatalytic process exhibiting a higher rate constant compared to the sum of the photocatalytic and sonocatalytic processes [120]. Sonophotocatalytic process was more effective in degrading RhB compared to photocatalytic and sonocatalytic processes due to the presence of more reactive radicals and the increased active surface area of the ZnO/CNT photocatalyst. Togther, these findings highlight the potential of sonophotocatalysis as a promising approach for the efficient degradation of organic dyes in wastewater.

During the sonophotocatalytic process, the combination of ultrasonic wave, light, and photocatalyst can lead to synergistic effect that enhances the degradation of organic pollutants in wastewater. The synergistic effect is attributed to several factors, including the increased production of reactive radicals and the improved mass transfer of the pollutants to the photocatalyst surface. One of the key advantages of sonophotocatalysis is the increased production of reactive radicals, such as •OH, which is highly effective in breaking down organic pollutants. Ultrasonic wave can induce cavitation, which generates high-energy bubbles that collapse and release shockwave and heat, leading to the formation of reactive radicals. Similarly, when a photocatalyst is illuminated with light, electrons are excited, leading to the production of reactive radicals. The combination of ultrasonic wave and light in sonophotocatalysis can lead to a higher production of reactive radicals, as the ultrasonic wave can promote the separation of electron-hole pairs, which are the precursors of reactive radicals, while also enhancing the mass transfer of the pollutants to the photocatalyst surface. Another factor that contributes to the synergistic effect of sonophotocatalysis is the improved mass transfer of the pollutants to the photocatalyst surface. In traditional photocatalysis, the efficiency of pollutant degradation is often limited by the mass transfer of the pollutants from the bulk solution to the photocatalyst surface. The use of ultrasonic wave in sonophotocatalysis can enhance the mass transfer of the pollutants by promoting the formation of micro-scale streams and turbulence, which increase the contact between the pollutants and the photocatalyst surface. In summary, the synergistic effect of sonophotocatalysis in the degradation of organic pollutants can be attributed to the increased production of reactive radicals and the improved mass transfer of the pollutants to the photocatalyst surface.

4. Degradation of Dyes Using TiO$_2$-Based Semiconductor Catalysts

TiO$_2$ has been widely examined among numerous photocatalysts because of its chemical stability, non-toxicity, strong oxidation ability, low cost, high catalytic activity, and photo corrosion resistance. It has been the focus of research in the field of photocatalysis and is one of the most commonly used and most promising photocatalysts [121–123]. The photocatalytic activity of anatase TiO$_2$ is limited to ultraviolet light with wavelength shorter than 387 nm due to its wide band gap of 3.23 eV. As the energy of UV light accounts for only 4% of the total energy of sunlight, TiO$_2$ cannot effieiciently utilize sunlight, which seriously limits its application in photocatalysis [18]. In practical applications, researchers have modified TiO$_2$ to enhance its catalytic activity. There are several primary methods for TiO$_2$ modification, such as noble metal deposition, compound semiconductor, ion doping, and photosensitization. The primary objective of modification is to expand the light-absorption of TiO$_2$ to the visible light spectrum and inhibit the recombination of electron-hole pairs [124]. Additionally, the incorporation of other materials into the TiO$_2$ photocatalyst can enhance its performance. For example, graphene oxide (GO) has been used as a support material for TiO$_2$ nanoparticles to form GO-TiO$_2$ composites. The incorporation of GO can improve the adsorption capacity and photocatalytic activity of TiO$_2$ by increasing the specific surface area and promoting the separation of photogenerated electron-hole pairs [125]. GO also has excellent electrical conductivity, which can facilitate the transfer of electrons and improve the efficiency of photocatalytic reactions. Moreover, metal ions, such as Fe, Cu, and Ag, can be doped into the TiO$_2$ lattice to form metal-doped TiO$_2$ photocatalysts. The incorporation of metal ions can modify the band gap of TiO$_2$ and enhance its photocatalytic activity [126–128]. The metal ions can also act as active sites for the adsorption and degradation of organic dyes [129]. Therefore, the combination of TiO$_2$ with other materials can enhance the its photocatalystic performance and broaden its application in the treatment of organic dye wastewater.

TiO$_2$-based catalysts have shown promising performance in the degradation of organic dyes in various processes, including photocatalysis, sonocatalysis, and sonophotocatalysis.

The efficiency of these processes depends largely on the generation of free radicals, such as •OH, •O_2^-. Table 1 summarizes the performance of TiO_2-based catalysts in the degradation of organic dyes in recent studies.

Table 1. Summary of performance of TiO_2 based catalysts in the degradation of organic dyes.

TiO_2-Based Catalyst	Dye	Catalytic Conditions	Experiment Conditions	Result (Kinetic Constant (k) or Degradation Efficiency (%))	Ref.
ZnO/graphene/TiO_2 (ZGT)	Methylene blue	Bath sonicator Power = 750 W Frequency = 20 kHz	[Catalyst] = 1.00 g/L [Pollutant] = 20 mg/L	1.97×10^{-2} min^{-1}	[130]
N/Ti^{3+} TiO_2/BiOBr0.3	Methylene blue, rhodamine B	Bath sonicator Power = 180 W Frequency = 30 kHz	[Catalyst] = 7.5 mg [Pollutant] = 5 mg/L Time = 50 min	98.2%	[131]
Er^{3+}: YAlO$_3$/TiO_2-ZnO	Acid red B	Bath sonicator Power = 50 W Frequency = 40 kHz	[Catalyst] = 1.0 g/L [Pollutant] = 10 mg/L Time = 60 min	76.84%	[58]
RGO-TiO_{2-x}	Methylene blue	Light power = 150 W	[Catalyst] = 20 mg [Pollutant] = 5 ppm	0.075 min^{-1}	[132]
Black-TiO_2/CoTiO$_3$	Rhodamine B, methylene blue, and methyl orange	Light power = 50 W	[Catalyst] = 100 mg [Pollutant] = 5 ppm Time = 60 min	99%	[133]
Au-TiO_2	Patent blue V	Light power = 570 W/m^2	[Catalyst] = 23 g/L [Pollutant] = 7 mg/L Time = 180 min	93%	[88]
TiO_2_Ag_Graphene	Black 5	Bath sonicator Power = 30 W/L Frequency = 40 kHz UV light power = 5 W	[Catalyst] = 0.03 g [Pollutant] = 5 mg/L	0.05 min^{-1}	[134]
NT-TBWx	Methylene blue (MB)	Bath sonicator Power = 180 W Frequency = 35 kHz UV light power = 100 mW/cm^2	[Catalyst] = 7.5 mg [Pollutant] = 5 mg/L Time = 50 min	99%	[135]
CNTs/TiO_2	methyl orange (MO)	Bath sonicator Power = 50 W Frequency = 20kHz UV light Power = 30 W	[Catalyst] = 50 mg [Pollutant] = 25 ppm	0.01118 min^{-1}	[113]

Nuengmatcha et al. showed that the ZnO/graphene/TiO_2 hybrid catalyst prepared using solvothermal method was more efficient at degrading ZGT dye compared to the indiviudal components [130]. The high surface area of graphene allows for better dispersion of ZnO and TiO_2, leading to increased absorption of ultrasonic irradiation and the generation of more electron-hole pairs. Yao et al. synthesized TiO_2/BiOBr heterojunctions with N/Ti^{3+} co-doping using one-step in situ hydrothermal method and demonstrated that they exhibited higher sonocatalytic activity in degrading methylene blue compared to pristine TiO_2 [131]. Specifically, NT-TB$_{0.3}$ exhibited the highest degradation efficiency of 98.2% after 50 min of ultrasound irradiation. The improved catalytic activity was attributed to the formation of a heterojunction between TiO_2/BiOBr, which enhances the separation of electron-hole pairs. These studies highlight the potential of hybrid and composite catalysts in enhancing the performance of sonocatalytic and photocatalytic reactions and provide insight into the mechanisms underlying their improved activity.

Sriramoju et al. synthesized RGO-TiO_{2-x} nanocomposites using one-step in situ hydrothermal method and observed that these nanocomposites displayed exceptional photocatalytic degradation performance against diverse organic dyes when exposed to UV-visible irradiation [132]. The rate constants for Rhodamine-B, methylene blue, and rose red dye were 0.083 min^{-1}, 0.075 min^{-1}, and 0.093 min^{-1}, respectively. The superior photocatalytic performance observed in TiO_{2-x} samples was linked to the presence of highly conductive RGO, which improves the mobility of photo-generated charge carriers and reduces electron-hole pair recombination. The oxygen vacancy/Ti^{3+} was also identified as an important contributor to the enhanced photocatalytic activity. Mousavi and colleagues developed a Z-scheme heterojunction photocatalyst consisting of Black-TiO_2 and $CoTiO_3$, which exhibits visible-light responsiveness and is capable of decomposing a variety of organic dyes, includimg rhodamine B, methylene blue, and methyl orange [133]. The much higher photocatalytic activity of B-TiO_2/CTO nanocomposites compared to B-TiO_2 and CTO is attributed to the improved generation, separation, and transportation of charge carriers. Moreover, the combination of B-TiO_2 with CTO increased the specific surface area of the nanocomposites, which increases the active sites on the catalyst surface and the generation of photo-generated electron-hole pairs. These studies highlight the potential of constructing hybrid nanocomposites of TiO_2 to enhance the photocatalytic removal of organic dyes.

Lozano et al. synthesized a novel Ag-graphene oxide/TiO_2 catalyst and showed that it effectively degraded Black 5 and orange II dyes in a sonophotocatalysis system under ultrasonic and UV irradiation [134]. Ultrasound and UV light were observed to have significant synergistic effect on the degradation of organic dyes in the presence of the catalyst. Sun et al. investigated the sonophotocatalytic removal of organic pollutants in water using N/Ti^{3+}-doped biphasic TiO_2/Bi_2WO_6 heterojunctions, and found that the catalytic activity of NT-TBWx in the sonophotocatalytic system for the degradation of methyl blue was much higher than that in the photocatalytic and sonocatalytic systems [135]. Compared to TiO_2 and NT-TiO_2, the NT-TBWx heterojunctions exhibited superior sonophotocatalytic activity. The improved sonophotocatalytic efficiency of the NT-TBWx composites is likely due to the synergistic effect of photocatalysis and sonocatalysis, as well as the N/Ti^{3+} co-doping and heterophase junctions.

The performance of sonocatalytic, photocatalytic, and sonophotocatalytic processes is influenced by a range of factors, including catalyst dose, solution pH, and the type and concentration of organic dyes. These factors must be optimized to improve the overall dye removal efficiency in practical applications. For instance, pH plays a critical role in determining the surface charge potential of the catalyst, which can significantly impact its interaction with the organic dye molecules. Under acidic conditions, the surface of TiO_2 is positively charged, allowing the adsorption of negatively charged dye molecules, thus increases the efficiency of the photocatalytic degradation. On the other hand, the surface of TiO_2 becomes negatively charged under alkaline conditions, which reduces photocatalytic activityin the degradation of negatively charged dyes. The catalyst dose is another crucial factor that affects the performance of TiO_2-based catalysts. The amount of catalyst used affects the number of active sites available for the adsorption of dye molecules, which directly influences the overal degradation rate. However, high catalyst doses may cause light shielding in the solution and reduce photocatalytic performance. In addition, the type and concentration of organic dyes also play important roles in the efficiency of TiO_2-based catalysts. The adsorption of dye molecules on the surface of TiO_2 is influenced by the size, structure, and chemical composition of the dye molecules, which affect the overall degradation rate. High concentrations of organic dyes can lead to increased light scattering and lower photocatalytic activity.

Taken together, photocatalytic, sonocatalytic, and sonophotocatalytic activity of TiO_2-based catalysts is influenced by various factors, such as pH, catalyst dose, and the type and concentration of organic dyes. In practical applications, it is necessary to optimize these factors to achieve efficient treatment of organic pollutants in water and wastewater. Addressing the challenges of high cost and limited efficacy under visible light is crucial for the widespread adoption of TiO_2-based catalysts in water and wastewater treatment.

5. Further Research Trends

While the general mechanism of sonophotocatalysis has been relativly well understood, there are several challenges that need to be addressed to make it a practical and effective method for dye decolorization. One of the major challenges is the scale-up of the sonophotocatalytic process from laboratory to industrial scale, as the reaction conditions and equipment used in the laboratory may not be suitable for large-scale applications. Thus, it is necessary to develop and optimize the sonophotocatalytic process for industrial applications, which may require innovative catalyst and reactor design, as well as novel ways of supplying the ultrasound and light energy.

The cost-effectiveness of the sonophotocatalytic process is a crucial aspect that needs to be considered. Despite the tremendous photocatalytic activity of noble metal/TiO_2 systems, their practical utilization is remarkably constrained due to the high cost as well as limited accessibility for precious metals. This hinders the widespread application of sonophotocatalysis for dye decolorization, necessitating the development of cost-effective catalysts with high activity and stability, such as non-noble metal-based catalysts or composites of TiO_2 with other materials. To enhance the properties of TiO_2-based heterojunction photocatalysts, more efficient synthesis techniques must be explored to produce catalysts with tailored morphologies and compositions. However, it is challenging to mass produce high-quality, homogeneous TiO_2-based heterostructure photocatalysts. Therefore, the design as well as performance for TiO_2-based heterojunctions must be further improved, which requires better understanding of the photocatalytic reaction mechanism. Additional investigation is required to explore both the thermodynamics and kinetics of surface catalytic processes, as well as the mechanism of charge carrier transfer. To enable the effective utilization of TiO_2-based heterojunction photocatalysts in natural environments with sunlight, it is crucial to extend the excitation wavelength of photocatalysts, particularly by broadening their light-response from UV to visible light, which can enhance their solar conversion efficiency. Additionally, the effect of environmental factors, such as temperature, pH, and the presence of other pollutants, on the sonophotocatalytic process need to be investigated. Changes in these factors can affect the performance of sonophotocatalytic processes, necessitating the optimization of treatment conditions to aqequate decolorization efficiency. In brief, the cost-effectiveness, synthesis techniques, photocatalytic reaction mechanism, and environmental factors are critical aspects that must be considered to enhance the practical application of sonophotocatalytic processes for dye decolorization.

TiO_2 is only responsive to UV light, which accounts for a small portion of the solar spectrum. Therefore, there is a need for the development of visible-light-responsive TiO_2-based catalysts to expand their applications in water and wastewater treatment. Several strategies have been proposed to improve the visible-light responsiveness of TiO_2-based catalysts, such as doping with transition metals, modifying with carbon materials, and forming heterojunctions with other semiconductors.

In addition to the technical challenges of sonophotocatalytic treatment of organic pollutants, more in-depth understanding of the mechanism of sonophotocatalytic degradation of dyes is essential. Investigating the interactions between the catalyst, organic dye, and environmental factors is critical for developing an effective and efficient sonophotocatalytic process. Characterization of the reaction pathways and intermediates of dyes enables the prediction of the toxicity and environmental impact of the degradation products. Therefore, future research should focus on developing detailed mechanistic models that can predict the reaction pathways and intermediates in sonophotocatalytic degradation of dyes. This requires a combination of experimental and theoretical approaches to elucidate the complex interplay between the catalyst, organic dye, and environmental factors.

Besides the above technical and mechanistic challenges for the sonophotocatalytic treatment of organic pollutants, there are potential risks associated with the process that must be assessed. The release of nanoparticles from catalyst breakdown may have adverse effects on human health and the ecosystem. This should be thoroughly explored, and appropriate measures should be taken to mitigate their impact on the ecosystem and human health. Nanoparticle release can be minimized by optimizing the sonophotocatalytic process and designing catalysts with minimal nanoparticle release. Overall, a comprehensive risk assessment of the sonophotocatalytic process is essential to ensure that it is safe and sustainable for practical applications.

6. Conclusions

The technology of sonophotocatalysis has become an important method for treating organic pollutants in water and wastewater. Sonophotocatalysis has significant synergistic effect, resulting in faster pollutant removal compared to sonocatalysis and photocatalysis. With significant improvements in terms of efficiency and treatment time, sonophotocatalysis has the potential to be a practical and effective method for dye decolorization,. However, several challenges need to be addressed, including scale-up, cost-effectiveness, optimization of process conditions, mechanistic understanding, and risk assessment, to ensure that sonophotocatalysis can be widely applied in the treatment of dye wastewater.

The synergistic effect of sonophotocatalysis offers unique opportunity to overcome some of the limitations of other treatment technologies, including sonocatalysis and photocatalysis. Therefore, further research in this field could lead to the development of new and efficient water treatment technologies that can address a wide range of environmental problems.

Author Contributions: G.W.: Conceptualization, Writing—original draft, Writing—review & editing; H.C.: Conceptualization, Resources, Supervision, Writing—original draft, Writing—review & editing. All authors have read and agreed to the published version of the manuscript.

Funding: This research was funded by the Natural Science Foundation of China (Grant Nos.: U2006212 and 41725015).

Institutional Review Board Statement: Not applicable.

Informed Consent Statement: Not applicable.

Data Availability Statement: No data are associated with this article.

Acknowledgments: This work was supported in parts by the Natural Science Foundation of China (Grant Nos. 41725015 and U2006212).

Conflicts of Interest: The authors declare no conflict of interest.

References

1. Yang, W.; Ding, K.; Chen, J.; Wang, H.; Deng, X. Synergistic Multisystem Photocatalytic Degradation of Anionic and Cationic Dyes Using Graphitic Phase Carbon Nitride. *Molecules* **2023**, *28*, 2796. [CrossRef] [PubMed]
2. Wei, X.; Feng, H.; Li, L.; Gong, J.; Jiang, K.; Xue, S.; Chu, P.K. Synthesis of tetragonal prismatic γ-In$_2$Se$_3$ nanostructures with predominantly {110} facets and photocatalytic degradation of tetracycline. *Appl. Catal. B-Environ.* **2020**, *260*, 118218. [CrossRef]
3. Zou, X.; Zhang, J.; Zhao, X.; Zhang, Z. MoS$_2$/RGO composites for photocatalytic degradation of ranitidine and elimination of NDMA formation potential under visible light. *Chem. Eng. J.* **2020**, *383*, 123084. [CrossRef]
4. Huang, D.; Wang, H.; Wu, Y. Photocatalytic Aerobic Oxidation of Biomass-Derived 5-HMF to DFF over MIL-53(Fe)/g-C$_3$N$_4$ Composite. *Molecules* **2022**, *27*, 8537. [CrossRef] [PubMed]
5. Heidari, S.; Haghighi, M.; Shabani, M. Sunlight-activated BiOCl/BiOBr–Bi$_{24}$O$_{31}$Br$_{10}$ photocatalyst for the removal of pharmaceutical compounds. *J. Clean. Prod.* **2020**, *259*, 120679. [CrossRef]
6. Zhou, Y.; Yu, M.; Liang, H.; Chen, J.; Xu, L.; Niu, J. Novel dual-effective Z-scheme heterojunction with g-C$_3$N$_4$, Ti$_3$C$_2$ MXene and black phosphorus for improving visible light-induced degradation of ciprofloxacin. *Appl. Catal. B-Environ.* **2021**, *291*, 120105. [CrossRef]
7. Liu, K.; Tong, Z.; Muhammad, Y.; Huang, G.; Zhang, H.; Wang, Z.; Zhu, Y.; Tang, R. Synthesis of sodium dodecyl sulfate modified BiOBr/magnetic bentonite photocatalyst with Three-dimensional parterre like structure for the enhanced photodegradation of tetracycline and ciprofloxacin. *Chem. Eng. J.* **2020**, *388*, 124374. [CrossRef]
8. Zhang, M.; Lai, C.; Li, B.; Huang, D.; Liu, S.; Qin, L.; Yi, H.; Fu, Y.; Xu, F.; Li, M. Ultrathin oxygen-vacancy abundant WO$_3$ decorated monolayer Bi$_2$WO$_6$ nanosheet: A 2D/2D heterojunction for the degradation of Ciprofloxacin under visible and NIR light irradiation. *J. Colloid Interface Sci.* **2019**, *556*, 557–567. [CrossRef]
9. Irshad, A.; Warsi, M.F.; Agboola, P.O.; Dastgeer, G.; Shahid, M. Sol-gel assisted Ag doped NiAl$_2$O$_4$ nanomaterials and their nanocomposites with g-C$_3$N$_4$ nanosheets for the removal of organic effluents. *J. Alloys Compd.* **2022**, *902*, 163805. [CrossRef]
10. Gao, P.; Cui, J.; Deng, Y. Direct regeneration of ion exchange resins with sulfate radical-based advanced oxidation for enabling a cyclic adsorption–regeneration treatment approach to aqueous perfluorooctanoic acid (PFOA). *Chem. Eng. J.* **2021**, *405*, 126698. [CrossRef]
11. Wu, J.; Wang, T.; Wang, J.; Zhang, Y.; Pan, W.-P. A novel modified method for the efficient removal of Pb and Cd from wastewater by biochar: Enhanced the ion exchange and precipitation capacity. *Sci. Total Environ.* **2021**, *754*, 142150. [CrossRef] [PubMed]

12. Vapnik, H.; Elbert, J.; Su, X. Redox-copolymers for the recovery of rare earth elements by electrochemically regenerated ion-exchange. *J. Mater. Chem. A* **2021**, *9*, 20068–20077. [CrossRef]
13. Chen, J.; Li, Y.; Li, M.; Shi, J.; Wang, L.; Luo, S.; Liu, H. Chemical Flocculation-Based Green Algae Materials for Photobiological Hydrogen Production. *ACS Appl. Bio. Mater.* **2022**, *5*, 897–903. [CrossRef] [PubMed]
14. Kurniawan, S.B.; Imron, M.F.; Sługocki, Ł.; Nowakowski, K.; Ahmad, A.; Najiya, D.; Abdullah, S.R.S.; Othman, A.R.; Purwanti, I.F.; Hasan, H.A. Assessing the effect of multiple variables on the production of bioflocculant by Serratia marcescens: Flocculating activity, kinetics, toxicity, and flocculation mechanism. *Sci. Total Environ.* **2022**, *836*, 155564. [CrossRef]
15. Wang, X.; Wang, D.; Xu, J.; Fu, J.; Zheng, G.; Zhou, L. Modified chemical mineralization-alkali neutralization technology: Mineralization behavior at high iron concentrations and its application in sulfur acid spent pickling solution. *Water Res.* **2022**, *218*, 118513. [CrossRef]
16. Xie, L.; Du, T.; Wang, J.; Ma, Y.; Ni, Y.; Liu, Z.; Zhang, L.; Yang, C.; Wang, J. Recent advances on heterojunction-based photocatalysts for the degradation of persistent organic pollutants. *Chem. Eng. J.* **2021**, *426*, 130617. [CrossRef]
17. Wang, L.; Bahnemann, D.W.; Bian, L.; Dong, G.; Zhao, J.; Wang, C. Two-dimensional layered zinc silicate nanosheets with excellent photocatalytic performance for organic pollutant degradation and CO_2 conversion. *Angew. Chem. Int. Ed.* **2019**, *131*, 8187–8192. [CrossRef]
18. Qutub, N.; Singh, P.; Sabir, S.; Sagadevan, S.; Oh, W.-C. Enhanced photocatalytic degradation of Acid Blue dye using CdS/TiO_2 nanocomposite. *Sci. Rep.* **2022**, *12*, 5759. [CrossRef]
19. Han, B.; Xie, A.; Yu, Q.; Huang, F.; Shen, Y.; Zhu, L. Synthesis of $PbSO_4$ crystals by hydrogel template on postprocessing strategy for secondary pollution. *Appl. Surf. Sci.* **2012**, *261*, 623–627. [CrossRef]
20. Bao, S.; Li, K.; Ning, P.; Peng, J.; Jin, X.; Tang, L. Highly effective removal of mercury and lead ions from wastewater by mercaptoamine-functionalised silica-coated magnetic nano-adsorbents: Behaviours and mechanisms. *Appl. Surf. Sci.* **2017**, *393*, 457–466. [CrossRef]
21. Xin, S.; Zeng, Z.; Zhou, X.; Luo, W.; Shi, X.; Wang, Q.; Deng, H.; Du, Y. Recyclable Saccharomyces cerevisiae loaded nanofibrous mats with sandwich structure constructing via bio-electrospraying for heavy metal removal. *J. Hazard. Mater.* **2017**, *324*, 365–372. [CrossRef] [PubMed]
22. Mamba, G.; Mishra, A. Advances in magnetically separable photocatalysts: Smart, recyclable materials for water pollution mitigation. *Catalysts* **2016**, *6*, 79. [CrossRef]
23. Wang, G.; Cheng, H. Facile synthesis of a novel recyclable dual Z-scheme $WO_3/NiFe_2O_4/BiOBr$ composite with broad-spectrum response and enhanced sonocatalytic performance for levofloxacin removal in aqueous solution. *Chem. Eng. J.* **2023**, *461*, 141941. [CrossRef]
24. Vasseghian, Y.; Dragoi, E.-N.; Almomani, F. A comprehensive review on MXenes as new nanomaterials for degradation of hazardous pollutants: Deployment as heterogeneous sonocatalysis. *Chemosphere* **2022**, *287*, 132387. [CrossRef] [PubMed]
25. Nas, M.S. $AgFe_2O_4$/MWCNT nanoparticles as novel catalyst combined adsorption-sonocatalytic for the degradation of methylene blue under ultrasonic irradiation. *J. Environ. Chem. Eng.* **2021**, *9*, 105207. [CrossRef]
26. Dulta, K.; Koşarsoy Ağçeli, G.; Chauhan, P.; Jasrotia, R.; Chauhan, P.; Ighalo, J.O. Multifunctional CuO nanoparticles with enhanced photocatalytic dye degradation and antibacterial activity. *Sustain. Environ. Res.* **2022**, *32*, 1–15. [CrossRef]
27. Ramamoorthy, S.; Das, S.; Balan, R.; Lekshmi, I. TiO_2-ZrO_2 nanocomposite with tetragonal zirconia phase and photocatalytic degradation of Alizarin Yellow GG azo dye under natural sunlight. *Mater. Today Proc.* **2021**, *47*, 4641–4646. [CrossRef]
28. Wani, S.I.; Ganie, A.S. Ag_2O incorporated ZnO-TiO_2 nanocomposite: Ionic conductivity and photocatalytic degradation of an organic dye. *Inorg. Chem. Commun.* **2021**, *128*, 108567. [CrossRef]
29. Vellingiri, K.; Vikrant, K.; Kumar, V.; Kim, K.-H. Advances in thermocatalytic and photocatalytic techniques for the room/low temperature oxidative removal of formaldehyde in air. *Chem. Eng. J.* **2020**, *399*, 125759. [CrossRef]
30. Wang, J.; Zhang, T.; Jiang, S.; Ma, X.; Shao, X.; Liu, Y.; Wang, D.; Li, X.; Li, B. Controllable self-assembly of BiOI/oxidized mesocarbon microbeads core-shell composites: A novel hierarchical structure facilitated photocatalytic activities. *Chem. Eng. Sci.* **2020**, *221*, 115653. [CrossRef]
31. Lei, X.; Ouyang, C.; Huang, K. A first-principles investigation of Janus MoSSe as a catalyst for photocatalytic water-splitting. *Appl. Surf. Sci.* **2021**, *537*, 147919. [CrossRef]
32. Majumder, S.; Chatterjee, S.; Basnet, P.; Mukherjee, J. ZnO based nanomaterials for photocatalytic degradation of aqueous pharmaceutical waste solutions–A contemporary review. *Environ. Nanotechnol. Monit. Manag.* **2020**, *14*, 100386. [CrossRef]
33. Costarramone, N.; Kartheuser, B.; Pecheyran, C.; Pigot, T.; Lacombe, S. Efficiency and harmfulness of air-purifying photocatalytic commercial devices: From standardized chamber tests to nanoparticles release. *Catal. Today* **2015**, *252*, 35–40. [CrossRef]
34. Cushing, S.K.; Li, J.; Meng, F.; Senty, T.R.; Suri, S.; Zhi, M.; Li, M.; Bristow, A.D.; Wu, N. Photocatalytic activity enhanced by plasmonic resonant energy transfer from metal to semiconductor. *J. Am. Chem. Soc.* **2012**, *134*, 15033–15041. [CrossRef]
35. Sheikh, M.; Pazirofteh, M.; Dehghani, M.; Asghari, M.; Rezakazemi, M.; Valderrama, C.; Cortina, J.-L. Application of ZnO nanostructures in ceramic and polymeric membranes for water and wastewater technologies: A review. *Chem. Eng. J.* **2020**, *391*, 123475. [CrossRef]
36. Zhou, D.; Wu, S.; Cheng, G.; Che, C.-M. A gold (iii)–TADF emitter as a sensitizer for high-color-purity and efficient deep-blue solution-processed OLEDs. *J. Mater. Chem. C* **2022**, *10*, 4590–4596. [CrossRef]

37. Nemati, F.; Nikkhah, S.H.; Elhampour, A. An environmental friendly approach for the catalyst-free synthesis of highly substituted pyrazoles promoted by ultrasonic radiation. *Chin. Chem. Lett.* **2015**, *26*, 1397–1399. [CrossRef]
38. Ali El-Remaily, M.A.E.A.A.; El-Dabea, T.; Alsawat, M.; Mahmoud, M.H.; Alfi, A.A.; El-Metwaly, N.; Abu-Dief, A.M. Development of new thiazole complexes as powerful catalysts for synthesis of pyrazole-4-carbonitrile derivatives under ultrasonic irradiation condition supported by DFT studies. *ACS Omega* **2021**, *6*, 21071–21086. [CrossRef]
39. Wojcieszyńska, D.; Łagoda, K.; Guzik, U. Diclofenac Biodegradation by Microorganisms and with Immobilised Systems—A Review. *Catalysts* **2023**, *13*, 412.
40. Zhu, Z.-H.; Liu, Y.; Song, C.; Hu, Y.; Feng, G.; Tang, B.Z. Porphyrin-Based Two-Dimensional Layered Metal–Organic Framework with Sono-/Photocatalytic Activity for Water Decontamination. *ACS Nano* **2021**, *16*, 1346–1357. [CrossRef]
41. Guo, L.; Chen, Y.; Ren, Z.; Li, X.; Zhang, Q.; Wu, J.; Li, Y.; Liu, W.; Li, P.; Fu, Y. Morphology engineering of type-II heterojunction nanoarrays for improved sonophotocatalytic capability. *Ultrason. Sonochem.* **2021**, *81*, 105849. [CrossRef] [PubMed]
42. Hosseini, M.; Kahkha, M.R.R.; Fakhri, A.; Tahami, S.; Lariche, M.J. Degradation of macrolide antibiotics via sono or photo coupled with Fenton methods in the presence of ZnS quantum dots decorated SnO_2 nanosheets. *J. Photochem. Photobiol. B* **2018**, *185*, 24–31. [CrossRef] [PubMed]
43. Preeyanghaa, M.; Vinesh, V.; Neppolian, B. Construction of S-scheme 1D/2D rod-like g-C_3N_4/V_2O_5 heterostructure with enhanced sonophotocatalytic degradation for Tetracycline antibiotics. *Chemosphere* **2022**, *287*, 132380. [CrossRef]
44. Liu, J.; Ma, N.; Wu, W.; He, Q. Recent progress on photocatalytic heterostructures with full solar spectral responses. *Chem. Eng. J.* **2020**, *393*, 124719. [CrossRef]
45. Rodríguez-González, V.; Obregón, S.; Patrón-Soberano, O.A.; Terashima, C.; Fujishima, A. An approach to the photocatalytic mechanism in the TiO_2-nanomaterials microorganism interface for the control of infectious processes. *Appl. Catal. B-Environ.* **2020**, *270*, 118853. [CrossRef] [PubMed]
46. Wen, X.-J.; Shen, C.-H.; Fei, Z.-H.; Fang, D.; Liu, Z.-T.; Dai, J.-T.; Niu, C.-G. Recent developments on AgI based heterojunction photocatalytic systems in photocatalytic application. *Chem. Eng. J.* **2020**, *383*, 123083. [CrossRef]
47. Theerthagiri, J.; Lee, S.J.; Karuppasamy, K.; Arulmani, S.; Veeralakshmi, S.; Ashokkumar, M.; Choi, M.Y. Application of advanced materials in sonophotocatalytic processes for the remediation of environmental pollutants. *J. Hazard. Mater.* **2021**, *412*, 125245. [CrossRef]
48. Qiu, P.; Park, B.; Choi, J.; Thokchom, B.; Pandit, A.B.; Khim, J. A review on heterogeneous sonocatalyst for treatment of organic pollutants in aqueous phase based on catalytic mechanism. *Ultrason. Sonochem.* **2018**, *45*, 29–49. [CrossRef] [PubMed]
49. Liu, P.; Wu, Z.; Abramova, A.V.; Cravotto, G. Sonochemical processes for the degradation of antibiotics in aqueous solutions: A review. *Ultrason. Sonochem.* **2021**, *74*, 105566. [CrossRef]
50. He, Y.; Ma, Z.; Junior, L.B. Distinctive binary g-C_3N_4/MoS_2 heterojunctions with highly efficient ultrasonic catalytic degradation for levofloxacin and methylene blue. *Ceram. Int.* **2020**, *46*, 12364–12372. [CrossRef]
51. Waheed, I.F.; Al-Janabi, O.Y.T.; Foot, P.J. Novel $MgFe_2O_4$-CuO/GO heterojunction magnetic nanocomposite: Synthesis, characterization, and batch photocatalytic degradation of methylene blue dye. *J. Mol. Liq.* **2022**, *357*, 119084. [CrossRef]
52. Xu, X.; Xu, X.; Wang, T.; Xu, M.; Yang, H.; Hou, J.; Cao, D.; Wang, Q. Construction of Z-scheme CdS/Ag/TiO_2 NTs photocatalysts for photocatalytic dye degradation and hydrogen evolution. *Spectrochim. Acta A* **2022**, *276*, 121215. [CrossRef]
53. Abazari, R.; Sanati, S.; Morsali, A.; Kirillov, A.M. Instantaneous sonophotocatalytic degradation of tetracycline over NU-1000@$ZnIn_2S_4$ core–shell nanorods as a robust and eco-friendly catalyst. *Inorg. Chem.* **2021**, *60*, 9660–9672. [CrossRef] [PubMed]
54. Hoo, D.Y.; Low, Z.L.; Low, D.Y.S.; Tang, S.Y.; Manickam, S.; Tan, K.W.; Ban, Z.H. Ultrasonic cavitation: An effective cleaner and greener intensification technology in the extraction and surface modification of nanocellulose. *Ultrason. Sonochem.* **2022**, *90*, 106176. [CrossRef] [PubMed]
55. Moftakhari Anasori Movahed, S.; Calgaro, L.; Marcomini, A. Trends and characteristics of employing cavitation technology for water and wastewater treatment with a focus on hydrodynamic and ultrasonic cavitation over the past two decades: A Scientometric analysis. *Sci. Total Environ.* **2023**, *858*, 159802. [CrossRef]
56. He, L.-L.; Zhu, Y.; Qi, Q.; Li, X.-Y.; Bai, J.-Y.; Xiang, Z.; Wang, X. Synthesis of $CaMoO_4$ microspheres with enhanced sonocatalytic performance for the removal of Acid Orange 7 in the aqueous environment. *Sep. Purif. Technol.* **2021**, *276*, 119370. [CrossRef]
57. Wang, G.; Ma, X.; Liu, J.; Qin, L.; Li, B.; Hu, Y.; Cheng, H. Design and performance of a novel direct Z-scheme $NiGa_2O_4$/CeO_2 nanocomposite with enhanced sonocatalytic activity. *Sci. Total Environ.* **2020**, *741*, 140192. [CrossRef]
58. Gao, J.; Jiang, R.; Wang, J.; Kang, P.; Wang, B.; Li, Y.; Li, K.; Zhang, X. The investigation of sonocatalytic activity of Er^{3+}: $YAlO_3$/TiO_2-ZnO composite in azo dyes degradation. *Ultrason. Sonochem.* **2011**, *18*, 541–548. [CrossRef]
59. Gao, H.; Pei, F.; Hu, G.; Liu, W.; Meng, A.; Wang, H.; Shao, H.; Li, W. The influence of pressure on the acoustic cavitation in saturated CO_2-expanded N, N-dimethylformamide. *Ultrason. Sonochem.* **2022**, *83*, 105934. [CrossRef]
60. Kozmus, G.; Zevnik, J.; Hočevar, M.; Dular, M.; Petkovšek, M. Characterization of cavitation under ultrasonic horn tip—Proposition of an acoustic cavitation parameter. *Ultrason. Sonochem.* **2022**, *89*, 106159. [CrossRef]
61. Yao, C.; Zhao, S.; Liu, L.; Liu, Z.; Chen, G. Ultrasonic emulsification: Basic characteristics, cavitation, mechanism, devices and application. *Front. Chem. Sci. Eng.* **2022**, *16*, 1560–1583. [CrossRef]
62. Zhang, H.; Qiao, J.; Li, G.; Li, S.; Wang, G.; Wang, J.; Song, Y. Preparation of Ce^{4+}-doped $BaZrO_3$ by hydrothermal method and application in dual-frequent sonocatalytic degradation of norfloxacin in aqueous solution. *Ultrason. Sonochem.* **2018**, *42*, 356–367. [CrossRef] [PubMed]

63. Wang, G.; Li, S.; Ma, X.; Qiao, J.; Li, G.; Zhang, H.; Wang, J.; Song, Y. A novel Z-scheme sonocatalyst system, Er^{3+}:$Y_3Al_5O_{12}$@Ni($Fe_{0.05}Ga_{0.95}$)$_2O_4$-Au-$BiVO_4$, and application in sonocatalytic degradation of sulfanilamide. *Ultrason. Sonochem.* **2018**, *45*, 150–166. [CrossRef] [PubMed]
64. Huang, Y.; Wang, G.; Zhang, H.; Li, G.; Fang, D.; Wang, J.; Song, Y. Hydrothermal-precipitation preparation of CdS@(Er^{3+}:$Y_3Al_5O_{12}$/ZrO_2) coated composite and sonocatalytic degradation of caffeine. *Ultrason. Sonochem.* **2017**, *37*, 222–234. [CrossRef] [PubMed]
65. Wang, G.; Huang, Y.; Li, G.; Zhang, H.; Wang, Y.; Li, B.; Wang, J.; Song, Y. Preparation of a novel sonocatalyst, Au/$NiGa_2O_4$-Au-Bi_2O_3 nanocomposite, and application in sonocatalytic degradation of organic pollutants. *Ultrason. Sonochem.* **2017**, *38*, 335–346. [CrossRef]
66. Abdurahman, M.H.; Abdullah, A.Z.; Shoparwe, N.F. A comprehensive review on sonocatalytic, photocatalytic, and sonophotocatalytic processes for the degradation of antibiotics in water: Synergistic mechanism and degradation pathway. *Chem. Eng. J.* **2021**, *413*, 127412. [CrossRef]
67. Hu, Y.; Wei, J.; Shen, Y.; Chen, S.; Chen, X. Barrier-breaking effects of ultrasonic cavitation for drug delivery and biomarker release. *Ultrason. Sonochem.* **2023**, *94*, 106346. [CrossRef]
68. Li, S.; Wang, G.; Qiao, J.; Zhou, Y.; Ma, X.; Zhang, H.; Li, G.; Wang, J.; Song, Y. Sonocatalytic degradation of norfloxacin in aqueous solution caused by a novel Z-scheme sonocatalyst, mMBIP-MWCNT-In_2O_3 composite. *J. Mol. Liq.* **2018**, *254*, 166–176. [CrossRef]
69. Hassandoost, R.; Kotb, A.; Movafagh, Z.; Esmat, M.; Guegan, R.; Endo, S.; Jevasuwan, W.; Fukata, N.; Sugahara, Y.; Khataee, A.; et al. Nanoarchitecturing bimetallic manganese cobaltite spinels for sonocatalytic degradation of oxytetracycline. *Chem. Eng. J.* **2022**, *431*, 133851. [CrossRef]
70. Afzal, M.Z.; Zu, P.; Zhang, C.-M.; Guan, J.; Song, C.; Sun, X.-F.; Wang, S.-G. Sonocatalytic degradation of ciprofloxacin using hydrogel beads of TiO_2 incorporated biochar and chitosan. *J. Hazard. Mater.* **2022**, *434*, 128879. [CrossRef]
71. Jorfi, S.; Pourfadakari, S.; Kakavandi, B. A new approach in sono-photocatalytic degradation of recalcitrant textile wastewater using MgO@ Zeolite nanostructure under UVA irradiation. *Chem. Eng. J.* **2018**, *343*, 95–107. [CrossRef]
72. Isari, A.A.; Mehregan, M.; Mehregan, S.; Hayati, F.; Kalantary, R.R.; Kakavandi, B. Sono-photocatalytic degradation of tetracycline and pharmaceutical wastewater using WO_3/CNT heterojunction nanocomposite under US and visible light irradiations: A novel hybrid system. *J. Hazard. Mater.* **2020**, *390*, 122050. [CrossRef] [PubMed]
73. Liu, Y.-C.; Wang, J.-Q.; Wang, Y.; Chen, C.-L.; Wang, X.; Xiang, Z. Sonocatalytic degradation of ciprofloxacin by BiOBr/$BiFeO_3$. *Appl. Catal. A* **2022**, *643*, 118776. [CrossRef]
74. Xu, L.; Liu, N.-P.; An, H.-L.; Ju, W.-T.; Liu, B.; Wang, X.-F.; Wang, X. Preparation of Ag_3PO_4/$CoWO_4$ S-scheme heterojunction and study on sonocatalytic degradation of tetracycline. *Ultrason. Sonochem.* **2022**, *89*, 106147. [CrossRef] [PubMed]
75. Pang, Y.L.; Koe, A.Z.Y.; Chan, Y.Y.; Lim, S.; Chong, W.C. Enhanced Sonocatalytic Performance of Non-Metal Graphitic Carbon Nitride (g-C_3N_4)/Coconut Shell Husk Derived-Carbon Composite. *Sustainability* **2022**, *14*, 3244. [CrossRef]
76. Sun, M.; Lin, X.; Meng, X.; Liu, W.; Ding, Z. Ultrasound-driven ferroelectric polarization of TiO_2/$Bi_{0.5}Na_{0.5}TiO_3$ heterojunctions for improved sonocatalytic activity. *J. Alloys Compd.* **2022**, *892*, 162065. [CrossRef]
77. Li, S.; Zhang, M.; Ma, X.; Qiao, J.; Zhang, H.; Wang, J.; Song, Y. Preparation of ortho-symmetric double (OSD) Z-scheme SnO_2\CdSe/Bi_2O_3 sonocatalyst by ultrasonic-assisted isoelectric point method for effective degradation of organic pollutants. *J. Ind. Eng. Chem.* **2019**, *72*, 157–169. [CrossRef]
78. Lu, L.; Wang, T.; Fang, C.; Song, L.; Qian, C.; Lv, Z.; Fang, Y.; Liu, X.; Yu, X.; Xu, X.; et al. Oncolytic Impediment/Promotion Balance Disruption by Sonosensitizer-Free Nanoplatforms Unfreezes Autophagy-Induced Resistance to Sonocatalytic Therapy. *ACS Appl. Mater. Interfaces* **2022**, *14*, 36462–36472. [CrossRef]
79. Haddadi, S.; Khataee, A.; Arefi-Oskoui, S.; Vahid, B.; Orooji, Y.; Yoon, Y. Titanium-based MAX-phase with sonocatalytic activity for degradation of oxytetracycline antibiotic. *Ultrason. Sonochem.* **2023**, *92*, 106255. [CrossRef]
80. Wang, G.; Dou, K.; Cao, H.; Du, R.; Liu, J.; Tsidaeva, N.; Wang, W. Designing Z-scheme CdS/WS_2 heterojunctions with enhanced photocatalytic degradation of organic dyes and photoreduction of Cr (VI): Experiments, DFT calculations and mechanism. *Sep. Purif. Technol.* **2022**, *291*, 120976. [CrossRef]
81. Dharman, R.K.; Shejale, K.P.; Kim, S.Y. Efficient sonocatalytic degradation of heavy metal and organic pollutants using CuS/MoS_2 nanocomposites. *Chemosphere* **2022**, *305*, 135415. [CrossRef] [PubMed]
82. Zhou, Q.; Ma, S.; Zhan, S. Superior photocatalytic disinfection effect of Ag-3D ordered mesoporous CeO_2 under visible light. *Appl. Catal. B-Environ.* **2018**, *224*, 27–37. [CrossRef]
83. Wan, L.; Zhou, Q.; Wang, X.; Wood, T.E.; Wang, L.; Duchesne, P.N.; Guo, J.; Yan, X.; Xia, M.; Li, Y.F. Cu_2O nanocubes with mixed oxidation-state facets for (photo) catalytic hydrogenation of carbon dioxide. *Nat. Catal.* **2019**, *2*, 889–898. [CrossRef]
84. Xu, L.; Wu, X.-Q.; Li, C.-Y.; Liu, N.-P.; An, H.-L.; Ju, W.-T.; Lu, W.; Liu, B.; Wang, X.-F.; Wang, Y.; et al. Sonocatalytic degradation of tetracycline by BiOBr/$FeWO_4$ nanomaterials and enhancement of sonocatalytic effect. *J. Clean. Prod.* **2023**, *394*, 136275. [CrossRef]
85. Xiang, W.; Ji, Q.; Xu, C.; Guo, Y.; Liu, Y.; Sun, D.; Zhou, W.; Xu, Z.; Qi, C.; Yang, S. Accelerated photocatalytic degradation of iohexol over Co_3O_4/g-C_3N_4/$Bi_2O_2CO_3$ of pn/nn dual heterojunction under simulated sunlight by persulfate. *Appl. Catal. B Environ.* **2021**, *285*, 119847. [CrossRef]
86. Qiu, J.; Li, M.; Xu, J.; Zhang, X.-F.; Yao, J. Bismuth sulfide bridged hierarchical Bi_2S_3/BiOCl@$ZnIn_2S_4$ for efficient photocatalytic Cr (VI) reduction. *J. Hazard. Mater.* **2020**, *389*, 121858. [CrossRef]

87. Liu, J.; Wang, G.; Li, B.; Ma, X.; Hu, Y.; Cheng, H. A high-efficiency mediator-free Z-scheme Bi_2MoO_6/AgI heterojunction with enhanced photocatalytic performance. *Sci. Total Environ.* **2021**, *784*, 147227. [CrossRef]
88. Vaiano, V.; Iervolino, G.; Sannino, D.; Murcia, J.J.; Hidalgo, M.C.; Ciambelli, P.; Navío, J.A. Photocatalytic removal of patent blue V dye on Au-TiO_2 and Pt-TiO_2 catalysts. *Appl. Catal. B-Environ.* **2016**, *188*, 134–146. [CrossRef]
89. Li, S.; Zhang, M.; Qu, Z.; Cui, X.; Liu, Z.; Piao, C.; Li, S.; Wang, J.; Song, Y. Fabrication of highly active Z-scheme Ag/g-C_3N_4-Ag-Ag_3PO_4 (1 1 0) photocatalyst photocatalyst for visible light photocatalytic degradation of levofloxacin with simultaneous hydrogen production. *Chem. Eng. J.* **2020**, *382*, 122394. [CrossRef]
90. Zhang, D.; Yang, Z.; Hao, J.; Zhang, T.; Sun, Q.; Wang, Y. Boosted charge transfer in dual Z-scheme $BiVO_4$@ $ZnIn_2S_4$/$Bi_2Sn_2O_7$ heterojunctions: Towards superior photocatalytic properties for organic pollutant degradation. *Chemosphere* **2021**, *276*, 130226. [CrossRef]
91. Wang, G.; Ma, X.; Wang, C.; Li, S.; Qiao, J.; Zhang, H.; Li, G.; Wang, J.; Song, Y. Highly efficient visible-light driven photocatalytic hydrogen evolution over Er^{3+}: $YAlO_3$/Ta_2O_5/rGO/$MoSe_2$ nanocomposite. *J. Mol. Liq.* **2018**, *260*, 375–385. [CrossRef]
92. Wang, G.; Ma, X.; Wei, S.; Li, S.; Qiao, J.; Wang, J.; Song, Y. Highly efficient visible-light driven photocatalytic hydrogen production from a novel Z-scheme Er^{3+}: $YAlO_3$/Ta_2O_5-V^{5+} | | Fe^{3+}-TiO_2/Au coated composite. *J. Power Sources* **2018**, *373*, 161–171. [CrossRef]
93. Zhao, G.; Ding, J.; Zhou, F.; Chen, X.; Wei, L.; Gao, Q.; Wang, K.; Zhao, Q. Construction of a visible-light-driven magnetic dual Z-scheme $BiVO_4$/g-C_3N_4/$NiFe_2O_4$ photocatalyst for effective removal of ofloxacin: Mechanisms and degradation pathway. *Chem. Eng. J.* **2021**, *405*, 126704. [CrossRef]
94. Molla, A.; Kim, A.Y.; Woo, J.C.; Cho, H.S.; Youk, J.H. Study on preparation methodology of zero-valent iron decorated on graphene oxide for highly efficient sonocatalytic dye degradation. *J. Environ. Chem. Eng.* **2022**, *10*, 107214. [CrossRef]
95. Wang, X.; He, X.-S.; Li, C.-Y.; Liu, S.-L.; Lu, W.; Xiang, Z.; Wang, Y. Sonocatalytic removal of tetracycline in the presence of S-scheme Cu_2O/$BiFeO_3$ heterojunction: Operating parameters, mechanisms, degradation pathways and toxicological evaluation. *J. Water Process Eng.* **2023**, *51*, 103345. [CrossRef]
96. Dharman, R.K.; Palanisamy, G.; Oh, T.H. Sonocatalytic degradation of ciprofloxacin and organic pollutant by 1T/2H phase MoS_2 in Polyvinylidene fluoride nanocomposite membrane. *Chemosphere* **2022**, *308*, 136571. [CrossRef]
97. Akdağ, S.; Sadeghi Rad, T.; Keyikoğlu, R.; Orooji, Y.; Yoon, Y.; Khataee, A. Peroxydisulfate-assisted sonocatalytic degradation of metribuzin by La-doped ZnFe layered double hydroxide. *Ultrason. Sonochem.* **2022**, *91*, 106236. [CrossRef]
98. Liu, C.; Mao, S.; Wang, H.; Wu, Y.; Wang, F.; Xia, M.; Chen, Q. Peroxymonosulfate-assisted for facilitating photocatalytic degradation performance of 2D/2D WO_3/BiOBr S-scheme heterojunction. *Chem. Eng. J.* **2022**, *430*, 132806. [CrossRef]
99. de Jesús Ruíz-Baltazar, Á. Sonochemical activation-assisted biosynthesis of Au/Fe_3O_4 nanoparticles and sonocatalytic degradation of methyl orange. *Ultrason. Sonochem.* **2021**, *73*, 105521. [CrossRef]
100. Zhang, J.; Zhao, Y.; Zhang, K.; Zada, A.; Qi, K. Sonocatalytic degradation of tetracycline hydrochloride with $CoFe_2O_4$/g-C_3N_4 composite. *Ultrason. Sonochem.* **2023**, *94*, 106325. [CrossRef]
101. Wang, X.; Yu, S.; Li, Z.-H.; He, L.-L.; Liu, Q.-L.; Hu, M.-Y.; Xu, L.; Wang, X.-F.; Xiang, Z. Fabrication Z-scheme heterojunction of Ag_2O/$ZnWO_4$ with enhanced sonocatalytic performances for meloxicam decomposition: Increasing adsorption and generation of reactive species. *Chem. Eng. J.* **2021**, *405*, 126922. [CrossRef]
102. Sadeghi Rad, T.; Ansarian, Z.; Khataee, A.; Vahid, B.; Doustkhah, E. N-doped graphitic carbon as a nanoporous MOF-derived nanoarchitecture for the efficient sonocatalytic degradation process. *Sep. Purif. Technol.* **2021**, *256*, 117811. [CrossRef]
103. Gote, Y.M.; Sinhmar, P.S.; Gogate, P.R. Sonocatalytic Degradation of Chrysoidine R Dye Using Ultrasonically Synthesized $NiFe_2O_4$ Catalyst. *Catalysts* **2023**, *13*, 597. [CrossRef]
104. Joseph, C.G.; Puma, G.L.; Bono, A.; Krishnaiah, D. Sonophotocatalysis in advanced oxidation process: A short review. *Ultrason. Sonochem.* **2009**, *16*, 583–589. [CrossRef] [PubMed]
105. Malika, M.; Sonawane, S.S. The sono-photocatalytic performance of a Fe_2O_3 coated TiO_2 based hybrid nanofluid under visible light via RSM. *Colloids Surf. A* **2022**, *641*, 128545. [CrossRef]
106. Mosleh, S.; Rahimi, M.R.; Ghaedi, M.; Asfaram, A.; Jannesar, R.; Sadeghfar, F. A rapid and efficient sonophotocatalytic process for degradation of pollutants: Statistical modeling and kinetics study. *J. Mol. Liq.* **2018**, *261*, 291–302. [CrossRef]
107. Karim, A.V.; Shriwastav, A. Degradation of amoxicillin with sono, photo, and sonophotocatalytic oxidation under low-frequency ultrasound and visible light. *Environ. Res.* **2021**, *200*, 111515. [CrossRef]
108. Dinesh, G.K.; Anandan, S.; Sivasankar, T. Sonophotocatalytic treatment of Bismarck Brown G dye and real textile effluent using synthesized novel Fe (0)-doped TiO_2 catalyst. *RSC Adv.* **2015**, *5*, 10440–10451. [CrossRef]
109. Al-Musawi, T.J.; Rajiv, P.; Mengelizadeh, N.; Mohammed, I.A.; Balarak, D. Development of sonophotocatalytic process for degradation of acid orange 7 dye by using titanium dioxide nanoparticles/graphene oxide nanocomposite as a catalyst. *J. Environ. Manag.* **2021**, *292*, 112777. [CrossRef]
110. Rameshbabu, R.; Kumar, N.; Pecchi, G.; Delgado, E.J.; Karthikeyan, C.; Mangalaraja, R. Ultrasound-assisted synthesis of rGO supported NiO-TiO_2 nanocomposite: An efficient superior sonophotocatalyst under diffused sunlight. *J. Environ. Chem. Eng.* **2022**, *10*, 107701. [CrossRef]
111. Gokul, P.; Vinoth, R.; Neppolian, B.; Anandhakumar, S. Binary metal oxide nanoparticle incorporated composite multilayer thin films for sono-photocatalytic degradation of organic pollutants. *Appl. Surf. Sci.* **2017**, *418*, 119–127. [CrossRef]
112. Ding, Z.; Sun, M.; Liu, W.; Sun, W.; Meng, X.; Zheng, Y. Ultrasonically synthesized N-TiO_2/Ti_3C_2 composites: Enhancing sonophotocatalytic activity for pollutant degradation and nitrogen fixation. *Sep. Purif. Technol.* **2021**, *276*, 119287. [CrossRef]

113. Wang, S.; Gong, Q.; Liang, J. Sonophotocatalytic degradation of methyl orange by carbon nanotube/TiO$_2$ in aqueous solutions. *Ultrason. Sonochem.* **2009**, *16*, 205–208. [CrossRef]
114. Ahmad, R.; Ahmad, Z.; Khan, A.U.; Mastoi, N.R.; Aslam, M.; Kim, J. Photocatalytic systems as an advanced environmental remediation: Recent developments, limitations and new avenues for applications. *J. Environ. Chem. Eng.* **2016**, *4*, 4143–4164. [CrossRef]
115. Sathishkumar, P.; Mangalaraja, R.V.; Mansilla, H.D.; Gracia-Pinilla, M.; Anandan, S. Sonophotocatalytic (42 kHz) degradation of Simazine in the presence of Au–TiO$_2$ nanocatalysts. *Appl. Catal. B-Environ.* **2014**, *160*, 692–700. [CrossRef]
116. Hapeshi, E.; Fotiou, I.; Fatta-Kassinos, D. Sonophotocatalytic treatment of ofloxacin in secondary treated effluent and elucidation of its transformation products. *Chem. Eng. J.* **2013**, *224*, 96–105. [CrossRef]
117. Mosleh, S.; Rahimi, M.; Ghaedi, M.; Dashtian, K. Sonophotocatalytic degradation of trypan blue and vesuvine dyes in the presence of blue light active photocatalyst of Ag$_3$PO$_4$/Bi$_2$S$_3$-HKUST-1-MOF: Central composite optimization and synergistic effect study. *Ultrason. Sonochem.* **2016**, *32*, 387–397. [CrossRef]
118. Babu, S.G.; Karthik, P.; John, M.C.; Lakhera, S.K.; Ashokkumar, M.; Khim, J.; Neppolian, B. Synergistic effect of sono-photocatalytic process for the degradation of organic pollutants using CuO-TiO$_2$/rGO. *Ultrason. Sonochem.* **2019**, *50*, 218–223. [CrossRef]
119. Benomara, A.; Guenfoud, F.; Mokhtari, M.; Boudjemaa, A. Sonolytic, sonocatalytic and sonophotocatalytic degradation of a methyl violet 2B using iron-based catalyst. *React. Kinet. Mech. Catal.* **2021**, *132*, 513–528. [CrossRef]
120. Ahmad, M.; Ahmed, E.; Hong, Z.; Ahmed, W.; Elhissi, A.; Khalid, N. Photocatalytic, sonocatalytic and sonophotocatalytic degradation of Rhodamine B using ZnO/CNTs composites photocatalysts. *Ultrason. Sonochem.* **2014**, *21*, 761–773. [CrossRef]
121. Abdullah, A.Z.; Ling, P.Y. Heat treatment effects on the characteristics and sonocatalytic performance of TiO$_2$ in the degradation of organic dyes in aqueous solution. *J. Hazard. Mater.* **2010**, *173*, 159–167. [CrossRef] [PubMed]
122. Mahanta, U.; Khandelwal, M.; Deshpande, A.S. TiO$_2$@SiO$_2$ nanoparticles for methylene blue removal and photocatalytic degradation under natural sunlight and low-power UV light. *Appl. Surf. Sci.* **2022**, *576*, 151745. [CrossRef]
123. Rajagopal, S.; Paramasivam, B.; Muniyasamy, K. Photocatalytic removal of cationic and anionic dyes in the textile wastewater by H$_2$O$_2$ assisted TiO$_2$ and micro-cellulose composites. *Sep. Purif. Technol.* **2020**, *252*, 117444. [CrossRef]
124. Li, S.; Wang, J.; Xia, Y.; Li, P.; Wu, Y.; Yang, K.; Song, Y.; Jiang, S.; Zhang, T.; Li, B. Boosted electron-transfer by coupling Ag and Z-scheme heterostructures in CdSe-Ag-WO$_3$-Ag for excellent photocatalytic H$_2$ evolution with simultaneous degradation. *Chem. Eng. J.* **2021**, *417*, 129298. [CrossRef]
125. Hunge, Y.M.; Yadav, A.A.; Dhodamani, A.G.; Suzuki, N.; Terashima, C.; Fujishima, A.; Mathe, V.L. Enhanced photocatalytic performance of ultrasound treated GO/TiO$_2$ composite for photocatalytic degradation of salicylic acid under sunlight illumination. *Ultrason. Sonochem.* **2020**, *61*, 104849. [CrossRef]
126. Ribao, P.; Corredor, J.; Rivero, M.J.; Ortiz, I. Role of reactive oxygen species on the activity of noble metal-doped TiO$_2$ photocatalysts. *J. Hazard. Mater.* **2019**, *372*, 45–51. [CrossRef]
127. Gogoi, D.; Namdeo, A.; Golder, A.K.; Peela, N.R. Ag-doped TiO$_2$ photocatalysts with effective charge transfer for highly efficient hydrogen production through water splitting. *Int. J. Hydrogen Energy* **2020**, *45*, 2729–2744. [CrossRef]
128. Wang, R.; Tang, T.; Wei, Y.; Dang, D.; Huang, K.; Chen, X.; Yin, H.; Tao, X.; Lin, Z.; Dang, Z.; et al. Photocatalytic debromination of polybrominated diphenyl ethers (PBDEs) on metal doped TiO$_2$ nanocomposites: Mechanisms and pathways. *Environ. Int.* **2019**, *127*, 5–12. [CrossRef]
129. Zhang, H.; Tang, P.; Yang, K.; Wang, Q.; Feng, W.; Tang, Y. PAA/TiO$_2$@C composite hydrogels with hierarchical pore structures as high efficiency adsorbents for heavy metal ions and organic dyes removal. *Desalination* **2023**, *558*, 116620. [CrossRef]
130. Nuengmatcha, P.; Chanthai, S.; Mahachai, R.; Oh, W.-C. Sonocatalytic performance of ZnO/graphene/TiO$_2$ nanocomposite for degradation of dye pollutants (methylene blue, texbrite BAC-L, texbrite BBU-L and texbrite NFW-L) under ultrasonic irradiation. *Dyes Pigm.* **2016**, *134*, 487–497. [CrossRef]
131. Yuan, Y.; Sun, M.; Yuan, X.; Zhu, Y.; Lin, X.; Anandan, S. One-step hydrothermal synthesis of N/Ti^{3+} co-doping multiphasic TiO$_2$/BiOBr heterojunctions towards enhanced sonocatalytic performance. *Ultrason. Sonochem.* **2018**, *49*, 69–78. [CrossRef]
132. Sriramoju, J.B.; Muniyappa, M.; Marilingaiah, N.R.; Sabbanahalli, C.; Shetty, M.; Mudike, R.; Chitrabanu, C.; Shivaramu, P.D.; Nagaraju, G.; Rangappa, K.S. Carbon-based TiO$_{2-x}$ heterostructure nanocomposites for enhanced photocatalytic degradation of dye molecules. *Ceram. Int.* **2021**, *47*, 10314–10321. [CrossRef]
133. Mousavi, M.; Ghasemi, J.B. Novel visible-light-responsive Black-TiO$_2$/CoTiO$_3$ Z-scheme heterojunction photocatalyst with efficient photocatalytic performance for the degradation of different organic dyes and tetracycline. *J. Taiwan Inst. Chem. Eng.* **2021**, *121*, 168–183. [CrossRef]
134. May-Lozano, M.; Lopez-Medina, R.; Escamilla, V.M.; Rivadeneyra-Romero, G.; Alonzo-Garcia, A.; Morales-Mora, M.; González-Díaz, M.; Martinez-Degadillo, S. Intensification of the Orange II and Black 5 degradation by sonophotocatalysis using Ag-graphene oxide/TiO$_2$ systems. *Chem. Eng. Process.* **2020**, *158*, 108175. [CrossRef]
135. Sun, M.; Yao, Y.; Ding, W.; Anandan, S. N/Ti^{3+} co-doping biphasic TiO$_2$/Bi$_2$WO$_6$ heterojunctions: Hydrothermal fabrication and sonophotocatalytic degradation of organic pollutants. *J. Alloys Compd.* **2020**, *820*, 153172. [CrossRef]

Disclaimer/Publisher's Note: The statements, opinions and data contained in all publications are solely those of the individual author(s) and contributor(s) and not of MDPI and/or the editor(s). MDPI and/or the editor(s) disclaim responsibility for any injury to people or property resulting from any ideas, methods, instructions or products referred to in the content.

Article

Photocatalytic Hydrogen Production by the Sensitization of Sn(IV)-Porphyrin Embedded in a Nafion Matrix Coated on TiO$_2$

Sung-Hyun Kim and Hee-Joon Kim *

Department of Chemistry and Bioscience, Kumoh National Institute of Technology, Gumi 39177, Korea; kashik@naver.com
* Correspondence: hjk@kumoh.ac.kr; Tel.: +82-54-478-7822

Abstract: Efficient utilization of visible light for photocatalytic hydrogen production is one of the most important issues to address. This report describes a facile approach to immobilize visible-light sensitizers on TiO$_2$ surfaces. To effectively utilize the sensitization of Sn(IV) porphyrin species for photocatalytic hydrogen production, perfluorosulfonate polymer (Nafion) matrix coated-TiO$_2$ was fabricated. Nafion coated-TiO$_2$ readily adsorbed *trans*-diaqua[*meso*-tetrakis(4-pyridinium)porphyrinato]tin(IV) cation [(TPy^{H}P)Sn(OH$_2$)$_2$]$^{6+}$ via an ion-exchange process. The uptake of [(TPy^{H}P)Sn(OH$_2$)$_2$]$^{6+}$ in an aqueous solution completed within 30 min, as determined by UV-vis spectroscopy. The existence of Sn(IV) porphyrin species embedded in the Nafion matrix coated on TiO$_2$ was confirmed by zeta potential measurements, UV-vis absorption spectroscopy, TEM combined with energy dispersive X-ray spectroscopy, and thermogravimetric analysis. Sn(IV)-porphyrin cationic species embedded in the Nafion matrix were successfully used as visible-light sensitizer for photochemical hydrogen generation. This photocatalytic system performed 45% better than the uncoated TiO$_2$ system. In addition, the performance at pH 7 was superior to that at pH 3 or 9. This work revealed that Nafion matrix coated-TiO$_2$ can efficiently produce hydrogen with a consistent performance by utilizing a freshly supplied cationic Sn(IV)-porphyrin sensitizer in a neutral solution.

Keywords: hydrogen production; photosensitization; Sn(IV)-porphyrin; Nafion; photocatalyst

Citation: Kim, S.-H.; Kim, H.-J. Photocatalytic Hydrogen Production by the Sensitization of Sn(IV)-Porphyrin Embedded in a Nafion Matrix Coated on TiO$_2$. *Molecules* **2022**, *27*, 3770. https://doi.org/10.3390/molecules27123770

Academic Editors: Hongda Li, Mohammed Baalousha and Victor A. Nadtochenko

Received: 26 May 2022
Accepted: 9 June 2022
Published: 11 June 2022

Publisher's Note: MDPI stays neutral with regard to jurisdictional claims in published maps and institutional affiliations.

Copyright: © 2022 by the authors. Licensee MDPI, Basel, Switzerland. This article is an open access article distributed under the terms and conditions of the Creative Commons Attribution (CC BY) license (https://creativecommons.org/licenses/by/4.0/).

1. Introduction

Photochemical generation of hydrogen has been intensively studied as a means of converting solar energy into chemical energy [1–6]. Solar energy is predominantly in the visible region, therefore, efficient utilization of visible light is one of the most important issues to address. During natural photosynthesis, the absorption of visible light by chlorophyll sensitizers initiates the light-harvesting process, followed by charge separation and transfer, which proceeds through redox reactions. Porphyrins and metalloporphyrins have been extensively explored in the context of light harvesting and photoinduced electron/energy transfer processes [7–9], because of their similarity to chlorophyll sensitizers in natural photosynthesis. They have also been extensively investigated for their photochemical properties in environmental photocatalysis [10], hydrogen production [11,12], and solar cell [13] applications.

Among the metalloporphyrins, Sn(IV)-porphyrin is particularly noteworthy as a photosensitizer or photocatalyst for the development of various photocatalytic systems. Sn(IV)-porphyrin has an intrinsically strong oxidation ability owing to the high charge of the Sn(IV) center; consequently, the excited state of SnP has a high affinity for electrons that initiate photooxidative reactions. The excited Sn(IV)-porphyrin exhibits a high photochemical activity for the oxidative degradation of organic pollutants under visible light [14–19]. Sn(IV)-porphyrin complex-based nanoparticles have been also used in photochemical hydrogen production [20]. A water-soluble Sn(IV)-porphyrin complex, *trans*-diaqua[*meso*-tetrakis(4-pyridinium)porphyrinato]tin(IV) hexanitrate [(TPy^{H}P)Sn(OH$_2$)$_2$](NO$_3$)$_6$, has been investigated as a visible light sensitizer of platinized TiO_2 nanoparticles for the production of hydrogen [21]. Although the Sn(IV)-porphyrin sensitizer was not bound to TiO_2, hydrogen was successfully generated under visible light over a wide pH range (pH 3–11). Efficient visible light sensitization generally requires strong chemical bonding between the semiconductor oxide (TiO_2) and sensitizer molecule, which results in significant electronic coupling between the semiconductor conduction band and the sensitizer's excited orbital. Therefore, molecular sensitizers with anchors such as carboxylate, phosphonate, and catechol groups are fixed on the surface of the semiconductor oxide [22–24]. In contrast, this study revealed that the sufficiently long lifetime of photogenerated π-radical anions of Sn(IV)-porphyrin ($SnP^{\bullet-}$) enables the diffusion of $SnP^{\bullet-}$ to the TiO_2 surface in the bulk solution. The disadvantage of the typical molecular dye-sensitized TiO_2 system, in which the chemical modification of the sensitizer for anchoring is essential and the hydrogen production is limited to acidic conditions, was addressed in this system. However, the efficiency of the electron transfer between the porphyrin sensitizer and redox mediator remains to be improved.

An alternative method of immobilizing sensitizing molecules on the surface of TiO_2 has been achieved using a polymer matrix [25,26]. Nafion, an anionic perfluorinated polymer, has hydrophilic pores (~4 nm) surrounded by sulfonate anion groups ($-SO_3^-$) capable of exchanging cationic species. Additionally, it is chemically and photochemically inert. $[(TPy^HP)Sn(OH_2)_2]^{6+}$, which does not bind to the TiO_2 surface, can be embedded into the Nafion-coated TiO_2 surface through ion exchange, which facilitates more efficient production of hydrogen under visible light.

Unlike Ru-based sensitizers, Sn(IV)-porphyrins can be developed and used as practical sensitizers for solar energy conversion because they are inexpensive, have low toxicity, and are rich in certain elements. To improve the efficiency of visible-light-sensitive hydrogen generation using porphyrin sensitizers, we investigated a photocatalytic hydrogen generation system that incorporated Sn(IV)-porphyrin cations into a perfluorosulfonate polymer (Nafion) matrix coated on platinized TiO_2 nanoparticles (Scheme 1).

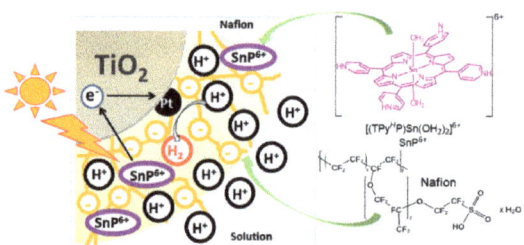

Scheme 1. Illustration for the photochemical hydrogen generation by sensitization of Sn(IV)-porphyrin embedded in Nafion matrix coated on TiO_2.

2. Results and Discussion

2.1. Fabrication of Photocatalyst

Scheme 2 illustrates the fabrication of photocatalysts used in this study. Platinized TiO_2 nanoparticles (**1**) were prepared by the chemical reduction of H_2PtCl_6 with $NaBH_4$. The TEM image, EDS spectrum, and elemental mapping images obtained by STEM showed the presence of Pt particles on the TiO_2 surface (Figure S1). The Nafion polymer was easily coated onto the Pt-TiO_2 surface by the drop-casting method using a commercial Nafion solution. Nafion-coated Pt-TiO_2 (**2**) was characterized by TEM, EDS, X-ray photoelectron spectroscopy (XPS), and FT-IR techniques. The EDS spectrum indicated the presence of F and S elements in the Nafion layer of **2** (Figure S2). The XPS spectra as shown in Figure 1 clearly confirmed the presence of F (1 s binding energy of 687 eV) in **2**, but not in **1**. The Ti 2p binding energies (464 and 458 eV) of both **2** and **1** were identical to those of pure TiO_2. In the IR spectra comparing **2** and **1** (Figure S3), the C–F vibration band was observed at

1239 cm^{-1} only from **2** further supporting the presence of the Nafion layer. Therefore, all the characterization data prove the successful fabrication of **2**.

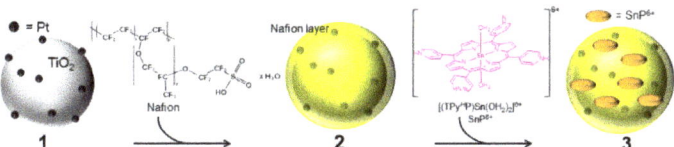

Scheme 2. Fabrication of photocatalysts used in this study.

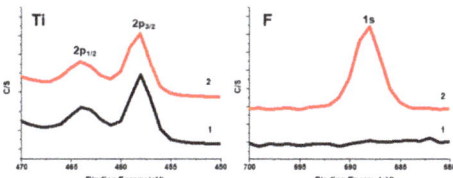

Figure 1. XPS spectra of platinized TiO$_2$ (**1**) and Nafion-coated Pt-TiO$_2$ (**2**) in the Ti 2p (left) and F 1s band regions.

2.2. Embedding of Sn(IV)-Porphyrin Cations into Nafion-Coated Photocatalyst

The adsorption of ionic surfactant polymers, such as Nafion, on the TiO$_2$ surface can modify the surface charge drastically. The zeta potential of suspended **2** at pH 7, −28.5 mV, was measured to be more negative than that of **1**, −8.8 mV, indicating that the anionic character due to the coated Nafion layer was significantly manifested on the surface of particles of **1**. **2** was expected to readily adsorb certain cationic species through an ion-exchange process in the Nafion matrix. The uptake of [(TPy^{H}P)Sn(OH$_2$)$_2$]$^{6+}$ (SnP^{6+}, water-soluble and highly charged Sn(IV)-porphyrin cation) by **2** suspended in an aqueous solution was monitored using UV-vis spectroscopy. Figure 2 shows that the absorption in the Soret band of SnP^{6+} decreased gradually, and the uptake was completed within 30 min. In contrast, **1** (uncoated Nafion polymer) did not adsorb SnP^{6+} at all, evident by the unchanged absorption spectra. The inset in Figure 2 also shows the uptake of SnP^{6+} quantitatively over time.

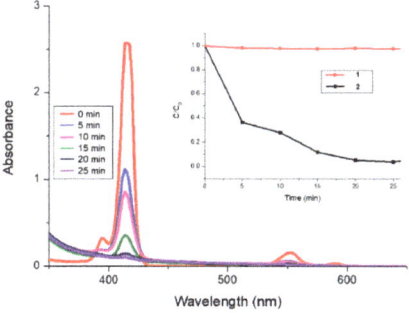

Figure 2. UV-vis absorption spectra of SnP^{6+} remaining after the equilibrated adsorption on Nafion-coated Pt-TiO$_2$ (**2**). The inset depicts the comparison of absorption changes in Soret bands of SnP^{6+} according to adsorption progress with time on **1** and **2**.

Isolated **2** containing SnP^{6+} (**3**) was further characterized. In the zeta potential measurement, **3** showed +20.9 mV at pH 7, which strongly implies SnP^{6+} was sufficiently incorporated into the Nafion matrix of **2**. In contrast, that of **1** was measured to be −14.5 mV when SnP^{6+} was present in aqueous solution, which indicates that **1** itself does not adsorb SnP^{6+} in aqueous solution. Figure 3 shows the UV-vis absorption spectra for each nanoparticle measured in the solid state. When compared to **1** and **2**, the spectrum of **3** exhibited strong absorption bands from SnP^{6+} at 414, 511, and 550 nm in the visible light region. The TEM image and the EDS spectrum for **3** further proves the existence of key elements such as Sn, S, F, Pt, and Ti constituting **3** (Figure S4).

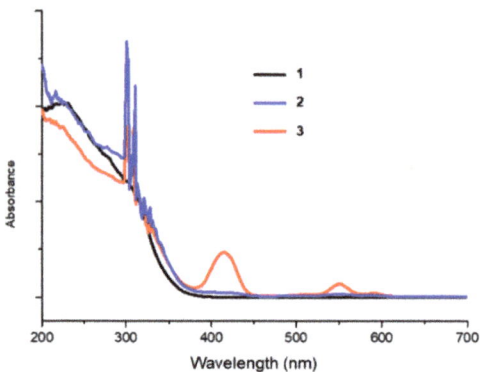

Figure 3. UV-vis absorption spectra of platinized TiO$_2$ (**1**), Nafion-coated Pt-TiO$_2$ (**2**), and SnP^{6+}-embedded Nafion/Pt-TiO$_2$ (**3**) measured in solid state.

Finally, TGA measurements were taken to determine the content of SnP^{6+} in **3** (Figure 4). In the TGA diagram of [(TPy^{H}P)Sn(OH$_2$)$_2$](NO$_3$)$_6$, the removal of solvent molecules such as water occurred up to ~150 °C, decomposition of nitrate anions occurred at 200–350 °C, followed by the degradation of the Sn(IV)-porphyrin. In addition, the TGA plot of the Nafion sample showed that Nafion gradually lost water molecules and began to decompose rapidly at 300 °C, eventually losing 97 wt.% at 500 °C. **2** and **3** exhibited similar behavior in TGA, where the final plateau was achieved above 500 °C for both samples. Based on the difference of weight loss between **2** and **3**, the content of SnP^{6+} was estimated to be about 2.4 wt.% in bulk **3**.

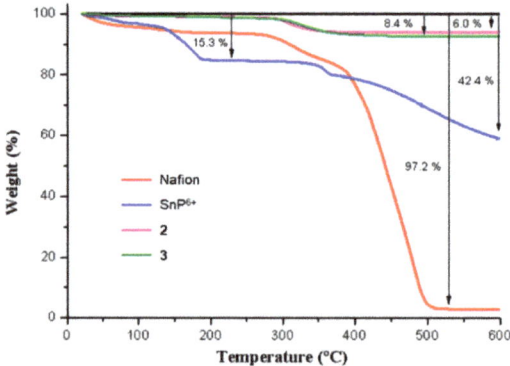

Figure 4. TGA thermogram of SnP^{6+}-embedded Nafion/Pt-TiO$_2$ (**3**).

2.3. Photocatalytic Hydrogen Generation

Photocatalytic hydrogen generation was first investigated under visible light irradiation in an aqueous suspension containing **3** (2.4 wt.% of SnP^{6+}, 1.5 g/L) as the photocatalyst and EDTA (1 mM) as the electron donor, but without an additional SnP^{6+} sensitizer. As shown in Figure 5, the amount of hydrogen generated continuously increased to 22 μmol after 4 h. It was clearly demonstrated that the SnP^{6+} species embedded in the Nafion matrix functioned successfully as a visible light sensitizer for photochemical hydrogen generation. The decrease in hydrogen production after 4 h may be due to the irreversible conversion of Sn(IV)-porphyrin to Sn(IV)-chlorin, a reduced form of the pyrrole ring [21].

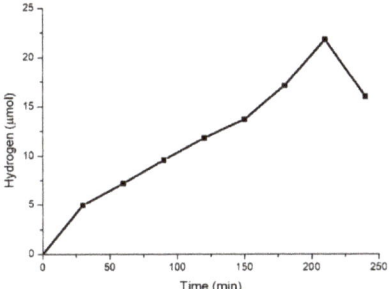

Figure 5. Time profile of H$_2$ production using the photocatalyst of SnP^{6+}-embedded Nafion/Pt-TiO$_2$ (**3**) in the absence of SnP^{6+} in the solution under visible light irradiation.

The effect of SnP^{6+} concentration on hydrogen generation in a suspension of **1** was examined, as displayed in Figure 6. While the amount of hydrogen generated was negligible at 0.01 mM concentration, increased remarkably as the concentration of the sensitizer increased to 0.1 mM. This means that the sensitizer in the solution must exceed a certain critical concentration to enter the Nafion matrix. On the other hand, at higher concentrations (0.5 and 1.0 mM), the hydrogen production increased sharply, but was prematurely saturated at approximately 30–45 min irradiation, and the amount produced was substantially less than that 0.1 mM concentration after further irradiation. This revealed that a large excess of the SnP^{6+} sensitizer can initially enhance the rate of incorporation into the Nafion matrix, but the surplus in solution has little effect on hydrogen production. Photocatalyst **2** itself does not contain SnP^{6+} sensitizer, so it is crucial to uptake SnP^{6+} sensitizer from solution at an initial stage. The uptake rate and efficiency probably depend on the concentration and mass transfer of SnP^{6+}. Consequently, the optimal concentration of the SnP^{6+} sensitizer could efficiently incorporate the sensitizer into the Nafion matrix to subsequently promote photosensitized hydrogen generation.

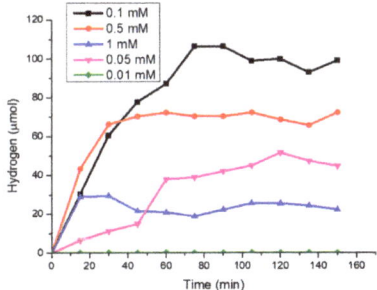

Figure 6. Comparison of visible light-irradiated hydrogen production of Nafion-coated Pt-TiO$_2$ (**2**) by varying the concentration of SnP^{6+}.

The adsorption of SnP^{6+} on TiO$_2$ is not required for photocatalytic H$_2$ production, hence, we further compared the performance of **2** with that of **1** in the presence of 0.1 mM SnP^{6+}. As shown in Figure 7, 193 and 133 μmol of hydrogen were generated by the photocatalysts **2** and **1**, respectively, after 2 h of irradiation. **2** exhibited a 45% better performance than uncoated TiO$_2$. SnP^{6+} sensitizers embedded into the Nafion matrix coated on the surface of TiO_2 facilitate the electron transfer process between the sensitizer and redox mediator (TiO_2) when compared to free SnP^{6+} sensitizers in solution. The higher local concentration of H^+ in the Nafion matrix also contributed significantly. The H^+ population on the Nafion polymer-coated TiO_2 surface increased considerably owing to the presence of the sulfonate groups in the Nafion polymer. It is well known that the pH of Nafion is much lower than that of the aqueous bulk phase. The protons trapped in the Nafion matrix of **2** could then be readily photochemically reduced to form hydrogen. Accordingly, the performance of **2** for the photocatalytic H_2 generation was enhanced compared to that of **1**.

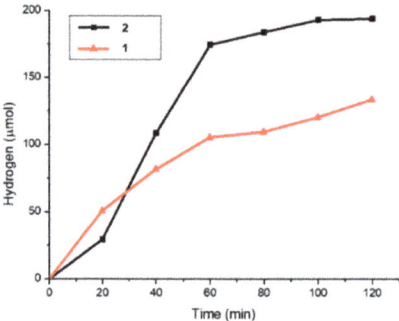

Figure 7. Comparison of visible light-irradiated hydrogen production in the presence of 0.1 mM SnP^{6+} of platinized TiO_2 (**1**) and Nafion-coated Pt-TiO_2 (**2**).

In a previous report [21], the unbound SnP^{6+}-sensitized TiO_2 system was found to successfully generate hydrogen under visible-light irradiation over a wide pH range (pH 3–11). Here, we investigated photocatalytic hydrogen generation with SnP^{6+}-sensitized **2** at different pH values to evaluate the effect of pH on the performance and stability of the photocatalyst. Figure 8 shows the performance of H_2 generation sensitized by SnP^{6+} in **2** at three different pH values (3, 7, and 9).

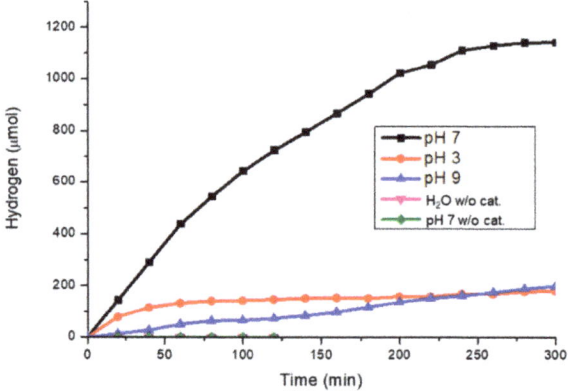

Figure 8. Comparison of visible light-irradiated hydrogen production of Nafion-coated Pt-TiO_2 (**2**) at different pH values in the presence of 0.1 mM SnP^{6+}.

The performance at pH 7 was superior to that at pH 3 or 9, where a similar performance was observed. At pH 3, hydrogen production gradually increased in the initial stage but almost ceased after 1 h. The photocatalytic production of hydrogen through the sensitization of Sn(IV)-porphyrin species is affected by the action of the corresponding π-radical anion species. Sn(IV)-chlorin, a reduced form of the pyrrole ring, is irreversibly formed by a bimolecular reaction between π-radical anions. The favorable formation of Sn(IV)-chlorin at an acidic pH inhibits the electron transfer process from the π-radical anions to the TiO_2 or platinum catalyst, thereby reducing hydrogen production performance. It was also observed that the Nafion matrix coated on the TiO_2 peeled off at pH 9 and above. This exfoliation can explain why hydrogen production under basic conditions did not increase as much as that at neutral pH over time. Therefore, it can be concluded that Nafion matrix coated-TiO_2 can efficiently produce hydrogen with a consistent performance by utilizing a freshly supplied cationic Sn(IV)-porphyrin sensitizer in a neutral solution.

3. Materials and Methods

Trans-diaqua [5,10,15,20-tetrakis(4-pyridinium)porphyrinato]tin(IV) hexanitrate, [(TPyHP)Sn(OH$_2$)$_2$](NO$_3$)$_6$, was prepared using a reported procedure [27]. TiO_2 nanoparticles (Degussa P25) were used as received. Nafion was purchased from Aldrich as a 5 wt. % solution in a mixture of alcohol and water. Chloroplatinic acid ($H_2PtCl_6 \cdot 6H_2O$) (Aldrich, St. Louis, MO, USA), methanol (Aldrich), and ethylenediaminetetraacetic acid (EDTA, Aldrich) were used as received. $HClO_4$ and NaOH were used to adjust the pH of aqueous suspensions. Ultrapure deionized water (18 MΩ·cm) and was prepared using the Barnstead purification system. Transmission electron microscopy (TEM), TEM-energy dispersive X-ray spectroscopy (TEM-EDS), and scanning transmission electron microscopy (S-TEM) images were obtained using a JEOL/JEM 2100 instrument. The zeta potentials of the catalyst particles in the aqueous suspension were measured using an electrophoretic light-scattering spectrophotometer (ELSZ-2, Otsuka, Osaka, Japan). The surface atomic composition was determined using X-ray photoelectron spectroscopy (XPS, ULVAC-PHI/Quantera). UV-visible spectra were recorded using a UV-vis spectrophotometer (UV-3600, Shimadzu, Tokyo, Japan). FT-IR spectra were recorded in the range of 4000–400 cm^{-1} on a Bruker Vertex 80v. Thermogravimetric analyses (TGA) were carried out on a TA Instruments/Auto-TGA Q502 instrument heated from room temperature to 600 °C at a ramp rate of 5 °C/min under nitrogen.

3.1. Preparation of Photocatalyst

3.1.1. Platinized TiO_2 (**1**)

A 2.0 g sample of TiO_2 nanoparticles were immersed in water with 100 mL of $H_2PtCl_6 \cdot H_2O$ (0.1 M) while being continuously stirred for 2 h. Then, 50 mL of $NaBH_4$ (1.0 M $NaBH_4$ in methanol) was added quickly and stirred continuously for 2 h. The photocatalyst color changed from white to black with increasing Pt loading. The powder was washed repeatedly with distilled water. The suspension was centrifuged and decanted. The residue was then dried overnight at 90 °C, and this yielded (2.2 g of **1**).

3.1.2. Nafion-Coated Pt-TiO_2 (**2**)

An aliquot of Nafion solution (2 mL) in H_2O/MeOH was added to **1** (1.0 g), and the mixture was mixed thoroughly. The suspension was centrifuged and decanted. The residue was washed with H_2O/MeOH, and dried overnight at 90 °C, and this yielded (1.2 g of **2**).

3.1.3. Sn(IV)-porphyrin cations-embedded Nafion/Pt-TiO_2 (**3**)

An aliquot of 0.1 mL of a 1.0 mM [(TPy^{H}P)Sn(OH$_2$)$_2$](NO$_3$)$_6$ solution in H_2O was added to **2** (0.1 g), and the reaction mixture was vigorously stirred for 1 h. The suspension was centrifuged and decanted. The residue was dried overnight at 90 °C, and this yielded (0.1 g of **3**).

3.2. Photocatalytic Hydrogen Generation

2 (7.5 mg, 1.5 g/L) was suspended in an aqueous solution of SnP^{6+} (0.1 mM) and EDTA (1 mM) in a glass reactor (20 mL, Wheaton, Stoke-on-Trent, UK). The mixture was vigorously stirred for 1 h to immobilize SnP^{6+} on the Nafion layer. The suspension was purged with N_2 for 1 h before illumination. A 150 W xenon arc lamp was used as the light source (LS 150, ABET-technologies, Milford, CT, USA). Light was passed through a 10-cm IR cut-off filter (λ > 900 nm, Edmund Optics, Barrington, IL, USA) and a UV cut-off filter (λ < 400 nm, Edmund Optics), and the headspace gas (15 mL) of the reactor was intermittently sampled and analyzed for hydrogen using a gas chromatograph (GC-2014, Shimadzu, Tokyo, Japan).

4. Conclusions

Perfluorosulfonate polymer (Nafion) matrix coated-TiO_2 was fabricated to effectively sensitize Sn(IV)-porphyrin species for photocatalytic hydrogen production. Nafion coated-TiO_2 readily adsorbed Sn(IV)-porphyrin cation species via an ion-exchange process. The presence of the Sn(IV)-porphyrin species embedded in the Nafion matrix coated on TiO_2 was confirmed using various instrumental techniques. Our investigation revealed that the Sn(IV)-porphyrin cationic species embedded in the Nafion matrix successfully functioned as a visible-light sensitizer for photocatalytic hydrogen generation. This photocatalytic system performed 45% better than the uncoated TiO_2 system. In addition, the performance at pH 7 is much better than that at pH 3 or 9. In conclusion, Nafion matrix coated-TiO_2 can efficiently produce hydrogen through the favorable uptake of cationic Sn(IV)-porphyrin sensitizer in a neutral solution. Our work makes an important contribution in the development of nanostructured photocatalysts that are more efficient and practical than Ru-based sensitization for visible-light-sensitized hydrogen production.

Supplementary Materials: The following supporting information can be downloaded at: https://www.mdpi.com/article/10.3390/molecules27123770/s1. Figure S1: Micrographs for platinized TiO_2 (1) showing (a) TEM image, (b) EDS spectrum, and (c) elemental mapping images by STEM for platinum (left, red), titanium (center, green) and oxygen (right, blue). Figure S2: Micrographs for Nafion-coated Pt-TiO_2 (2) showing (a) TEM image, and (b) EDS spectrum. Figure S3: FT-IR spectra of platinized TiO_2 (1) and Nafion-coated Pt-TiO_2 (2). Figure S4: Micrographs for SnP^{6+}-embedded Nafion/Pt-TiO_2 (3) showing (a) TEM image, and (b) EDS spectrum.

Author Contributions: Investigation, methodology, data curation, visualization, formal analysis, validation, and software, S.-H.K.; conceptualization, writing, review and editing, supervision, project administration, and funding acquisition, H.-J.K. All authors have read and agreed to the published version of the manuscript.

Funding: National Research Foundation of Korea (NRF) grant (No. 2017R1A2B2011585).

Institutional Review Board Statement: Not applicable.

Informed Consent Statement: Not applicable.

Data Availability Statement: Data are available in the article and Supplementary Materials.

Acknowledgments: This work was supported by the National Research Foundation of Korea (NRF) grant (No. 2017R1A2B2011585) funded by the Korean government (MSIT).

Conflicts of Interest: The authors declare no conflict of interest.

Sample Availability: Samples of the compounds are not available from the authors.

References

1. Wang, Y.; Suzuki, H.; Xie, J.; Tomita, O.; Martin, D.J.; Higashi, M.; Kong, D.; Abe, R.; Tang, J. Mimicking Natural Photosynthesis: Solar to Renewable H_2 Fuel Synthesis by Z-Scheme Water Splitting Systems. *Chem. Rev.* **2018**, *118*, 5201–5241. [CrossRef] [PubMed]
2. Balzani, V.; Credi, A.; Venturi, M. Photochemical Conversion of Solar Energy. *Chem. Sustain. Chem.* **2008**, *1*, 26–58. [CrossRef] [PubMed]
3. Chen, X.; Shen, S.; Guo, L.; Mao, S.S. Semiconductor-based Photocatalytic Hydrogen Generation. *Chem. Rev.* **2010**, *110*, 6503–6570. [CrossRef] [PubMed]
4. Andreiadis, E.S.; Chavarot-Kerlidou, M.; Fontecave, M.; Artero, V. Artificial Photosynthesis: From Molecular Catalysts for Light-driven Water Splitting to Photoelectrochemical Cells. *Photochem. Photobiol.* **2011**, *87*, 946–964. [CrossRef] [PubMed]

5. Young, K.J.; Martini, L.A.; Milot, R.L.; Snoeberger III, R.C.; Batista, V.S.; Schmuttenmaer, C.A.; Crabtree, R.H.; Brudvig, G.W. Light-driven water oxidation for solar fuels. *Coord. Chem. Rev.* **2012**, *256*, 2503–2520. [CrossRef] [PubMed]
6. Ismail, A.A.; Bahnemann, D.W. Photochemical splitting of water for hydrogen production by photocatalysis: A review. *Sol. Energy Mater. Sol. Cells* **2014**, *128*, 85–101. [CrossRef]
7. Imahori, H.; Umeyama, T.; Kurotobi, K.; Takanom, Y. Self-assembling porphyrins and phthalocyanines for photoinduced charge separation and charge transport. *Chem. Commun.* **2012**, *48*, 4032–4045. [CrossRef] [PubMed]
8. Bottari, G.; Trukhina, O.; Ince, M.; Torres, T. Towards artificial photosynthesis: Supramolecular, donor–acceptor, porphyrin- and phthalocyanine/carbon nanostructure ensembles. *Coord. Chem. Rev.* **2012**, *256*, 2453–2477. [CrossRef]
9. Jurow, M.; Schuckman, A.E.; Batteas, J.D.; Drain, C.M. Porphyrins as molecular electronic components of functional devices. *Coord. Chem. Rev.* **2010**, *254*, 2297–2310. [CrossRef]
10. Mahy, J.G.; Paez, C.A.; Carcel, C.; Bied, C.; Tatton, A.S.; Damblon, C.; Heinrichs, B.; Man, M.W.C.; Lambert, S.D. Porphyrin-based hybrid silica-titania as a visible-light photocatalyst. *J. Photochem. Photobiol. A* **2019**, *373*, 66–76. [CrossRef]
11. Ladomenoua, K.; Natali, M.; Iengo, E.; Charalampidis, G.; Scandola, F.; Coutsolelos, A.G. Photochemical hydrogen generation with porphyrin-based systems. *Coord. Chem. Rev.* **2015**, *304–305*, 38–54. [CrossRef]
12. Wang, L.; Fan, H.; Bai, F. Porphyrin-based photocatalysts for hydrogen production. *MRS Bull.* **2020**, *45*, 49–56. [CrossRef]
13. Mahmood, A.; Hu, J.-Y.; Xiao, B.; Tang, A.; Wang, X.; Zhou, E. Recent progress in porphyrin-based materials for organic solar cells. *J. Mater. Chem. A* **2018**, *6*, 16769–16797. [CrossRef]
14. Kim, H.; Kim, W.; Mackeyev, Y.; Lee, G.-S.; Kim, H.-J.; Tachikawa, T.; Hong, S.; Lee, S.; Kim, J.; Wilson, L.J.; et al. Selective Oxidative Degradation of Organic Pollutants by Singlet Oxygen Photosensitizing Systems: Tin Porphyrin versus C_{60} Aminofullerene Systems. *Environ. Sci. Technol.* **2012**, *46*, 9606–9613. [CrossRef] [PubMed]
15. Kim, W.; Park, J.; Jo, H.J.; Kim, H.-J.; Choi, W. Visible Light Photocatalysts Based on Homogeneous and Heterogenized Tin Porphyrins. *J. Phys. Chem. C* **2008**, *112*, 491–499. [CrossRef]
16. Shee, N.K.; Kim, M.K.; Kim, H.-J. Supramolecular porphyrin nanostructures based on coordination driven self-assembly and their visible light catalytic degradation of methylene blue dye. *Nanomaterials* **2020**, *10*, 2314. [CrossRef]
17. Shee, N.K.; Kim, H.-J. Self-assembled Nanomaterials Based on Complementary Sn(IV) and Zn(II)-porphyrins, and Their Photocatalytic Degradation for Rhodamine B Dye. *Molecules* **2021**, *26*, 3598. [CrossRef] [PubMed]
18. Shee, N.K.; Jo, H.J.; Kim, H.-J. Coordination framework materials fabricated by the self-assembly of Sn(IV) porphyrins with Ag(I) ions for the photocatalytic degradation of organic dyes in wastewater. *Inorg. Chem. Front.* **2022**, *9*, 1270–1280. [CrossRef]
19. Shee, N.K.; Kim, H.-J. Three Isomeric Zn(II)-Sn(IV)-Zn(II) Porphyrin-Triad-Based Supramolecular Nanoarchitectures for the Morphology-Dependent Photocatalytic Degradation of Methyl Orange. *ACS Omega* **2022**, *7*, 9775–9784. [CrossRef]
20. Li, C.; Park, K.-M.; Kim, H.-J. Ionic assembled hybrid nanoparticle consisting of tin(IV) porphyrin cations and polyoxomolybdate anions, and photocatalytic hydrogen production by its visible light sensitization. *Inorg. Chem. Comm.* **2015**, *60*, 8–11. [CrossRef]
21. Kim, W.; Tachikawa, T.; Majima, T.; Li, C.; Kim, H.-J.; Choi, W. Tin-porphyrin sensitized TiO_2 for the production of H_2 under visible light. *Energy Environ. Sci.* **2010**, *3*, 1789–1795. [CrossRef]
22. Bae, E.; Choi, W. Effect of the Anchoring Group (Carboxylate vs. Phosphonate) in Ru-Complex-Sensitized TiO_2 on Hydrogen Production under Visible Light. *J. Phys. Chem. B* **2006**, *110*, 14792–14799. [CrossRef] [PubMed]
23. Du, P.; Schneider, J.; Li, F.; Zhao, W.; Patel, U.; Castellano, F.N.; Eisenberg, R. Bi- and Terpyridyl Platinum(II) Chloro Complexes: Molecular Catalysts for the Photogeneration of Hydrogen from Water or Simply Precursors for Colloidal Platinum? *J. Am. Chem. Soc.* **2008**, *130*, 5056–5058. [CrossRef] [PubMed]
24. Ramakrishna, G.; Verma, S.; Jose, D.A.; Kumar, D.K.; Das, A.; Palit, D.K.; Ghosh, H.N. Interfacial Electron Transfer between the Photoexcited Porphyrin Molecule and TiO_2 Nanoparticles: Effect of Catecholate Binding. *J. Phys. Chem. B* **2006**, *110*, 9012–9021. [CrossRef] [PubMed]
25. Park, H.; Choi, W. Photocatalytic Reactivities of Nafion-Coated TiO_2 for the Degradation of Charged Organic Compounds under UV or Visible Light. *J. Phys. Chem. B* **2005**, *109*, 11667–11674. [CrossRef] [PubMed]
26. Park, H.; Choi, W. Visible-Light-Sensitized Production of Hydrogen Using Perfluorosulfonate Polymer-Coated TiO_2 Nanoparticles: An Alternative Approach to Sensitizer Anchoring. *Langmuir* **2006**, *22*, 2906–2911. [CrossRef] [PubMed]
27. Jo, H.J.; Kim, S.H.; Kim, H.-J. Supramolecular Assembly of Tin(IV) Porphyrin Cations Stabilized by Ionic Hydrogen-Bonding Interactions. *Bull. Korean Chem. Soc.* **2015**, *36*, 2348–2351. [CrossRef]

Article

A Three-Dimensional Melamine Sponge Modified with MnOx Mixed Graphitic Carbon Nitride for Photothermal Catalysis of Formaldehyde

Rongyang Yin, Pengfei Sun *, Lujun Cheng, Tingting Liu, Baocheng Zhou and Xiaoping Dong

Department of Chemistry, Key Laboratory of Surface & Interface Science of Polymer Materials of Zhejiang Province, Zhejiang Sci-Tech University, 928 Second Avenue, Xiasha Higher Education Zone, Hangzhou 310018, China
* Correspondence: sunpf@zju.edu.cn

Abstract: Much attention has been paid to developing effective visible light catalytic technologies for VOC oxidation without requiring extra energy. In this paper, a series of sponge-based catalysts with rich three-dimensional porosity are synthesized by combining MnOx and graphitic carbon nitride (GCN) with commercial melamine sponges (MS) coated with polydopamine (PDA), demonstrating excellent photothermal catalytic performance for formaldehyde (HCHO). The three-dimensional porous framework of MS can provide a good surface for material modification and a reliable interface for gas-solid interaction. The grown layer of PDA framework not only increases the near-infrared wavelength absorption for improving the light-to-heat conversion of catalysts, but also brings excellent adhesion for the subsequent addition of MnO_X and GCN. The efficient formaldehyde oxidation is attributed to the sufficient oxygen vacancies generated by co-loaded MnO_X and GCN, which is conducive to the activation of more O^{2-} in the oxidation process. As the surface temperature of catalyst rapidly increases to its maximum value at ca. 115 °C under visible light irradiation, the HCHO concentration drops from 160 ppm to 46 ppm within 20 min. The reaction mechanism is certified as a classical Mars-van Krevelen mechanism based on the photo-induced thermal catalysis process.

Keywords: photothermal conversion; graphitic carbon nitride; oxygen vacancy; manganese oxide; formaldehyde

Citation: Yin, R.; Sun, P.; Cheng, L.; Liu, T.; Zhou, B.; Dong, X. A Three-Dimensional Melamine Sponge Modified with MnOx Mixed Graphitic Carbon Nitride for Photothermal Catalysis of Formaldehyde. *Molecules* 2022, 27, 5216. https://doi.org/10.3390/molecules27165216

Academic Editors: Hongda Li, Mohammed Baalousha and Victor A. Nadtochenko

Received: 20 July 2022
Accepted: 11 August 2022
Published: 16 August 2022

Publisher's Note: MDPI stays neutral with regard to jurisdictional claims in published maps and institutional affiliations.

Copyright: © 2022 by the authors. Licensee MDPI, Basel, Switzerland. This article is an open access article distributed under the terms and conditions of the Creative Commons Attribution (CC BY) license (https://creativecommons.org/licenses/by/4.0/).

1. Introduction

HCHO has been extensively found in boards, paints, carpets, and wallpapers commonly used in interior decoration. According to the International Agency for Research on Cancer, formaldehyde can cause cancer in humans, especially nasopharyngeal cancer and leukemia [1,2]. With the general public's need for health and concern for a more efficient use of resources, it is of great importance to develop efficient and clean indoor HCHO degradation methods. At present, HCHO treatment methods mainly include absorption [3], biological incineration [4] and catalytic oxidation [5]. Among them, the activated carbon adsorption method is the most widely used; however, this requires regularly replacing activated carbon and recovering activated carbon wastes, thus resulting in higher elution costs. In contrast, catalytic oxidation converts exhaust gas into harmless carbon dioxide and water, which is more suitable for HCHO purification. In catalytic oxidation, synergistic photothermal catalysis technology is considered as an efficient and low-carbon HCHO removal technology that integrates the advantages of both photocatalysis and thermocatalysis. Under illumination, the material with photothermal effect does not directly change the state of the internal electrons after absorbing photons, but converts light energy into lattice thermal vibration, thus increasing the material's temperature. The catalysts can be activated using natural light sources without requiring additional energy.

Manganese oxide (MnOx), as a common transition metal oxide, has the above photothermal properties. Li Yuanzhi et al. reported MnOx nano-catalysts with various structures that have been proven to activate lattice oxygen with sunlight and efficiently remove various VOCs [6]. Wang Wenzhong also modified MnOx with cobalt oxide with significantly improved photothermal catalytic performance for the oxidation of VOCs, which was attributed to the changing lattice oxygen structure in MnOx [7]. Wang Zhongsen also synthesized two-dimensional nanosheets of MnOx by adding graphene oxide and polymeric carbon nitride, achieving excellent visible light oxidation efficiency of HCHO [2]. The photothermal effect of MnOx mainly originates from the non-radiative recombination of electron-hole pairs produced by the d-d transition of metal ions upon the absorption of photons. The surface temperature of MnOx increases even more efficiently after being modified by other photothermal materials such as black carbon or graphene, as black carbon-based materials are able to more sufficiently absorb the solar spectrum than metal oxide, ranging from ultraviolet light to the entire near-infrared band [8]. Thus, these materials endow MnOx with a stronger light-to-heat conversion ability, eventually inducing a thermal catalysis process with the Mars-van Krevelen (MvK) mechanism [9,10].

A three-dimensional porous composite MS/PDA/MnOx/GCN catalyst is synthesized by a simple method in this paper. A melamine sponge (MS) is used as a three-dimensional porous framework where a layer of polydopamine (PDA) is grown. Similar to carbon-based materials, black PDA has excellent light-to-heat conversion properties due to its strong ability to absorb near-infrared wavelengths [11]. Additionally, PDA exhibits excellent adhesion, which can firmly adhere subsequent MnOx nanosheets and graphitic carbon nitride (GCN) to further improve oxidation susceptibility and light-to-heat conversion. Compared with other photocatalysts without a fixed structure, this 3D skeleton structure exhibits better photoresponse performance. It provides a larger surface area for active phase loading and visible light absorption, which thus improves the removal efficiency of HCHO.

According to the experimental results, the synthesized catalyst exhibits a considerable HCHO photothermal catalytic removal efficiency with HCHO concentration dropping from 160 ppm to 46 ppm in 20 min. The morphology and surface elemental properties of the catalyst are also studied by SEM, HR-TEM, XRD, XPS, ESR, etc. Moreover, a surface reaction mechanism is proposed by using in situ DRIFTS and reactive oxygen species detection.

2. Experimental

2.1. Catalyst Synthesis

The flow chart of the synthesis process is shown in Figure 1: 50 mg dopamine hydrochloride was dispersed into 50 mL Tris-HCl solution (pH = 8.5) under agitation. Then, a blank melamine sponge (MS) with size = 2.5 cm^3 was immersed into the solution and stirred magnetically for 6 h, which was named MS/PDA. After being washed with distilled water, the MS/PDA was further immersed in 50 mL 0.05 mol/L KMnO$_4$ solution for 5 h at 60 °C and dried at 60 °C. The prepared sample was named MS/PDA/MnOx. Finally, the dried MS/PDA/MnOx was immersed into 0.2 g/L GCN suspension for 10 min by ultrasound. The obtained sample was washed by distilled water and dried at room temperature, which was named MS/PDA/MnOx/GCN. In order to investigate the influence of the loading sequence between GCN and MnOx on photothermal response performance, we also prepared the MS/PDA/GCN/MnOx catalyst by exchanging the synthesis order of the last two steps.

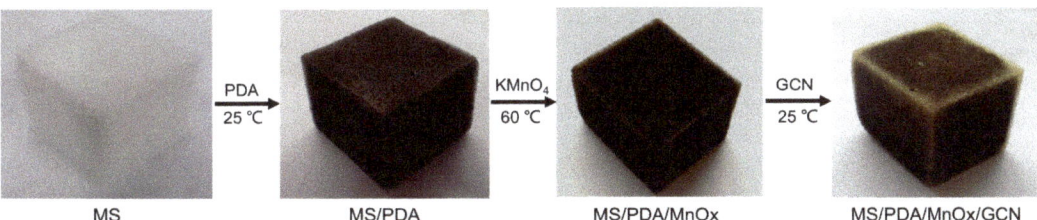

Figure 1. Schematic diagram of the sample preparation process.

2.2. Characterization

The catalyst was characterized by scanning electron microscopy (SEM) using a field emission scanning electron microscope (Model SU-8100, Hitachi Co., Tokyo, Japan) equipped with an energy dispersive X-ray spectrometer (EDS). X-ray diffraction (XRD) was tested by a powder diffractometer with Cu $K\alpha$ radiation (Model D/max RA, Rigaku Co., Tokyo, Japan). The chemical compositions and states were explored by X-ray photoelectron spectroscopy (XPS, Model AXIS UltraDLD, Kratos Co., Manchester, UK) using an Al $K\alpha$ monochromatic source. Electron spin resonance (ESR) spectra were tested using an electron paramagnetic resonance spectrometer (Model ESP 300 E, Bruker Co., Ettlingen, Germany) equipped with a xenon lamp as a light source (420–600 nm). UV-vis-NIR diffuse-reflectance spectra (DRS) were analyzed on a HITACHI U-4100 spectrophotometer. The changes in the surface temperature of different catalysts were characterized by a thermal image (Model H21PRO, Hikvision Co., Hangzhou, China).

2.3. Photothermal Catalytic Performance

The performance of the catalyst was tested in a 0.5 L glass container containing a 0.2 g sponge sample, with observation glass in the middle. After the gas-phase reactor was preheated to ca. 40 °C, 2.0 µL HCHO (37%) solution was injected into the gas-phase reactor till completely evaporated. At this time, the initial concentration of HCHO was approximately ca. 160 ppm. A 300 W xenon lamp (Model HSX-F300, NBet Co., Beijing, China) was used as light source. The HCHO removal activity under light irradiation was indirectly evaluated by measuring the absorbance obtained by phenol reagent spectrophotometry (Model UV-2600, Shimadzu Co., Kyoto, Japan). During measurements, 5 mL gas was sampled with a syringe and injected into 5 mL 0.05 mol/L phenol reagent (MBTH hydrochloride hydrate) solution. Subsequently, 0.4 mL 1 wt.% $NH_4Fe(SO_4)_2$ solution was added. After approximately 10 min, the absorbance of the above solution at 630 nm was measured. The changing surface temperature of the catalyst under light illuminations was also recorded by an infrared thermometer in the experimental process (Model H16, Hikvision Co., Hangzhou, China). A picture of the experimental set-up has been listed in the Supplementary Material.

2.4. Reaction Mechanism

The HCHO decomposition mechanism was clarified based on in situ diffuse reflectance infrared Fourier transform spectroscopy (in situ DRIFTS) using a Bruker Co., Model TENSOR II spectrometer. The reaction spectra of HCHO on the surface of the catalyst were acquired by accumulating 32 scans at a 4 cm^{-1} resolution under the irradiation of a xenon lamp. To confirm the formation of oxidized OH radical and O^{2-} radical during the reaction, the photoluminescence (PL) method with terephthalic acid (TA) oxidized by ·OH and blue tetrazolium (NBT) by O^{2-} was employed. Both reaction products were detected by changing the absorbance using a fluoroanalyzer (Model Fluoromax-4, HORIBA Co., Kyoto, Japan).

3. Results and Discussion

3.1. Photothermal Catalytic Performance of Formaldehyde

The photothermal activity of different sponge loaded catalysts was evaluated by HCHO degradation (the concentration of HCHO was ca. 160 ppm) using 300 W xenon lamps. As displayed in Figure 2A, the HCHO concentration of four catalysts decreased to varying degrees. For MS/PDA, the poor removal efficiency of HCHO was obviously due to the lack of reactive oxygen species caused by the absence of loaded metal oxide. The catalysts MS/PDA/MnOx and MS/PDA/GCN/MnOx exhibited similar HCHO removal efficiency. The HCHO concentration declined from 160 ppm to 46 ppm within 20 min, implying that neither the oxidation susceptibility of subsequently loaded MnOx nor the photothermal synergistic performance of the catalyst was improved by adding GCN. However, the catalyst MS/PDA/MnOx/GCN revealed significantly improved HCHO removal performance compared with MS/PDA/GCN/MnOx, suggesting that the loading sequence of GCN had a great effect on the performance of the photothermal co-catalytic system.

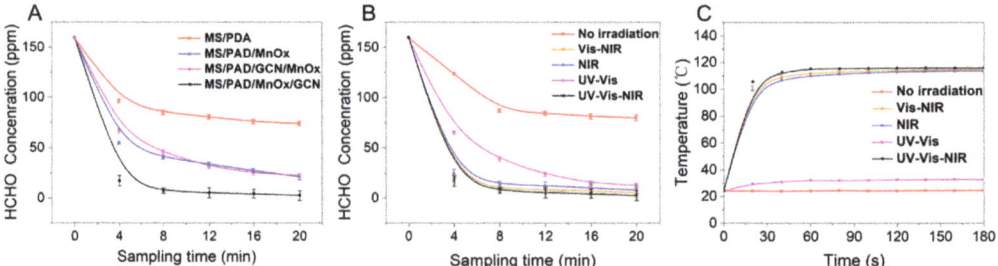

Figure 2. (**A**) Catalytic performance for HCHO degradation under the UV-Vis-NIR irradiation; (**B**) catalytic performance of catalyst MS/PDA/MnOx/GCN for HCHO degradation under irradiation with different wavelength ranges; (**C**) temperature change for catalyst MS/PDA/MnOx/GCN under irradiation with different wavelength ranges.

Figure 2B shows the degradation of HCHO under different light sources. Obviously, the degradation efficiency of HCHO without any light irradiation was not ideal. Compared with the rapid decline in the degradation efficiency of HCHO under other conditions, NIR light played a key role in the HCHO removal though UV and visible light also contributed to the photodegradation of HCHO. Figure 2C also exhibits the change in the surface temperature of the catalyst MS/PDA/MnOx/GCN under the condition of different light sources. The surface temperature of the catalyst MS/PDA/MnOx/GCN filtered by NIR light did not obviously change compared with that with the presence of NIR light. The latter temperatures all significantly increased to ca. 115 °C. Thus, we believe the degradation of HCHO on the manganese-based catalyst was mainly affected by increased temperature, which induced the well-known Mars-van Krevelen process that decomposes C_xH_y utilizing $\cdot O^{2-}$, OH and other oxidation groups [10,12].

The UV-Vis-NIR DRS was also tested to evaluate the photo-thermal effect of four different catalysts. As shown in Figure S1, four catalysts all exhibited considerable NIR absorption (wavelength ranging from 400 to 800 nm) due to the loaded PDA. These were the efficient NIR-absorbing materials as reported in the literature [11,13]. For MS/PDA/MnOx/GCN, the highest NIR absorption was found. The NIR absorption of the catalyst MS/PDA/MnOx/GCN was found to be the highest because its outermost surface was further wrapped by GCN, which further increased NIR light absorption [14]. The changes in the surface temperature of the four catalysts were tested by thermal imaging, as exhibited in Figures S2 and S3. This was consistent with the UV-Vis-NIR DRS result that the catalyst MS/PDA/MnOx/GCN displayed the highest temperature.

3.2. Morphological Characteristics

As shown in Figure 3, the SEM image of the catalyst MS/PDA/MnOx/GCN exhibited uniformly attached MnOx on the 3D structure of MS. The EDX mapping also confirmed that the scanning of the Mn and O elements was in accordance with the MS skeleton. The XRD result is displayed in Figure 4. It can be observed from the figure that all the samples revealed an obvious peak at 2θ = ca. 22°, which was ascribed to the surface diffraction of the melamine sponge [11,13]. With the increase in surface load, the intensity of the diffraction peak became weaker. The characteristic peaks of Mn loaded samples at 2θ = 36.5° and 65.5° were ascribed to the (100) and (110) planes of an Akhtenskite MnOx phase (JCPDS NO.30-0820) [11,13], indicating that MnOx was effectively formed on the surface of the melamine sponge merely by using potassium permanganate. However, the loaded PDA and GCN were not identified by XRD diffraction patterns mainly due to the fact that the loading of these two substances always presented good dispersion, as exhibited by the SEM image.

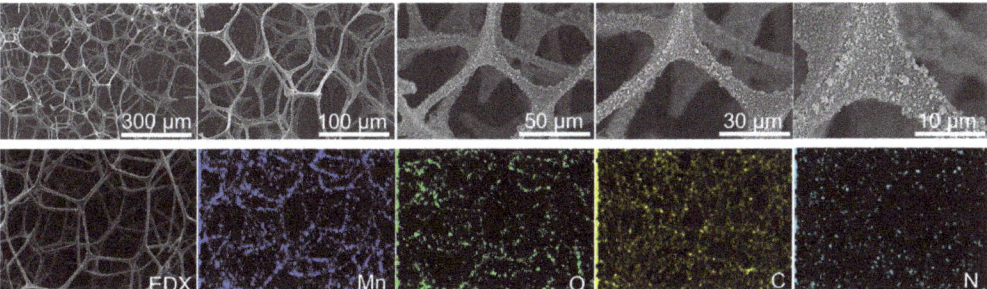

Figure 3. SEM image and EDX mapping of the catalyst MS/PDA/MnOx/GCN.

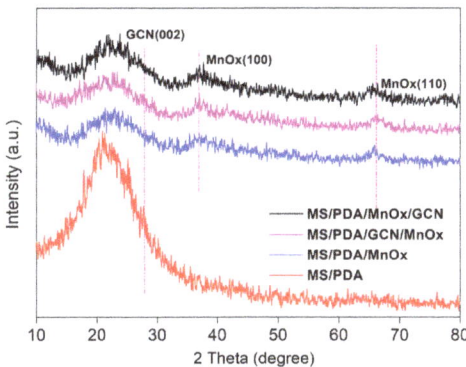

Figure 4. XRD profiles of the catalysts MS/PDA, MS/PDA/MnOx, MS/PDA/GCN/MnOx and MS/PDA/MnOx/GCN.

3.3. Surface Properties

In order to further understand the different surface properties of the catalysts MS/PDA/MnOx, MS/PDA/MnOx/GCN and MS/PDA/GCN/MnOx, XPS analysis was carried out. As shown in Figure 5A, Mn 2p spectrum showed two characteristic peaks ascribed to Mn 2p 3/2 (641.9 eV) and Mn 2p 1/2 (653.4 eV) [15,16], which suggests the presence of MnO_2. It should be noted that the splitting energy between these two characteristic peaks was 11.6 eV for MS/PDA/MnOx/GCN and 11.8 eV for MS/PDA/MnOx and MS/PDA/GCN/MnOx. The splitting energy of MS/PDA/MnOx/GCN decreased because of the enhanced π electron cloud density [16], which was probably originated from the reaction between the residual GCN on the external surface of MnOx and its functional groups.

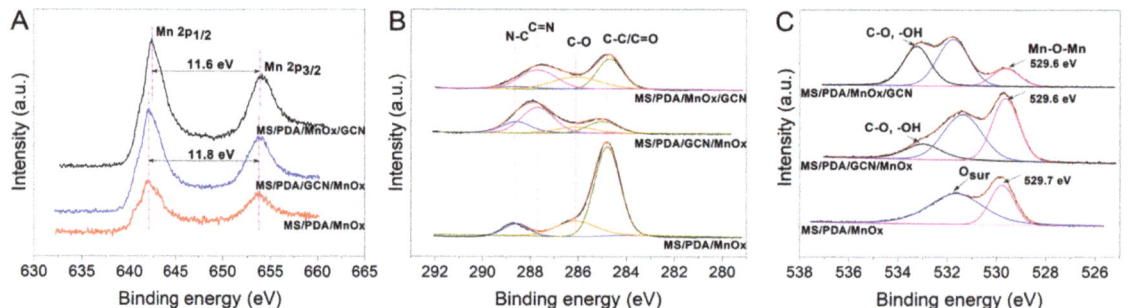

Figure 5. XPS spectra of the catalysts MS/PDA/MnOx, MS/PDA/GCN/MnOx and MS/PDA/MnOx/GCN: (**A**) Mn 2p; (**B**) C 1s; (**C**) O 1s.

The C 1s spectrum (Figure 5B) shows four peaks at 284.4, 286.2, 287.7 and 288.6 eV, respectively. The peak at 287.7 eV was ascribed to carbon in the melamine sponge conjugated system [11]. The peaks at 284.4 and 286.2 eV corresponded to sp2 graphite carbon in the PDA structure and C-O carbon in the benzene ring of the PDA molecules, respectively [11,17]. Additionally, the peak at 287.7 eV was ascribed to C=N to the GCN structure [18]. It was also observed that the order of GCN addition determined whether GCN reacted with PDA or MnOx. For MS/PDA/GCN/MnOx, PDA (peak at 284.4 eV) was found to be depleted compared with MS/PDA/MnOx/GCN and MS/PDA/MnOx. For the latter two catalysts, the PDA firstly reacted with manganese, resulting in a similar PDA structure. Especially for MS/PDA/MnOx/GCN, as GCN was loaded after MnOx, a tight bonding with MnOx as the splitting energy of Mn 2p shift was detected, as shown in Figure 5A.

The O 1s spectrum (Figure 5C) was divided into three peaks. The peak at 529.6 eV was ascribed to the lattice oxygen of Mn-O-Mn in MnOx [19], that at 531.3 eV was ascribed to surface chemisorbed oxygen species [20], and that at 532.8 eV was ascribed to hydroxyl groups such as Mn-OH or H-OH [21,22]. All of three Mn loaded catalysts exhibited chemisorbed oxygen species and lattice oxygen due to the predominant oxygen species formed on the surface MnOx. After the addition of GCN, the ionized oxygen species such as OH$^-$ emerged, indicating that the functional groups of GCN promoted the transformation of Mn-O to Mn-OH. As reported in the literature [23–25], GCN improves electrical conductivity and reduces the electron-transfer resistance of metal oxide and thus promotes the formation of Mn atoms and oxygen atoms into a covalent coordination bond or a hydrogen bond, such as Mn-OH.

Three fitted absorption bands can be observed in the N 1s spectrum in Figure S4. The peak around 398.6 eV resulted from N species in the C=N-C unit, while that at ca. 399.5 eV was assigned to the N-H groups in PDA and melamine sponge [11,26] and that around 400.2 eV to the N-(C)$_3$ in GCN [2,27]. For MS/PDA/GCN/MnOx and MS/PDA/MnOx/GCN, the peak around 404.2 eV ascribed to π electron cloud density was caused by doped GCN [25].

3.4. Reaction Mechanism

The photo-thermal-induced ROS were tested to further explore the reaction mechanism on the catalyst surface. The PL detection method was employed to confirm the generation of hydroxide radical with p-Phthalic acid (TA) as indicator. As shown in Figure 6A, the oxidative product TAOH had a sole fluorescence response at 425 nm [28]. It can be clearly found that the PL strength on the three Mn supported catalysts was significantly improved as the irradiation time increased, indicating that hydroxide was produced during the reaction. The formation of $\cdot O^{2-}$ radical was also tested by nitro blue tetrazolium (NBT) as the indicator [29]. As shown in Figure 6B, the concentration of NBT over the catalysts MS/PDA/GCN/MnOx and MS/PDA/MnOx did not decrease significantly, meaning that O_2 molecules were not activated on both catalysts. However, the generation of O^{2-} radical on MS/PDA/MnOx/GCN was obvious. The result implies that the outermost layer of GCN provides favorable conditions for the adsorption and activation of O_2 molecules on the reaction interface, which was also confirmed by the ESR result for oxygen vacancy determination. As shown in Figure 7, all samples showed a pair of steep peaks with a symmetric distribution in accordance with g = 2.002, an indication of electron trapping at oxygen vacancies [30]. It can be concluded from the ESR result of the catalyst MS/PDA/GCN that the addition of GCN provides a large number of oxygen vacancies for the catalyst system. Additionally, the catalysts MS/PDA/MnOx/GCN had more oxygen vacancies than MS/PDA/GCN/MnOx and MS/PDA/MnOx, as the loaded GCN on the outermost layer was partially retained. The formation of sufficient oxygen vacancies promoted the absorption and activation of O_2 atoms [31], leading to the formation of more $\cdot O^{2-}$ radicals over MS/PDA/MnOx/GCN.

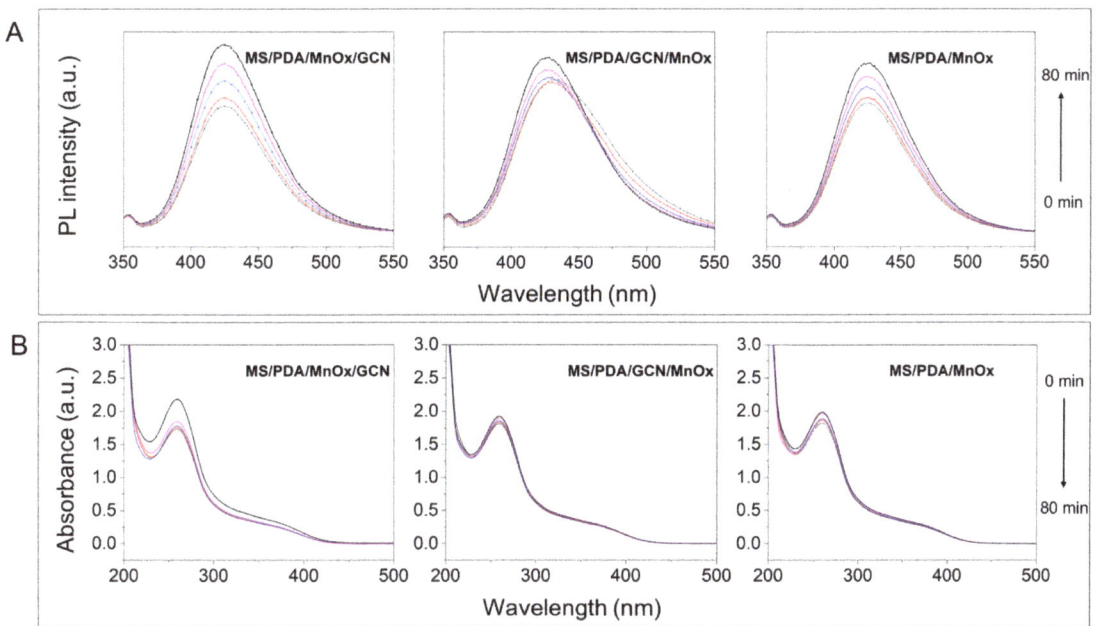

Figure 6. PL spectra of TA solution (**A**) and absorption spectra of NBT solution (**B**) with the catalysts MS/PDA/MnOx/GCN, MS/PDA/GCN/MnOx and MS/PDA/MnOx under xenon light irradiation.

In situ DRIFTS measurements were also conducted to obtain insight into the HCHO degradation mechanism. As shown in Figure 8, several vibration peaks were identified at approximately 3634, 2363, 2335, 1712, 1550, 1425 and 1243 cm^{-1}. It can be found that the peak at 1425 cm^{-1} represented the vibrations of dioxymethylene (DOM) [32,33]. Additionally, another group peak at 1712 and 1550 cm^{-1} represented formate (HCOO$^-$) species [32,33]. These are the primary degradation products of HCHCO oxidation. With the extension of reaction time, their corresponding vibration peaks gradually increased, indicating the absorption and accumulation of HCHO. The shoulder peaks at 2363 and 2335 cm^{-1} corresponded to CO_2 adsorption with a corresponding vibration peak at 1243 cm^{-1} (assigned to the CO_3^{2-} oxidized from HCOO$^-$) [9,34], implying the final mineralization of HCHO. There was also a very strong negative peak at 3634 cm^{-1}, which was assigned to hydroxyl vibration peak, representing a large consumption of Mn-OH during HCHO oxidation [35,36].

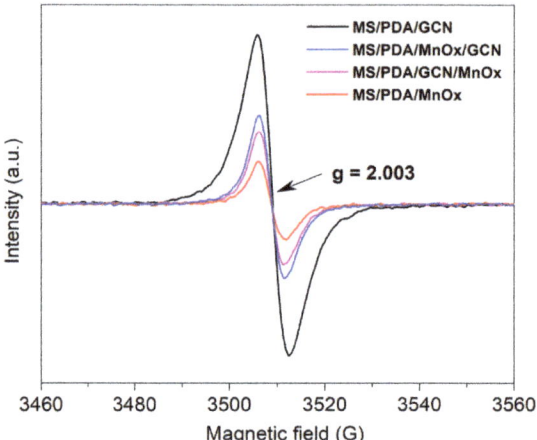

Figure 7. ESR test for the catalysts MS/PDA/GCN, MS/PDA/MnOx/GCN, MS/PDA/GCN/MnOx and MS/PDA/MnOx.

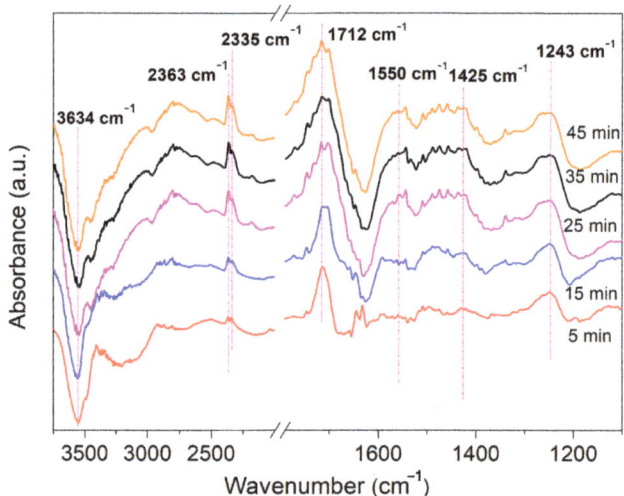

Figure 8. In situ DRIFTS of HCHO reaction on the catalyst MS/PDA/MnOx/GCN under xenon light irradiation.

Based on all the above, a plausible mechanism for HCHO oxidation over MS/PDA/MnOx/GCN is proposed as Figure 9. HCHO molecules were firstly adsorbed onto Mn-OH groups through hydrogen bonding [37,38], further utilizing adjacent Mn-O to form CO_2 and H_2O. Concurrently, O_2 molecules absorbed onto conterminal oxygen vacancies to activate O^{2-} and were involved in the formation of CO_2 and the replenishment of Mn-O.

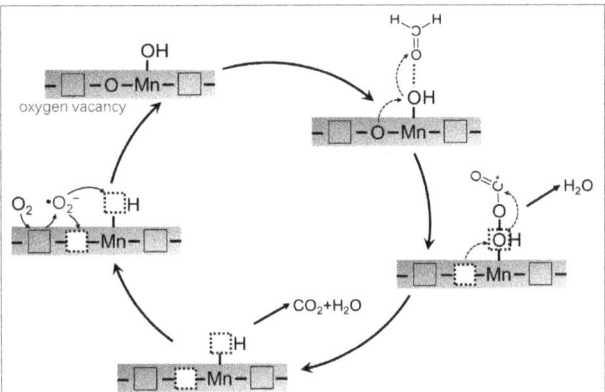

Figure 9. Probable degradation mechanism of HCHO over catalyst MS/PDA/MnOx/GCN.

3.5. Reusability

The cyclic performance of catalysts is very important for photocatalytic materials, which can be used to estimate whether the catalyst is easy to deactivate. The deactivation of a catalyst always arises due to the loss of intrinsic (per-site) activity, or a decrease in the number of active sites, or increasingly restricted access to the active sites [39]. Based on the above in situ DRIFTS analysis and mechanism elaboration results, we can speculate that the most likely cause of deactivation over this catalyst was only due to the accumulation of residue at the active site. As shown in Figure 10, HCHO was purified by the MS/PDA/MnOx/GCN catalyst five times. The results showed that the purification effect of HCHO was basically the same as that purified for the first time. Although it would be wrong to repeat the experiment to measure the stability of the catalyst at 100% conversion without studying under kinetically controlled conditions [39], we could not measure stability at the 50% or 90% conversion rates because the high sensitivity of this catalyst could achieve a 100% conversion rate of HCHO in a very short time. However, it is worth noting that the MS/PDA/MnOx/GCN catalyst exhibited excellent reuse performance for at least five tests, which indicates the potential to be developed as a commercial catalyst against indoor HCHO removal.

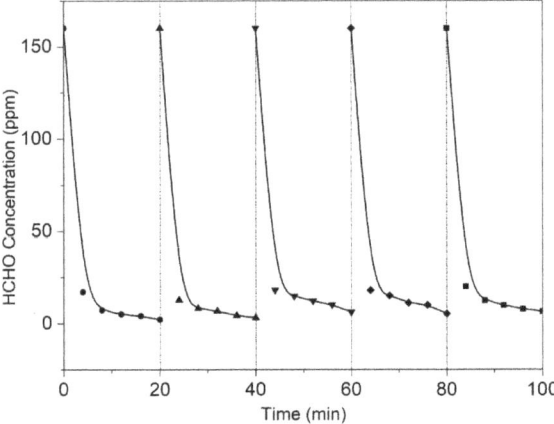

Figure 10. Cyclic performance test for HCHO removal.

4. Conclusions

To sum up, we designed an integral HCHO photocatalyst characterized by impressive light-to-heat conversion property and excellent reusability. The concentration of HCHO dropped from 160 ppm to 46 ppm within 20 min under the irradiation of visible light. The addition of GCN promoted the formation of oxygen vacancy, thus facilitating the activation of surface reactive oxygen species and maintaining the efficient oxidation of HCHO under a spontaneously rising surface temperature. The association of GCN with MnOx over common three-dimensional sponge materials offers great potential for commercial catalyst development aimed at indoor HCHO control without using noble metals, which also makes a significant contribution to the literature regarding catalysts designed to reduce environmental air pollution.

Supplementary Materials: The following supporting information can be downloaded at: https://www.mdpi.com/article/10.3390/molecules27165216/s1, Figure S1: UV–Vis-NIR DRS profiles of catalyst MS/PDA, MS/PDA/MnOx, MS/PDA/GCN/MnOx, and MS/PDA/MnOxGCN; Figure S2: Changes in surface temperature of catalyst MS/PDA, MS/PDA/MnOx, MS/PDA/GCN/MnOx, and MS/PDA/MnOxGCN over visible light irradiation; Figure S3: The thermal imagery of catalyst MS/PDA, MS/PDA/MnOx, MS/PDA/GCN/MnOx, and MS/PDA/MnOxGCN over visible light irradiation; Figure S4: N 1s XPS spectra of catalyst MS/PDA/MnOx, MS/PDA/GCN/MnOx and MS/PDA/MnOx/GCN; Figure S5: Photos of the experiment process.

Author Contributions: Conceptualization, X.D.; methodology, P.S. and L.C.; validation, P.S., L.C. and T.L.; formal analysis, P.S. and T.L.; investigation, R.Y.; writing—original draft preparation, R.Y.; writing—review and editing, P.S., B.Z. and X.D.; supervision, P.S., B.Z. and X.D. All authors have read and agreed to the published version of the manuscript.

Funding: This research was funded by the Natural Science Founding of China (51908491), the National Key Research and Development Program of China (2017YFE0127400) and Science Fund of Zhejiang Sci-Tech University (21062255-Y).

Institutional Review Board Statement: Not applicable.

Informed Consent Statement: Not applicable.

Data Availability Statement: Not applicable.

Conflicts of Interest: The authors declare no conflict of interest.

Sample Availability: Samples mentioned in paper are available from the authors.

References

1. Cogliano, V.J.; Grosse, Y.; Baan, R.A.; Straif, K.; Secretan, M.B.; Ghissassi, F.E.; Working Group for Volume 88. Meeting report: Summary of IARC monographs on formaldehyde, 2-butoxyethanol, and 1-tert-butoxy-2-propanol. *Environ. Health Perspect.* **2005**, *113*, 1205–1208. [CrossRef] [PubMed]
2. Wang, Z.; Yu, H.; Xiao, Y.; Zhang, L.; Guo, L.; Zhang, L.; Dong, X. Free-standing composite films of multiple 2D nanosheets: Synergetic photothermocatalysis/photocatalysis for efficient removal of formaldehyde under ambient condition. *Chem. Eng. J.* **2020**, *394*, 125014. [CrossRef]
3. Zhang, W.; Huang, T.; Ren, Y.; Wang, Y.; Yu, R.; Wang, J.; Tu, Q. Preparation of chitosan crosslinked with metal-organic framework (MOF-199)@aminated graphene oxide aerogel for the adsorption of formaldehyde gas and methyl orange. *Int. J. Biol. Macromol.* **2021**, *193*, 2243–2251. [CrossRef] [PubMed]
4. Rovina, K.; Vonnie, J.M.; Shaeera, S.N.; Yi, S.X.; Halid, N.F.A. Development of biodegradable hybrid polymer film for detection of formaldehyde in seafood products. *Sens. Bio-Sens. Res.* **2020**, *27*, 100310. [CrossRef]
5. Cao, K.; Dai, X.; Wu, Z.; Weng, X. Unveiling the importance of reactant mass transfer in environmental catalysis: Taking catalytic chlorobenzene oxidation as an example. *Chin. Chem. Lett.* **2021**, *32*, 1206–1209. [CrossRef]
6. Liu, F.; Zeng, M.; Li, Y.; Yang, Y.; Mao, M.; Zhao, X. UV-Vis-Infrared Light Driven Thermocatalytic Activity of Octahedral Layered Birnessite Nanoflowers Enhanced by a Novel Photoactivation. *Adv. Funct. Mater.* **2016**, *26*, 4518–4526. [CrossRef]
7. Zheng, Y.; Wang, W.; Jiang, D.; Zhang, L.; Li, X.; Wang, Z. Ultrathin mesoporous Co_3O_4 nanosheets with excellent photo-/thermo-catalytic activity. *J. Mater. Chem. A* **2016**, *4*, 105–112. [CrossRef]
8. Fu, L.; Wang, R.; Zhao, C.; Huo, J.; He, C.; Kim, K.-H.; Zhang, W. Construction of Cr-embedded graphyne electrocatalyst for highly selective reduction of CO_2 to CH_4: A DFT study. *Chem. Eng. J.* **2021**, *414*, 128857. [CrossRef]

9. Tang, X.; Chen, J.; Li, Y.; Li, Y.; Xu, Y.; Shen, W. Complete oxidation of formaldehyde over Ag/MnOx-CeO$_2$ catalysts. *Chem. Eng. J.* **2006**, *118*, 119–125. [CrossRef]
10. Zhang, J.; Li, Y.; Wang, L.; Zhang, C.; He, H. Catalytic oxidation of formaldehyde over manganese oxides with different crystal structures. *Catal. Sci. Technol.* **2015**, *5*, 2305–2313. [CrossRef]
11. Wang, Z.; Yu, H.; Xiao, Y.; Guo, L.; Zhang, L.; Dong, X. Polydopamine mediated modification of manganese oxide on melamine sponge for photothermocatalysis of gaseous formaldehyde. *J. Hazard. Mater.* **2021**, *407*, 124795. [CrossRef] [PubMed]
12. Mao, C.; Vannice, M. Formaldehyde Oxidation over Ag Catalysts. *J. Catal.* **1995**, *154*, 230–244. [CrossRef]
13. Sun, P.; Yu, H.; Liu, T.; Li, Y.; Wang, Z.; Xiao, Y.; Dong, X. Efficiently photothermal conversion in a MnOx-based monolithic photothermocatalyst for gaseous formaldehyde elimination. *Chin. Chem. Lett.* **2022**, *33*, 2564–2568. [CrossRef]
14. Liang, Q.; Li, Z.; Huang, Z.-H.; Kang, F.; Yang, Q.-H. Holey Graphitic Carbon Nitride Nanosheets with Carbon Vacancies for Highly Improved Photocatalytic Hydrogen Production. *Adv. Funct. Mater.* **2015**, *25*, 6885–6892. [CrossRef]
15. Sun, P.; Wang, W.; Dai, X.; Weng, X.; Wu, Z. Mechanism study on catalytic oxidation of chlorobenzene over Mn$_x$Ce$_{1-x}$O$_2$/H-ZSM5 catalysts under dry and humid conditions. *Appl. Catal. B Environ.* **2016**, *198*, 389–397. [CrossRef]
16. Lu, L.; Tian, H.; He, J.; Yang, Q. Graphene-MnO$_2$ hybrid nanostructure as a new catalyst for formaldehyde oxidation. *J. Phys. Chem. C* **2016**, *120*, 23660–23668. [CrossRef]
17. Li, L.; Chi, L.; Zhang, H.; Wu, S.; Wang, H.; Luo, Z.; Li, Y.; Li, Y. Fabrication of Ti-PDA nanoparticles with enhanced absorption and photocatalytic activities for hexavalent chromium Cr(VI) removal. *Appl. Surf. Sci.* **2022**, *580*, 152168. [CrossRef]
18. Gao, P.; Zhou, W.; Zhang, J.; Yang, Z.; Yuan, X.; Wang, Z. The effect of acidified graphite carbon nitride on the removal of pollutants by coupling filtration and photocatalysis. *Appl. Surf. Sci.* **2021**, *542*, 148675. [CrossRef]
19. Zhao, Q.; Chen, D.; Li, Y.; Zhang, G.; Zhang, F.; Fan, X. Rhodium complex immobilized on graphene oxide as an efficient and recyclable catalyst for hydrogenation of cyclohexene. *Nanoscale* **2013**, *5*, 882–885. [CrossRef]
20. Sun, P.; Zhai, S.; Chen, J.; Yuan, J.; Wu, Z.; Weng, X. Development of a multi-active center catalyst in mediating the catalytic destruction of chloroaromatic pollutants: A combined experimental and theoretical study. *Appl. Catal. B Environ.* **2020**, *272*, 119015. [CrossRef]
21. Li, L.; Yan, J.; Wang, T.; Zhao, Z.-J.; Zhang, J.; Gong, J.; Guan, N. Sub-10 nm rutile titanium dioxide nanoparticles for efficient visible-light-driven photocatalytic hydrogen production. *Nat. Commun.* **2015**, *6*, 5881. [CrossRef] [PubMed]
22. Wang, M.; Zhang, P.; Li, J.; Jiang, C. The effects of Mn loading on the structure and ozone decomposition activity of MnOx supported on activated carbon. *Chin. J. Catal.* **2014**, *35*, 335–341. [CrossRef]
23. Li, K.; Tong, Y.; Feng, D.; Chen, P. Electronic regulation of platinum species on metal nitrides realizes superior mass activity for hydrogen production. *J. Colloid Interface Sci.* **2022**, *622*, 410–418. [CrossRef] [PubMed]
24. Simaioforidou, A.; Georgiou, Y.; Bourlinos, A.; Louloudi, M. Molecular Mn-catalysts grafted on graphitic carbon nitride (gCN): The behavior of gCN as support matrix in oxidation reactions. *Polyhedron* **2018**, *153*, 41–50. [CrossRef]
25. Wang, Z.; Gu, Z.; Wang, F.; Hermawan, A.; Hirata, S.; Asakura, Y.; Hasegawa, T.; Zhu, J.; Inada, M.; Yin, S. An ultra-sensitive room temperature toluene sensor based on molten-salts modified carbon nitride. *Adv. Powder Technol.* **2021**, *32*, 4198–4209. [CrossRef]
26. Peng, S.; Yang, X.; Strong, J.; Sarkar, B.; Jiang, Q.; Peng, F.; Liu, D.; Wang, H. MnO$_2$-decorated N-doped carbon nanotube with boosted activity for low-temperature oxidation of formaldehyde. *J. Hazard. Mater.* **2020**, *396*, 122750. [CrossRef]
27. Yao, C.; Wang, R.; Wang, Z.; Lei, H.; Dong, X.; He, C. Highly dispersive and stable Fe^{3+} active sites on 2D graphitic carbon nitride nanosheets for efficient visible-light photocatalytic nitrogen fixation. *J. Mater. Chem. A* **2019**, *7*, 27547–27559. [CrossRef]
28. Lei, H.; Zhang, H.; Zou, Y.; Dong, X.; Jia, Y.; Wang, F. Synergetic photocatalysis/piezocatalysis of bismuth oxybromide for degradation of organic pollutants. *J. Alloy Compd.* **2019**, *809*, 151840. [CrossRef]
29. Wang, Z.; Yu, H.; Zhang, L.; Guo, L.; Dong, X. Photothermal conversion of graphene/layered manganese oxide 2D/2D composites for room-temperature catalytic purification of gaseous formaldehyde. *J. Taiwan Inst. Chem. Eng.* **2020**, *107*, 119–128. [CrossRef]
30. Han, Z.; Choi, C.; Hong, S.; Wu, T.S.; Soo, Y.L.; Jung, Y.; Qiu, J.; Sun, Z. Activated TiO$_2$ with tuned vacancy for efficient electrochemical nitrogen reduction. *Appl. Catal. B Environ.* **2019**, *257*, 117896. [CrossRef]
31. Ye, K.; Li, K.; Lu, Y.; Guo, Z.; Ni, N.; Liu, H.; Huang, Y.; Ji, H.; Wang, P. An overview of advanced methods for the characterization of oxygen vacancies in materials. *TrAC Trends Anal. Chem.* **2019**, *116*, 102–108. [CrossRef]
32. Chen, M.; Wang, H.; Chen, X.; Wang, F.; Qin, X.; Zhang, C.; He, H. High-performance of Cu-TiO$_2$ for photocatalytic oxidation of formaldehyde under visible light and the mechanism study. *Chem. Eng. J.* **2020**, *390*, 124481. [CrossRef]
33. Wang, J.; Li, J.; Jiang, C.; Zhou, P.; Zhang, P.; Yu, J. The effect of manganese vacancy in birnessite-type MnO$_2$ on room-temperature oxidation of formaldehyde in air. *Appl. Catal. B Environ.* **2017**, *204*, 147–155. [CrossRef]
34. Miao, L.; Xie, Y.; Xia, Y.; Zou, N.; Wang, J. Facile photo-driven strategy for the regeneration of a hierarchical C@MnO$_2$ sponge for the removal of indoor toluene. *Appl. Surf. Sci.* **2019**, *481*, 404–413. [CrossRef]
35. Fang, R.; Huang, H.; Ji, J.; He, M.; Feng, Q.; Zhan, Y.; Leung, D.Y. Efficient MnOx supported on coconut shell activated carbon for catalytic oxidation of indoor formaldehyde at room temperature. *Chem. Eng. J.* **2018**, *334*, 2050–2057. [CrossRef]
36. Wang, J.; Zhang, G.; Zhang, P. Layered birnessite-type MnO$_2$ with surface pits for enhanced catalytic formaldehyde oxidation activity. *J. Mater. Chem. A* **2017**, *5*, 5719–5725. [CrossRef]
37. Zhang, C.; Liu, F.; Zhai, Y.; Ariga, H.; Yi, N.; Liu, Y.; Asakura, K.; Flytzani-Stephanopoulos, M.; He, H. Alkali-metal-promoted Pt/TiO$_2$ opens a more efficient pathway to formaldehyde oxidation at ambient temperatures. *Angew. Chem. Int. Ed.* **2012**, *51*, 9628–9632. [CrossRef]

38. Wang, Y.; Zhu, X.; Crocker, M.; Chen, B.; Shi, C. A comparative study of the catalytic oxidation of HCHO and CO over $Mn_{0.75}Co_{2.25}O_4$ catalyst: The effect of moisture. *Appl. Catal. B Environ.* **2014**, *160*, 542–551. [CrossRef]
39. Scott, S.L. *A Matter of Life (Time) and Death*; ACS: Washington, DC, USA, 2018; pp. 8597–8599.

Article

Comparative Studies of g-C$_3$N$_4$ and C$_3$N$_3$S$_3$ Organic Semiconductors—Synthesis, Properties, and Application in the Catalytic Oxygen Reduction

Ewelina Wierzyńska [1], Marcin Pisarek [2], Tomasz Łęcki [1] and Magdalena Skompska [1,*]

[1] Faculty of Chemistry, University of Warsaw, Pasteura 1, 02-093 Warsaw, Poland
[2] Institute of Physical Chemistry, Polish Academy of Sciences, Kasprzaka 44/52, 01-224 Warsaw, Poland
* Correspondence: mskomps@chem.uw.edu.pl

Abstract: Exfoliated g-C$_3$N$_4$ is a well-known semiconductor utilized in heterogenous photocatalysis and water splitting. An improvement in light harvesting and separation of photogenerated charge carriers may be obtained by polymer doping with sulfur. In this work, we incorporate sulfur into the polymer chain by chemical polymerization of trithiocyanuric acid (C$_3$N$_3$S$_3$H$_3$) to obtain C$_3$N$_3$S$_3$. The XRD measurements and TEM images indicated that C$_3$N$_3$S$_3$, in contrast to g-C$_3$N$_4$, does not exist in the form of a graphitic structure and is not exfoliated into thin lamellas. However, both polymers have similar optical properties and positions of the conduction and valence bands. The comparative studies of electrochemical oxygen reduction and hydrogen evolution indicated that the overpotentials for the two processes were smaller for C$_3$N$_3$S$_3$ than for g-C$_3$N$_4$. The RDE experiments in the oxygen-saturated solutions of 0.1 M NaOH have shown that O$_2$ is electrochemically reduced via the serial pathway with two electrons involved in the first step. The spectroscopic experiments using NBT demonstrated that both polymers reveal high activity in the photocatalytic reduction of oxygen to superoxide anion radical by the photogenerated electrons.

Keywords: g-C$_3$N$_4$; C$_3$N$_3$S$_3$; band diagrams; oxygen reduction

1. Introduction

Graphitic carbon nitride (g-C$_3$N$_4$) is one of the most popular organic semiconductors utilized in heterogeneous photocatalysis [1–3]. It may be synthesized by thermal condensation of cyanamide, dicyandiamide, melamine, thiourea, and urea ([4] and the references therein). The band gap of g-C$_3$N$_4$, 2.7 eV, is small enough to absorb the photons from the visible light but sufficiently large to fulfill the thermodynamic requirements for water splitting. Namely, the bottom of the conduction band is located above the redox potential of the H$^+$/H$_2$ couple, while the edge of the valence band is more positive than the oxidation potential of water to O$_2$ [5]. Moreover, the microstructure of g-C$_3$N$_4$ with a large number of termination atoms and defects is beneficial for anchoring the active sites. Therefore, the bulk g-C$_3$N$_4$ is exfoliated in aqueous (acidic or basic solutions) or organic solvents to obtain ultrathin or monolayer nanosheets of more abundant surface active sites [6,7] and very high specific surface area.

Using g-C$_3$N$_4$ as a metal-free photocatalyst for water splitting to H$_2$ under visible light, with triethanolamine as the hole scavenger, has been first reported by Wang et al. [8]. However, an efficient and stable hydrogen evolution was possible after modification of g-C$_3$N$_4$ with a small amount (3 wt%) of Pt co-catalyst. It was also found that some improvement in the photocatalytic properties of g-C$_3$N$_4$ could be achieved by molecular doping. For example, the band gap energy of sulfur-doped g-C$_3$N$_4$ (S-g-C$_3$N$_4$) is reduced to 2.63 eV [9], which allows better utilization of the solar spectrum. The theoretical calculations predicted equal band gaps of g-C$_3$N$_4$ and S-g-C$_3$N$_4$, but the S-doped polymer has the impurity level located in the band gap. This ensures easy excitation of electrons from the valence band to the impurity level. Launching sulfur into the g-C_3N_4 structure also improves the charge carrier separation and prevents recombination. The electrons from sulfur-doped g-C_3N_4 have been utilized to reduce CO_2 to obtain CH_3OH, while the holes from the valence band were involved in water oxidation [9]. The hybrid system of TiO_2/S-g-C_3N_4 was used in the photocatalytic degradation of Congo Red [10]. The significant enhancement of photocatalytic activity was ascribed to the S-scheme mechanism and well-distributed 1D nanostructure of doped polymer.

Another possibility of sufur incorporation into the polymer chain is the synthesis from the monomer containing sulfur in the molecular structure. Therefore, in this work, we synthesized $C_3N_3S_3$ by chemical polymerization of trithiocyanuric acid ($C_3N_3S_3H_3$) and compared the properties of the obtained organic semiconductor with those of g-C_3N_4 formed by thermal condensation of urea. It has been shown that both polymers reveal ambipolar properties, i.e., behave as n-type or p-type semiconductors, depending on the polarization range. The band diagrams of both semiconductors were constructed taking into account the position of the Fermi level determined from photocurrent onset potential, and the valence band position was verified by VB X-ray photoelectron spectra. The formation of superoxide anion radicals by the electrons photogenerated in the conduction bands of both semiconductors was confirmed by the Nitro Blue Tetrazolium chloride (NBT) experiments. Finally, both polymers were used in the reactions of oxygen reduction and electrocatalytic hydrogen generation.

2. Results

2.1. Characterization of $C_3N_3S_3$

The successful synthesis of $C_3N_3S_3$ polymer by chemical oxidation of $C_3N_3S_3H_3$ with I_2 has been confirmed by FTIR spectra presented in Figure 1.

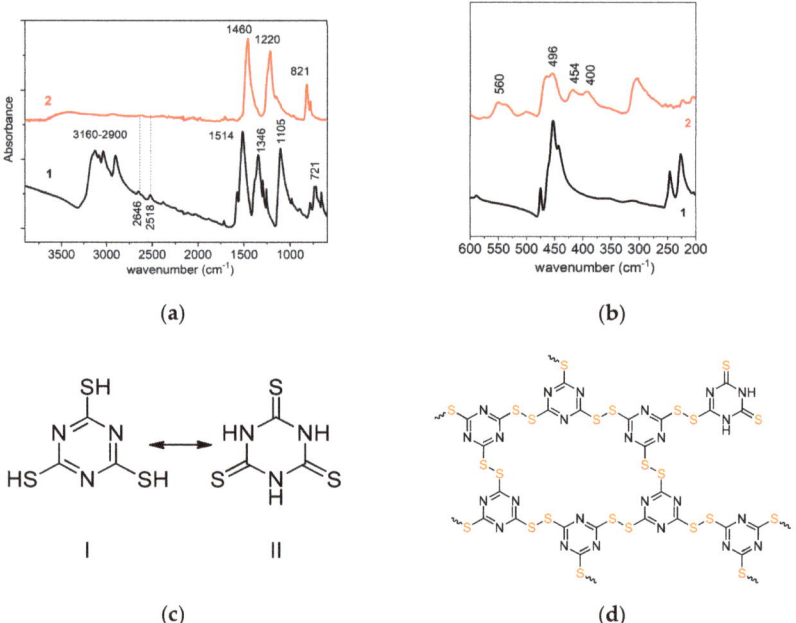

Figure 1. FTIR spectra of the monomer $C_3N_3S_3H_3$ (lines 1) and the polymer $C_3N_3S_3$ (lines 2) in the wavenumber range 3800 cm^{-1}–600 cm^{-1} (**a**) and 600 cm^{-1}–350 cm^{-1} (**b**); the chemical structures of the monomer in the tautomeric form of thiol (I) and thione (II) (**c**), and probable structure of the polymer $C_3N_3S_3$ (**d**).

In the spectrum of the monomer (line 1), one can observe several sharp peaks in the range 3160–2900 cm^{-1}, characteristic for N-H stretching vibrations in triazine groups in the monomer [11] (Figure 1c), which are not visible in the spectrum of the polymer (line 2). The peaks at 1514, 1346, 1105, and 721 cm^{-1} in the monomer spectrum are ascribed to the stretching vibrations of non-aromatic heterocycle thione (tautomeric structure II in Figure 1c). Specifically, the bands at 1514 and 1346 cm^{-1} may be ascribed to C=N and C-N stretching vibrations [12] in the triazine ring, while that at 1105 cm^{-1} is assigned to C=S stretching vibrations [13]. In the polymer, the peaks at 1514, 1346, and 720 cm^{-1} are shifted to 1460, 1220, and 821 cm^{-1}, respectively, while the trace of the C=S peak is observed in the band shoulder at about 1140 cm^{-1}. It likely origins from the thione groups in the external, terminal units of the polymer network. The formation of the polymer is supported by the presence of new peaks in the range 560–400 cm^{-1} (Figure 1b), ascribed to disulfide S-S linkages, and the disappearance of two weak peaks at 2518 and 2646 cm^{-1} corresponding to S-H stretching. The peak at about 450 cm^{-1} observed both in the spectrum of the monomer and polymer originates from NCN bending vibration [14].

The successful polymerization of thiocyanuric acid to $C_3N_3S_3$ has also been confirmed by XPS measurements. The main peaks in the survey spectrum presented in Figure S1 originate from C 1s, N 1s, and S 2p. Two additional peaks of very small intensities were detected at the binding energies 621.2 eV and 632.5 eV (Figure S2a), corresponding to I $3d_{5/2}$ and I $3d_{3/2}$, originate from the traces of I_2 oxidant (about 0.2 at.%) used for polymerization. The formation of S-S bonds in the polymer is manifested in the high-resolution XPS spectrum (Figure 2a) by two peaks at the binding energies 164.7 eV and 165.9 eV, ascribed to S $2p_{3/2}$ and S $2p_{1/2}$, respectively [15,16].

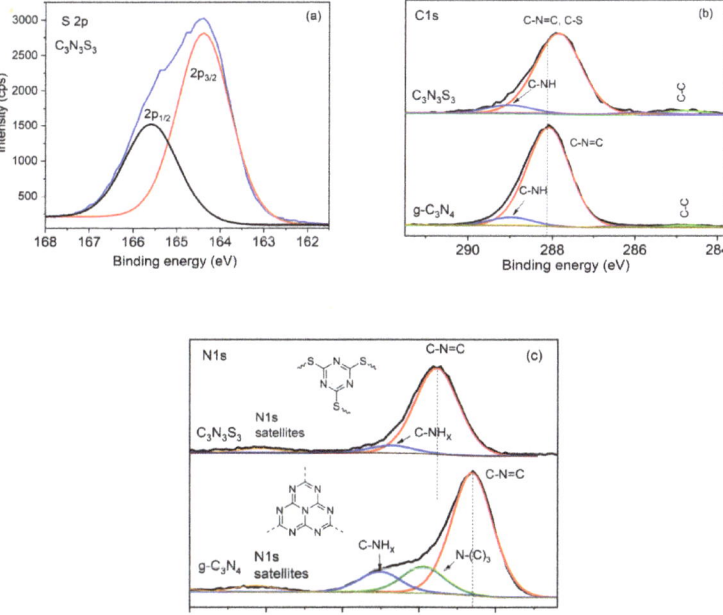

Figure 2. High-resolution XPS spectrum of S 2p of $C_3N_3S_3$ (**a**), and comparison of HR-XPS spectra of C 1s (**b**), N 1s (**c**) of $C_3N_3S_3$ and g-C_3N_4.

In the deconvoluted spectrum of $C_3N_3S_3$, the main C 1s peak at the binding energy 287.9 eV, corresponding to N-C=N and C-S bonds (Figure 2b), is shifted by about 0.2 eV towards lower energies with respect to the C 1s peak ascribed to sp^2-bonded carbon in N-

C=N, in the spectrum of g-C_3N_4, due to the presence of sulfur in the chemical environment of carbon. There is also a small component of C 1s peak in the spectra of both polymers at about 289 eV, which may be ascribed to sp^2 C bonded to the NH group (sp^2 C-NH) [17]. On the other hand, the signal at this binding energy may also originate from oxygen-carbon functional groups (C=O and C-O) due to the presence of surface carboxylic groups. However, the O 1s peaks of very small intensity were detected only in the spectrum of $C_3N_3S_3$ (see the survey and high-resolution spectra in Figures S1a and S2b) but not for g-C_3N_4 (Figure S1b).

The significant differences in the spectra of the two polymers are observed for the peak N 1s (Figure 2c). The peak N 1s for $C_3N_3S_3$ is narrow because it has only two components: the peak at 399.6 eV corresponding to sp^2-hybridized nitrogen in C-N=C, and a small peak at about 400.7 eV corresponding to C-NH_x, due to the presence of the external nitrogen atoms in the polymer network (Figure 1d). In contrast, the N 1s peak in the spectrum of g-C_3N_4, with a broad shoulder on the higher energy side, can be deconvoluted in three components. The main peak located at about 398.6 eV corresponds to sp^2-hybridized nitrogen (C-N=C), while the component at about 400 eV is ascribed to sp^3 tertiary/bridging N-$(C)_3$ nitrogens [18] (not present in the case of $C_3N_3S_3$). A weak peak at the binding energy ~401.0 eV is attributed to the nitrogen in terminal amino groups, C-NH_x. In the spectra of both polymers, there are also very small peaks at the binding energy of about 404.0 and 407.0 eV, which correspond to shake-up satellite peaks of N 1s. These data are in good agreement with the XPS observations and DFT calculations of Zhang and co-authors for different g-C_3N_4 models [19]. Such satellite signals can also be observed for carbon C 1s.

The atomic ratios obtained from XPS data are C:N = 1:1.27 for g-C_3N_4 (being very close to the theoretical ratio 1:1.3) and C:N:S = 1:0.84:1.1 for $C_3N_3S_3$ (see details in Tables S1 and S2 in SM).

The XRD patterns presented in Figure 3a indicate the significant differences in the crystallographic structures of $C_3N_3S_3$ and g-C_3N_4.

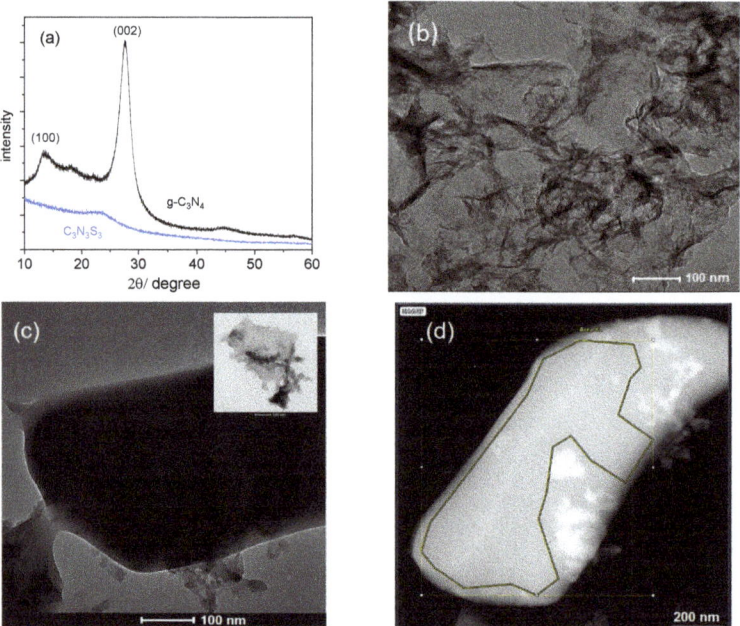

Figure 3. The XRD patterns of $C_3N_3S_3$ and g-C_3N_4 (**a**); TEM images of g-C_3N_4 (**b**) and $C_3N_3S_3$ (**c**) after sonication in DMSO, and HAADF image of $C_3N_3S_3$ with the indicated area of elemental analysis (**d**).

The XRD pattern of g-C$_3$N$_4$ is consistent with the reference diffractogram in the JCPDS (no. 87-1526) [20]. A high signal at 2θ of about 27° is indexed as (002) peak originating from the interplanar diffraction, confirming the graphitic-like structure of this polymer [8]. It corresponds to the interlayer distance of d = 3.36 Å, which is comparable to the packing in the crystalline graphite. The second peak at about 13° of markedly lower intensity, indexed as (100), is ascribed to in-planar repeated triazine units [21]. In contrast, the diffractogram of C$_3$S$_3$N$_3$ with only one broad peak at about 23°, similar to that reported in the literature [16], is typical for amorphous material. Different morphology of C$_3$S$_3$N$_3$ and g-C$_3$N$_4$ results in different behavior of the two polymers during sonication in DMSO used as the solvent. As visible in TEM images of the polymers sonicated for 10 h, the g-C$_3$N$_4$ underwent exfoliation into lamellas (Figure 3b), while C$_3$S$_3$N$_3$ still existed in the form of thicker flakes of micrometer length and smooth surface. However, in the latter case, one can observe the thin flakes or aggregated particles attached to the large and flat elements (see inset in Figure 3c).

The atomic ratio C:N:S obtained for the area marked in Figure 3d is about 1:0.8:1.1, which is practically the same as the ratio obtained from XPS measurements (1:0.84:1:1), and in both cases, the amount of N in C$_3$N$_3$S$_3$ is a little lower than the expected one, while the relative S content is a little too high. However, EDS elemental maps have shown that all elements are uniformly distributed over the whole sample (Figure S3).

2.2. Determination of the Band Diagrams of g-C$_3$N$_4$ and C$_3$N$_3$S$_3$

According to the UV-Vis absorption spectra presented in Figure 4a, the optical absorption edges of g-C$_3$N$_4$ and C$_3$N$_3$S$_3$ are at the same wavelength, 430 nm. This wavelength corresponds to the band gap energy 2.88 eV calculated from the equation E_g (eV) = 1240/λ(nm).

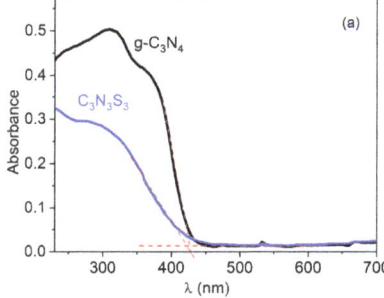
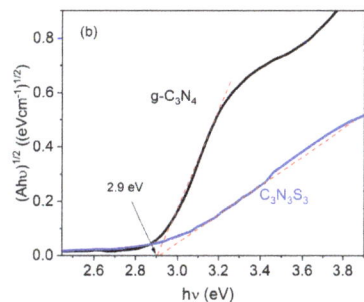
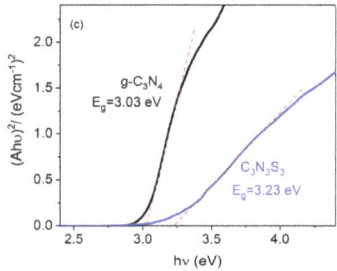

Figure 4. UV-Vis absorption spectra (**a**) and Tauc plots for C$_3$N$_3$S$_3$ and g-C$_3$N$_4$ for indirect (**b**) and direct (**c**) band transitions.

The band gap of both materials was also determined from the Tauc plot:

$$(\alpha h\nu)^{1/n} = A(h\nu - E_g) \quad (1)$$

where α is an absorption coefficient, h is the Planck constant, ν is the photon's frequency, and A is a constant. The value of factor n is equal to 1/2 for direct band transition and 2 for indirect transition [22]. The g-C_3N_4 is reported in the literature as an indirect semiconductor [23–25], while no data on the type of transition is available for $C_3N_3S_3$. Therefore, the Tauc plots were done for n = 2 (Figure 4b) as well as for n = 1/2 (Figure 4c). Since the value of E_g = 2.9 eV, obtained for both polymers from the plot $(\alpha h\nu)^{1/2}$ vs $h\nu$, is close to band gap energy determined directly from UV-Vis spectrum (Figure 4a), one can conclude that the transition in the $C_3N_3S_3$ is also indirect. The band gap energy obtained for g-C_3N_4 is a little higher than the values reported in the literature (2.6–2.85 eV) [6,8,26], which may be explained by lower temperature applied during the precursor condensation (500 °C) than most often applied in the literature (550–600 °C). It has been reported that the decrease of the band gap energy with the increase of the processing temperature is related to a gradual increase in the polymerization degree [27]. On the other hand, the increase of the synthesis temperature above 550 °C may lead to some distortion of the (100) crystal plane and a decrease in the photocatalytic ability of g-C_3N_4 [20].

The experimental data on $C_3N_3S_3$ are very scarce, but according to DFT calculations, the band gap of this polymer may vary from 3.77 eV to 1.9, depending on the polymerization degree [16].

The edge of the conduction band of g-C_3N_4 is often estimated in the literature from the Butler and Ginley relationship [28]:

$$E_{CB}[vs\ SHE] = \frac{1}{e}\left(\chi - \frac{1}{2}E_g - E^e\right)\ [eV] \quad (2)$$

where E^e is the energy of free electrons with respect to SHE (0 V vs. SHE corresponds to 4.5 eV), E_g is the band energy, and χ is the semiconductor electronegativity, which is defined as the geometric mean of the electronegativities of the constituent neutral atoms $\chi(M)$. However, according to Praus [29], this method is not relevant for the determination of the band edges of g-C_3N_4 since this material is not characterized by a single value of electronegativity due to possible structural defects, layer distortions, and the presence of natural impurities, such as oxygen, etc. It has been shown that the χ obtained for g-C_3N_4 (6.91 eV) leads to the band edges E_{CB} and E_{VB} far from those obtained from experimental values [29]. Recently, we have reported the same problem with the application of the Butler and Ginley approach in the determination of correct potentials of the band edges of $BiVO_4$ [30].

Experimentally, the conduction band of the semiconductors is most often determined from the flat band potential (E_{fb}) obtained from the Mott–Schottky plot [31]. However, the main assumption made in the derivation of the Mott–Schottky equation, such as a perfectly planar semiconductor surface or a homogeneous distribution of electronic defects, is not fulfilled in the case of lamellar g-C_3N_4. Alternatively, the value of E_{fb} may be determined from the onset potential of photocurrent by analysis of the voltammograms in the dark and under illumination [32,33], and this method gave very reliable results for Fe_2O_3 [33] and $BiVO_4$ [30].

The cyclic voltammograms of FTO/g-C_3N_4 and FTO/$C_3N_3S_3$ electrodes obtained in the solution of 0.1 M Na_2SO_4, presented in Figure 5a, indicate that both polymers exhibit an ambipolar behavior, i.e., as a p-type semiconductor with negative photocurrents under cathodic bias, and n-type with positive photocurrents in the range of anodic potentials. This type of behavior is typical for organic semiconductors, as well as two-dimensional layered materials, such as metal dichalcogenides [34]. It has also been reported for g-C_3N_4 as amphoteric behavior [35,36]. As visible in Figure 5a, the anodic photocurrent densities for both polymers studied were markedly lower than the reduction photocurrents. This may be explained by considering the electrode–solution interface reactions. In the range of positive potentials, the photogenerated holes may be involved in water oxidation. However, this reaction is rather complex because it requires the transfer of four electrons to oxidize two H_2O molecules with the removal of four protons to form the O=O bond [37]. In contrast, the cathodic current for both polymers was high, also without illumination, since the electrons delivered to the electrode/solution interface may reduce the oxygen dissolved in the solution. This reaction will be discussed in more detail in Section 2.3. Therefore, in order to determine the photocurrent onset potential the voltammetric measurements were performed in deaerated solutions (to minimize the dark current). The potential of the electrodes was scanned from 0.2 V vs Ag/AgCl towards the negative values, at a low scan rate of 1 mV s^{-1} to diminish the capacitance current. The light was chopped at the frequency of about 0.15 Hz.

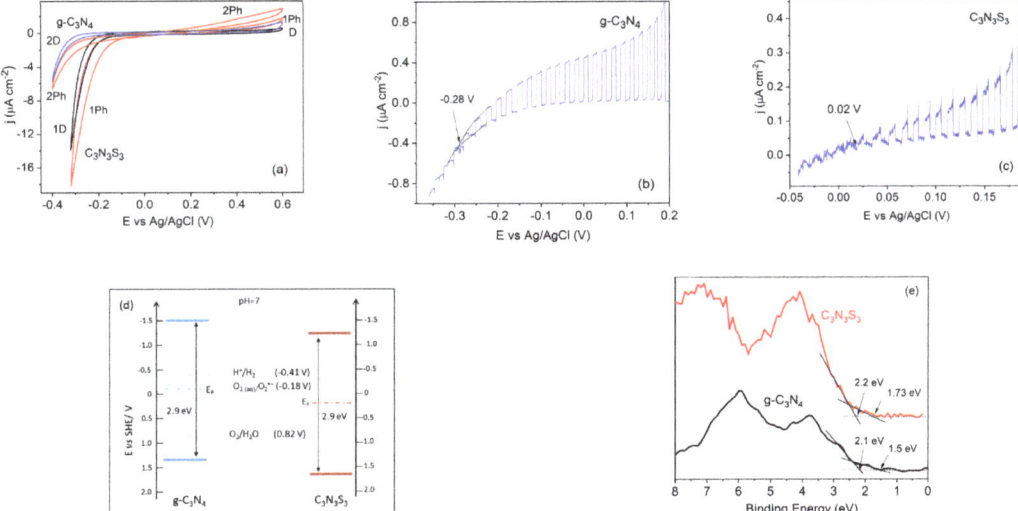

Figure 5. Cyclic voltammograms for $C_3N_3S_3$ (lines 1) and g-C_3N_4 (lines 2) in the dark (D) and under illumination (Ph) in the solution of 0.1 M Na_2SO_4 (**a**); linear sweep voltammograms (LSVs) for g-C_3N_4 (**b**) and $C_3N_3S_3$ (**c**) at the sweep rate of 1 mV s^{-1} under chopped illumination in a deaerated solution of 0.1 M Na_2SO_4; the band energy diagrams of g-C_3N_4 and $C_3N_3S_3$ constructed on the base of electrochemical data with the redox potentials of the solution species at pH 7 (**d**), and VB XP spectra of $C_3N_3S_3$ and g-C_3N_4 (**e**).

As visible in Figure 5b,c, the photocurrents reached the dark currents at the potentials -0.28 V vs. Ag/AgCl (i.e., -0.07 V vs. SHE) for g-C_3N_4 and 0.02 V vs. Ag/AgCl (i.e., 0.23 V vs. SHE) for $C_3N_3S_3$. The photocurrent onset potentials were assumed as the Fermi levels (E_F) of these two materials [35]. Taking into account that the Fermi level of the ambipolar semiconductor is located approximately at the middle of the forbidden band [38] and that the optical band gap of the g-C_3N_4 and $C_3N_3S_3$ determined from the UV-Vis spectra is about 2.9 eV, the estimated values of E_{VB} are of about 1.38 V vs. NHE for g-C_3N_4 and 1.68 V vs. NHE for $C_3N_3S_3$, while the E_{CB} is of about -1.52 V vs. NHE for g-C_3N_4 and -1.22 V vs. NHE for $C_3N_3S_3$, as presented in the band diagram in Figure 5d.

The valence band positions for both polymers were also determined from VB XP spectra presented in Figure 5e. The main absorption onsets are located at 2.1 eV for g-C_3N_4 and 2.2 eV for $C_3N_3S_3$; these values are higher than those obtained from electrochemical experiments. It is worth noting that both XPS spectra also have characteristic tails that cross the energy axes at lower binding energies (at 1.5 eV for g-C_3N_4 and 1.73 eV for $C_3N_3S_3$). Similar features have often been observed in XPS spectra of g-C_3N_4, but in general, they were ignored. However, Kang et al. ascribed this tail to the absence of long-range atomic order in the polymer matrix, which results in the dangling bonds, and they determined the valence band position of g-C_3N_4 from the tail end [39]. Thus, it is difficult to decide which approach is more appropriate. It should also be taken into account that the state of the semiconductor surface in the electrolyte solution is different than that in a vacuum due to the adsorption of the species at the polymer/solution interface [40]. However, irrespective of the method of determination, the E_{CB} of both polymers is located above the redox potential of H^+/H_2, while the E_{VB} is below the oxidation potential of H_2O to O_2, which means that g-C_3N_4 and $C_3S_3N_3$ are good candidates for photocatalytic hydrogen and oxygen evolution. The photocatalytic ability of g-C_3N_4 in the reaction of water splitting to produce H_2 has been reported for the first time by Wang et al. [8], but there are several drawbacks, such as limited visible light harvesting efficiency and the fast recombination rate of the photogenerated electron-hole pairs, limiting the hydrogen evolution efficiency on the pristine polymer [41]. A similar problem has been reported for applying g-C_3N_4 alone in the photocatalytic degradation of dye pollutants. Therefore, g-C_3N_4 is usually combined with other semiconductors, such as TiO_2, ZnO, WO_3, Bi_2WO_6, and MoS_2, to reduce the charge recombination (see [2,41,42] and references therein). According to the literature, the g-C_3N_4-based heterostructures may also be used for the pyrocatalytic decomposition of dyes [43].

The electrons excited to the conduction band may also be involved in the reduction of oxygen to superoxide anion radical since the redox potential $O_2/O_2^{-\bullet}$ (−0.18 V vs. SHE) [44] is located much below the conduction band edges of both polymers (as presented in the band diagram in Figure 5d). The generation of superoxide radicals under the illumination of the g-C_3N_4 and $C_3S_3N_3$ deposited on FTO has been monitored by the changes in the UV-Vis absorption spectra (Figure 6) that occurred due to the reaction of $O_2^{-\bullet}$ with NBT (as described in Section 3.7). It was found that the rate of transformation of NBT into diformazan was the same for both polymers (inset in Figure 6), which suggests the same ability of both materials in photocatalytic generation of superoxide anion radicals.

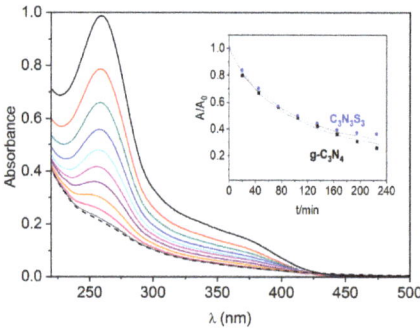

Figure 6. Evolution of UV-Vis spectra of NBT solution under irradiation with a diode 365 nm in the presence of FTO/g-C_3N_4 and the change of relative absorbance (A/A_0) in the peak at 260 nm in a function of time under illumination in the presence of g-C_3N_4 and $C_3N_3S_3$ (inset).

2.3. Electrochemical Oxygen Reduction and Hydrogen Evolution at the g-C_3N_4 and $C_3N_3S_3$-Modified Electrodes

In order to confirm that the cathodic currents recorded in the negative potential range on FTO modified with g-C_3N_4 and $C_3N_3S_3$ (presented in Figure 5a) result from oxygen reduction reaction (ORR), the comparative experiments were performed in the presence of oxygen and in a deaerated solution of 0.1 M Na_2SO_4. As visible in Figure 7a, deaeration of the solution led to a significant decrease of the cathodic currents for both polymers (dashed lines), confirming that the oxygen reduction is the main electrode reaction at the semiconductor/solution interface in the potential range from 0 V to −0.4 V vs. Ag/AgCl.

It is also worth noting that the process starts at a less negative potential at the $C_3N_3S_3$ (in the dark and under illumination) than at g-C_3N_4 (Figure 5a).

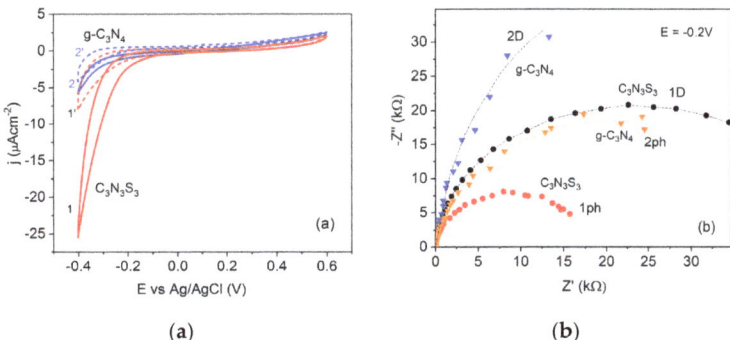

(a)   (b)

Figure 7. Cyclic voltammograms of $C_3N_3S_3$ (lines 1) and g-C_3N_4 (lines 2) in the solution of 0.1 M Na_2SO_4 in the dark, in the presence of oxygen (lines 1,2) and after deaeration (lines 1' and 2') (**a**), and EIS curves in the dark (D) and under illumination (ph) recorded at a constant potential of -0.2 V vs. Ag/AgCl in the frequency range 10^5–0.1 Hz (**b**).

This is also confirmed by the smaller arc radii in the EIS plots obtained for $C_3S_3N_3$ than that for g-C_3N_4 both in the dark and under illumination at the potential -0.2 V vs. Ag/AgCl (Figure 7b). The exact mechanism of oxygen reduction in metals and non-metallic catalysts is still a matter of extensive studies and discussions in the literature [45]. In general, two alternative reaction pathways are considered: a direct four-electron O_2 reduction or a "serial pathway" with two successive two-electron steps. In acidic solutions, the product of the direct reaction is water, while the two-electron process leads to the formation of H_2O_2, which may be further reduced to water in a two-electron reaction. In an alkaline solution, the O_2 is reduced in a direct four-electron reaction to OH^- anions, while in the serial pathway, O_2 is reduced to peroxide ion (HO_2^-), which may be followed by either further reduction to OH^- or disproportionation to OH^- and O_2 [46,47]. It has also been postulated that the first step of O_2 reduction on glassy carbon (GC) electrodes in neutral and mild alkaline (pH $\leq$ 10) solutions is the formation of superoxide anion radical ($O_2^{-\bullet}$) [48,49].

In order to determine the O_2 reduction pathway at the g-C_3N_4 and $C_3N_3S_3$ electrodes, linear sweep voltammetric (LSV) experiments using a rotating disc electrode (RDE) were performed. The polymers were deposited on the glassy carbon electrodes by drop cast and polarized in the potential range from 0.1 V to -0.6 V vs. Ag/AgCl at the scan rate of 2 mV s^{-1}, and different rotation rates in the solution of 0.1 M Na_2SO_4. However, no limiting current was observed in the voltammograms recorded in this solution (Figure S4), and therefore, the investigations were carried out in 0.1 M NaOH. For both polymers, the shape of the voltammograms was similar, with a broad wave at about -0.5 V vs. Ag/AgCl, followed by the continuously increasing cathodic current at more positive potentials, as illustrated in Figure 8a. After O_2 saturation of the solution by oxygen bubbling for 15 min, the limiting current density increased about 4 times (Figure 8b).

The number of electrons (n) involved in the oxygen reduction was determined from the slope of the Levich–Koutecky plot (j_{lim}^{-1} vs. $\omega^{-1/2}$), presented in the inset in Figure 8b, according to the relationship [50]:

$$\frac{1}{j_{lim}} = \frac{1}{j_k} + \frac{1}{j_D} = \frac{1}{nFkc} + \frac{1}{0.62nFD^{2/3}\nu^{-1/6}c} \cdot \frac{1}{\omega^{\frac{1}{2}}} \quad (3)$$

where j_k and j_D are the current densities controlled by the reaction kinetics and diffusion process, respectively, F is the Faraday constant, c is a concentration of oxygen in water

($1.26 \cdot 10^{-6}$ mol cm^{-3} for O$_2$-saturated solution), D is the oxygen diffusion coefficient ($D_{O_2} = 1.96 \cdot 10^{-5}$ cm^2 s^{-1}), and ν is a kinematic viscosity of the solution (0.01 cm^2 s^{-1}) (all data from [46]).

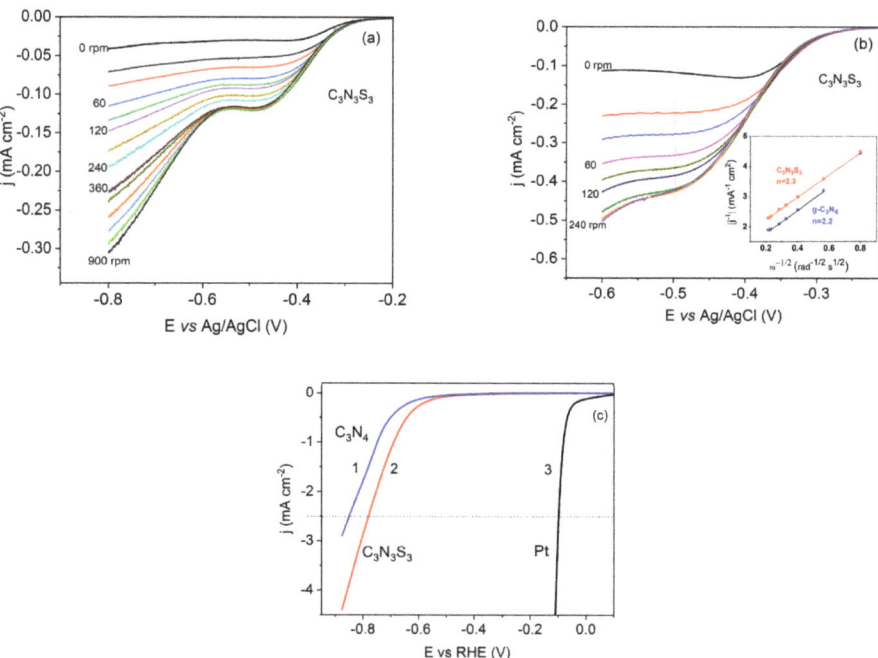

Figure 8. LSVs obtained on GC/C$_3$N$_3$S$_3$ RDE in 0.1 M NaOH at the sweep rate of 2 mVs^{-1} and different rotation rates (**a**), and in the same solution after saturation with O$_2$ (**b**); inset: Levich–Koutecky plots obtained at C$_3$N$_3$S$_3$ and g-C$_3$N$_4$-modified electrodes at the potential −0.5 V vs. Ag/AgCl, comparison of the voltammograms for FTO/g-C$_3$N$_4$ (line 1), FTO/C$_3$N$_3$S$_3$ (line 2) and Pt (line 3) electrodes in 0.5 M H$_2$SO$_4$ at the scan rate 50 mV/s (**c**).

The obtained number of electrons, 2.2–2.3 per one O$_2$ molecule, suggests the serial ORR pathway for both electrodes (modified with g-C$_3$N$_4$ and C$_3$N$_3$S$_3$), with the formation of HO_2^- intermediate in the first two-electron step in the alkaline solution.

Both polymers deposited on the FTO were also used in the preliminary studies of the electrochemical hydrogen evolution in the solution of 0.5 M H$_2$SO$_4$, deaerated before experiments to avoid the cathodic wave corresponding to the oxygen reduction. As visible in Figure 8c, the hydrogen evolution on g-C$_3$N$_4$ and C$_3$N$_3$S$_3$ occurs at a much higher overpotential than that at the Pt electrode. On the other hand, the process at C$_3$N$_3$S$_3$ starts a little earlier than at the g-C$_3$N$_4$ electrode. This may be ascribed to the presence of unsaturated dangling sulfur atoms, which act as the active sites for binding the H$^+$ ions. However, since the overpotential of hydrogen evolution at C$_3$N$_3$S$_3$ is large, further modifications, for example, by incorporation of the transition metal ions into the polymer matrix or/and combination with other semiconductors, such as MoS$_2$, are needed to improve the catalytic activity of C$_3$N$_3$S$_3$ or C$_3$N$_3$S$_3$-based hybrid system. Thus, the procedures of hybridization of the polymeric and inorganic semiconductor and immobilization of the hybrid system on the surface of FTO to obtain a stable photocatalyst should be developed.

3. Materials and Methods

3.1. Chemicals

All reagents were of analytical grade and used without further purification. Trithiocyanuric acid ($C_3N_3S_3H_3$), Nitro Blue Tetrazolium chloride (NBT), and dimethyl sulfoxide (DMSO) were purchased from Sigma-Aldrich (Darmstadt, Germany). Sodium hydroxide (NaOH, 99.8%), Na_2SO_4, H_2SO_4, KI, I_2, urea, and absolute ethanol were purchased from POCh S.A (Gliwice, Poland). The aqueous solutions were prepared using deionized water (DI, 18.2 MΩ cm) (Rephile, Shanghai, China). A conducting FTO (F-doped tin oxide) glass of a resistance 20 Ω square^{-1} was obtained from Dyenamo AB (Stockholm, Sweden).

3.2. Chemical Synthesis of $C_3N_3S_3$

Chemical synthesis of $C_3N_3S_3$ was carried out according to the procedure described in the literature [16]. Namely, 0.3 g of $C_3N_3S_3H_3$ (1.7 mmol) was dissolved in the solution of 0.6 mol/L NaOH under constant stirring for 24 h at room temperature. Then, the temperature of the monomer solution was diminished to 0 °C, and 10 mL of the saturated solution of KI containing 0.65 g (2.56 mmol) of I_2 was added to maintain the $C_3N_3S_3H_3$:I_2 (oxidant) ratio equal to 1:1.5. Then, the temperature gradually increased to the RT, and the reaction mixture was stirred for 24 h. The obtained yellowish precipitate was washed with deionized water and ethanol. The obtained product was dried at RT and stored in a desiccator. Before application, 6 mg of $C_3N_3S_3$ was added to 5 mL of DMSO and sonicated for 12 h. The obtained yellowish suspension was stored at the temperature 8 °C. DMSO was chosen as the solvent because of the long-term stability of the prepared suspension.

3.3. Chemical Synthesis of g-C_3N_4

The urea, used as a precursor of g-C_3N_4, was ground in the mortar and placed in a ceramic vessel covered with a lid. The powder was heated in the muffle furnace to a temperature of 500 °C with a ramp rate of 13 °C/min and kept for 1 h. Next, an excess of unreacted urea was washed out with distilled water. Then, 60 mg of obtained pale-yellow g-C_3N_4 (60 mg) was suspended in DMSO (50 mL) and sonicated for 10 h in an ultrasonic bath (150 W). The suspension was centrifuged for 10 min at 3000 rpm to remove the nonexfoliated polymer. The obtained white suspension of exfoliated g-C_3N_4 in DMSO was stored at the temperature of 8 °C.

3.4. Preparation of the Polymer-Modified Electrodes for Electrochemical Measurements

The FTO plates (of the size 2.5 × 1 cm) used as the substrates were washed with acetone by sonication for 15 min. Next, each plate was immersed in the solution of 3M NaOH for 30 s, rinsed with DI water, then immersed for 15 s in the solution of concentrated H_2SO_4, and again dipped in DI water. Finally, the FTO plates were dried and used for the deposition of the polymeric semiconductors.

The polymers' suspensions (600 µL in three portions, 200 µL each) were applied on the FTO substrates by drop casting. After each application, the samples were dried in air. The surface area of FTO covered with the polymers was about 1.5 cm^2.

A similar procedure was used to modify the glassy carbon (GC) disc electrode with the polymers. The electrode of a surface area of 0.07 cm^2 was polished by alumina slurry on the felt polishing pad, and then the polymer suspension was applied in three portions of 3 µL each. Roughly, the amount of the polymers applied on GC was about 150 µg cm^{-2}.

3.5. Characterization Methods

The details on the X-ray diffraction and UV-Vis measurements are provided in our recent paper [30] and the Supplementary Materials.

The FTIR spectra were recorded using a Nicolet iS 50 FTIR spectrometer (Thermo Fisher Scientific, Waltham, MA, USA) in the reflection mode in the wavenumber range 4000–350 cm^{-1}.

The chemical composition and chemical state of the prepared samples were characterized by X-ray photoelectron spectroscopy (XPS). The details on the equipment and data treatment are presented in ref. [51] and in the Supplementary Materials. The measured binding energies for individual elements were corrected in relation to the C1s carbon peak at 284.8 eV.

The binding energy for the valence band (VB) XP spectrum was calibrated using Au of the work function 4.5 eV. This value, typical for a thin polycrystalline gold film [52], practically meets the absolute potential of a standard hydrogen electrode (0 V vs. SHE, i.e., -4.44 ± 0.02 eV vs. vacuum [53]). Therefore, the VB maximum vs. SHE was determined by a linear extrapolation of low binding energy valence band emission edge [54].

3.6. Electrochemical Measurements

All electrochemical measurements were performed in a standard three-electrode cell with FTO/g-C_3N_4 or FTO/$C_3N_3S_3$ working electrode, Ag/AgCl (3 M KCl) reference electrode, and Pt plate counter electrode (see details in the Supplementary Materials). The measured potentials were recalculated to the SHE scale using the equation:

$$E(V, vs.\ SHE) = E(V, vs.\ Ag/AgCl) + 0.21 V \quad (4)$$

The EIS measurements were done at the ac voltage of the amplitude of 10 mV, in the frequency range 10^5–0.1 Hz. In the photoelectrochemical measurements, the working electrode was illuminated with a diode of the wavelength 365 nm.

3.7. Detection of Superoxide Radicals $O_2^{\bullet-}$

The samples of FTO/g-C_3N_4 or FTO/$C_3N_3S_3$ immersed in an aqueous solution of Nitro Blue Tetrazolium chloride (NBT) of concentration $8 \cdot 10^{-3}$ g L^{-1} was illuminated with a diode (365 nm), under constant stirring. The light intensity in the place of the photocatalyst was 100 mW cm^{-2}. The electrons photogenerated in the semiconductor reduce oxygen to superoxide radical ($O_2^{\bullet-}$), which is then involved in the reduction of NBT to diformazan, according to Scheme 1 [55]. In effect, the intensity of the main absorption peak of NBT in the UV-Vis spectrum, observed at the wavelength of 260 nm, decreases.

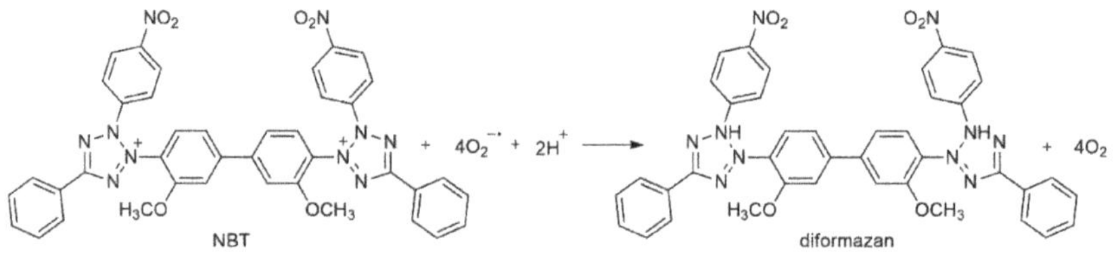

Scheme 1. The reaction of Nitro Blue Tetrazolium chloride (NBT) with superoxide radicals [55].

The reaction rates at different semiconductors were compared by plotting the changes of relative absorbance (A/A$_0$) at 260 nm as a function of time, where A$_0$ is the initial absorbance of the NBT solution.

4. Conclusions

In this work, we have shown that $C_3N_3S_3$, chemically synthesized from trithiocyanuric acid, reveals very similar electrochemical properties to g-C_3N_4. Both polymers exhibit ambipolar behavior, with the Fermi level located in the middle of the band gap. Although the bottom edge of the conduction band in $C_3N_3S_3$ is located a little below the CB of g-C_3N_4, the formation rate of superoxide anion radicals by the photogenerated electrons is the same in both polymers. Both polymers are good electrocatalysts in oxygen reduction, and the reaction starts at lower negative potentials at FTO/$C_3N_3S_3$ than that at the FTO/g-C_3N_4 electrode. The experiments performed using RDE in the solution of 0.1 M NaOH have shown that oxygen reduction occurs via the serial pathway with two successive two-electron steps. The presence of sulfur in the structure of $C_3N_3S_3$ is probably responsible for the lowering of hydrogen evolution overpotential at this polymer with respect to that at the g-C_3N_4-modified FTO electrode due to the presence of additional active sites for binding the H^+ ions. However, further improvement of $C_3N_3S_3$ for practical application in HER is necessary.

Supplementary Materials: The following supporting information can be downloaded at: https://www.mdpi.com/article/10.3390/molecules28062469/s1, Figure S1: Survey spectra of $C_3N_3S_3$ and g-C_3N_4; Figure S2: HR-XPS of I 3d and O 1s of $C_3N_3S_3$; Figure S3: Elemental maps of $C_3N_3S_3$; Figure S4: LSVs on RDE ($C_3N_3S_3$) for oxygen reduction; Table S1: XPS data for g-C_3N_4; Table S2: XPS data for $C_3N_3S_3$.

Author Contributions: E.W.: investigation, methodology, visualization, validation; M.P.: XPS investigation and analysis; T.Ł.: synthesis; M.S.: conceptualization, methodology, writing—original draft, supervision. All authors have read and agreed to the published version of the manuscript.

Funding: This research was funded by the National Science Centre of Poland under grant 2019/33/B/ST5/01720.

Institutional Review Board Statement: Not applicable.

Informed Consent Statement: Not applicable.

Data Availability Statement: Data available on request due to privacy restrictions.

Acknowledgments: The authors thank Kamil Sobczak from the Biological and Chemical Research Center, University of Warsaw, for TEM imaging and EDS analysis of the samples.

Conflicts of Interest: The authors declare no conflict of interest.

Sample Availability: Samples of the compounds are available from the authors.

References

1. Liao, G.; Gong, Y.; Zhang, L.; Gao, H.; Yang, G.J.; Fang, B. Semiconductor Polymeric Graphitic Carbon Nitride Photocatalysts: The "Holy Grail" for the Photocatalytic Hydrogen Evolution Reaction under Visible Light. *Energy Environ. Sci.* **2019**, *12*, 2080–2147. [CrossRef]
2. Mamba, G.; Mishra, A.K. Graphitic Carbon Nitride (g-C_3N_4) Nanocomposites: A New and Exciting Generation of Visible Light Driven Photocatalysts for Environmental Pollution Remediation. *Appl. Catal. B Environ.* **2016**, *198*, 347–377. [CrossRef]
3. Safaei, J.; Mohamed, N.A.; Mohamad Noh, M.F.; Soh, M.F.; Ludin, N.A.; Ibrahim, M.A.; Roslam Wan Isahak, W.N.; Mat Teridi, M.A. Graphitic Carbon Nitride (g-C_3N_4) Electrodes for Energy Conversion and Storage: A Review on Photoelectrochemical Water Splitting, Solar Cells and Supercapacitors. *J. Mater. Chem. A* **2018**, *6*, 22346–22380. [CrossRef]
4. Cao, S.; Low, J.; Yu, J.; Jaroniec, M. Polymeric Photocatalysts Based on Graphitic Carbon Nitride. *Adv. Mater.* **2015**, *27*, 2150–2176. [CrossRef] [PubMed]
5. Zhang, J.; Chen, X.; Takanabe, K.; Maeda, K.; Domen, K.; Epping, J.D.; Fu, X.; Antonieta, M.; Wang, X. Synthesis of a Carbon Nitride Structure for Visible-Light Catalysis by Copolymerization. *Angew. Chem. Int. Ed.* **2010**, *49*, 441–444. [CrossRef]
6. Han, Q.; Wang, B.; Gao, J.; Cheng, Z.; Zhao, Y.; Zhang, Z.; Qu, L. Atomically Thin Mesoporous Nanomesh of Graphitic C_3N_4 for High-Efficiency Photocatalytic Hydrogen Evolution. *ACS Nano* **2016**, *10*, 2745–2751. [CrossRef]
7. Xu, J.; Zhang, L.; Shi, R.; Zhu, Y. Chemical Exfoliation of Graphitic Carbon Nitride for Efficient Heterogeneous Photocatalysis. *J. Mater. Chem. A* **2013**, *1*, 14766–14772. [CrossRef]
8. Wang, X.; Maeda, K.; Thomas, A.; Takanabe, K.; Xin, G.; Carlsson, J.M.; Domen, K.; Antonietti, M. A Metal-Free Polymeric Photocatalyst for Hydrogen Production from Water under Visible Light. *Nat. Mater.* **2009**, *8*, 76–80. [CrossRef]
9. Wang, K.; Li, Q.; Liu, B.; Cheng, B.; Ho, W.; Yu, J. Sulfur-Doped g-C_3N_4 with Enhanced Photocatalytic CO2-Reduction Performance. *Appl. Catal. B Environ.* **2015**, *176–177*, 44–52. [CrossRef]
10. Wang, J.; Wang, G.; Cheng, B.; Yu, J.; Fan, J. Sulfur-Doped g-C_3N_4/TiO_2 S-Scheme Heterojunction Photocatalyst for Congo Red Photodegradation. *Chin. J. Catal.* **2020**, *42*, 56–68. [CrossRef]
11. Ko, D.; Lee, J.S.; Patel, H.A.; Jakobsen, M.H.; Hwang, Y.; Yavuz, C.T.; Hansen, H.C.B.; Andersen, H.R. Selective Removal of Heavy Metal Ions by Disulfide Linked Polymer Networks. *J. Hazard. Mater.* **2017**, *332*, 140–148. [CrossRef]
12. Socrates, G. *Infrared and Raman Characteristic Group Frequencies*; Wiley: Hoboken, NJ, USA, 2001.
13. Yin, J.; Xu, H. Degradation of Organic Dyes over Polymeric Photocatalyst $C_3N_3S_3$. In *2014 International Conference on Mechatronics, Electronic, Industrial and Control Engineering (MEIC-14)*; Atlantis Press: Paris, France, 2014; pp. 349–352. [CrossRef]

14. Drożdżewski, P.; Malik, M.; Kopel, P.; Bieńko, D.C. Normal Vibrations and Vibrational Spectra of Trithiocyanuric Acid in Its Natural, Deuterated, Anionic and Metal Coordinated Forms. *Polyhedron* **2022**, *220*, 115819. [CrossRef]
15. Xu, J.; Luo, L.; Xiao, G.; Zhang, Z.; Lin, H.; Wang, X.; Long, J. Layered $C_3N_3S_3$ Polymer/Graphene Hybrids as Metal-Free Catalysts for Selective Photocatalytic Oxidation of Benzylic Alcohols under Visible Light. *ACS Catal.* **2014**, *4*, 3302–3306. [CrossRef]
16. Zhang, Z.; Long, J.; Yang, L.; Chen, W.; Dai, W.; Fu, X.; Wang, X. Organic Semiconductor for Artificial Photosynthesis: Water Splitting into Hydrogen by a Bioinspired $C_3N_3S_3$ Polymer under Visible Light Irradiation. *Chem. Sci.* **2011**, *2*, 1826–1830. [CrossRef]
17. Li, X.; Hartley, G.; Ward, A.J.; Young, P.A.; Masters, A.F.; Maschmeyer, T. Hydrogenated Defects in Graphitic Carbon Nitride Nanosheets for Improved Photocatalytic Hydrogen Evolution. *J. Phys. Chem. C* **2015**, *119*, 14938–14946. [CrossRef]
18. Hong, Z.; Shen, B.; Chen, Y.; Lin, B.; Gao, B. Enhancement of Photocatalytic H2 Evolution over Nitrogen-Deficient Graphitic Carbon Nitride. *J. Mater. Chem. A* **2013**, *1*, 11754. [CrossRef]
19. Zhang, J.R.; Ma, Y.; Wang, S.Y.; Ding, J.; Gao, B.; Kan, E.; Hua, W. Accurate K-Edge X-Ray Photoelectron and Absorption Spectra of g-C_3N_4 Nanosheets by First-Principles Simulations and Reinterpretations. *Phys. Chem. Chem. Phys.* **2019**, *21*, 22819–22830. [CrossRef]
20. Ge, L. Synthesis and Photocatalytic Performance of Novel Metal-Free g-C 3N4 Photocatalysts. *Mater. Lett.* **2011**, *65*, 2652–2654. [CrossRef]
21. Zhang, J.; Sun, J.; Maeda, K.; Domen, K.; Liu, P.; Antonietti, M.; Fu, X.; Wang, X. Sulfur-Mediated Synthesis of Carbon Nitride: Band-Gap Engineering and Improved Functions for Photocatalysis. *Energy Environ. Sci.* **2011**, *4*, 675–678. [CrossRef]
22. Viezbicke, B.D.; Patel, S.; Davis, B.E.; Birnie, D.P. Evaluation of the Tauc Method for Optical Absorption Edge Determination: ZnO Thin Films as a Model System. *Phys. Status Solidi Basic Res.* **2015**, *252*, 1700–1710. [CrossRef]
23. Wang, Y.; Di, Y.; Antonietti, M.; Li, H.; Chen, X.; Wang, X. Excellent Visible-Light Photocatalysis of Fluorinated Polymeric Carbon Nitride Solids. *Chem. Mater.* **2010**, *22*, 5119–5121. [CrossRef]
24. Zheng, Y.; Zhang, Z.; Li, C. A Comparison of Graphitic Carbon Nitrides Synthesized from Different Precursors through Pyrolysis. *J. Photochem. Photobiol. A Chem.* **2017**, *332*, 32–44. [CrossRef]
25. Liu, G.; Wang, T.; Zhang, H.; Meng, X.; Hao, D.; Chang, K.; Li, P.; Kako, T.; Ye, J. Nature-Inspired Environmental "Phosphorylation" Boosts Photocatalytic H 2 Production over Carbon Nitride Nanosheets under Visible-Light Irradiation. *Angew. Chem.* **2015**, *127*, 13765–13769. [CrossRef]
26. Martin, D.J.; Qiu, K.; Shevlin, S.A.; Handoko, A.D.; Chen, X.; Guo, Z.; Tang, J. Highly Efficient Photocatalytic H2 Evolution from Water Using Visible Light and Structure-Controlled Graphitic Carbon Nitride. *Angew. Chem. Int. Ed.* **2014**, *53*, 9240–9245. [CrossRef]
27. Tyborski, T.; Merschjann, C.; Orthmann, S.; Yang, F.; Lux-Steiner, M.C.; Schedel-Niedrig, T. Tunable Optical Transition in Polymeric Carbon Nitrides Synthesized via Bulk Thermal Condensation. *J. Phys. Condens. Matter* **2012**, *24*, 162201. [CrossRef]
28. Butler, M.A.; Ginley, D.S. Prediction of Flatband Potentials at Semiconductor-Electrolyte Interfaces from Atomic Electronegativities. *J. Electrochem. Soc.* **1978**, *125*, 228–232. [CrossRef]
29. Praus, P. On Electronegativity of Graphitic Carbon Nitride. *Carbon* **2021**, *172*, 729–732. [CrossRef]
30. Łęcki, T.; Hamad, H.; Zarębska, K.; Wierzyńska, E.; Skompska, M. Mechanistic Insight into Photochemical and Photoelectrochemical Degradation of Organic Pollutants with the Use of $BiVO_4$ and $BiVO_4$/Co-Pi. *Electrochim. Acta* **2022**, *434*, 141292. [CrossRef]
31. Cardon, F.; Gomes, W.P. On the Determination of the Flat-Band Potential of a Semiconductor in Contact with a Metal or an Electrolyte from the Mott-Schottky Plot. *J. Phys. D Appl. Phys.* **1978**, *11*, L63. [CrossRef]
32. Beranek, R. (Photo)Electrochemical Methods for the Determination of the Band Edge Positions of TiO_2-Based Nanomaterials. *Adv. Phys. Chem.* **2011**, *2011*, 80–83. [CrossRef]
33. Hankin, A.; Bedoya-Lora, F.E.; Alexander, J.C.; Regoutz, A.; Kelsall, G.H. Flat Band Potential Determination: Avoiding the Pitfalls. *J. Mater. Chem. A* **2019**, *7*, 26162–26176. [CrossRef]
34. Ren, Y.; Yang, X.; Zhou, L.; Mao, J.Y.; Han, S.T.; Zhou, Y. Recent Advances in Ambipolar Transistors for Functional Applications. *Adv. Funct. Mater.* **2019**, *29*, 1902105. [CrossRef]
35. Jing, J.; Chen, Z.; Feng, C. Dramatically Enhanced Photoelectrochemical Properties and Transformed p/n Type of g-C_3N_4 Caused by K and I Co-Doping. *Electrochim. Acta* **2019**, *297*, 488–496. [CrossRef]
36. Jing, J.; Chen, Z.; Feng, C.; Sun, M.; Hou, J. Transforming G-C_3N_4 from Amphoteric to n-Type Semiconductor: The Important Role of p/n Type on Photoelectrochemical Cathodic Protection. *J. Alloys Compd.* **2021**, *851*, 156820. [CrossRef]
37. Kanan, M.W.; Nocera, D.G. In Situ Formation of an Oxygen-Evolving Catalyst in Neutral Water Containing Phosphate and Co2+. *Science* **2008**, *321*, 1072–1075. [CrossRef]
38. Zhang, Y.; Antonietti, M. Photocurrent Generation by Polymeric Carbon Nitride Solids: An Initial Step towards a Novel Photovoltaic System. *Chem.—Asian J.* **2010**, *5*, 1307–1311. [CrossRef]
39. Kang, Y.; Yang, Y.; Yin, L.C.; Kang, X.; Liu, G.; Cheng, H.M. An Amorphous Carbon Nitride Photocatalyst with Greatly Extended Visible-Light-Responsive Range for Photocatalytic Hydrogen Generation. *Adv. Mater.* **2015**, *27*, 4572–4577. [CrossRef] [PubMed]
40. Chun, W.J.; Ishikawa, A.; Fujisawa, H.; Takata, T.; Kondo, J.N.; Hara, M.; Kawai, M.; Matsumoto, Y.; Domen, K. Conduction and Valence Band Positions of Ta2O5, TaON, and Ta3N5 by UPS and Electrochemical Methods. *J. Phys. Chem. B* **2003**, *107*, 1798–1803. [CrossRef]

41. Ismael, M. A Review on Graphitic Carbon Nitride (g-C_3N_4) Based Nanocomposites: Synthesis, Categories, and Their Application in Photocatalysis. *J. Alloys Compd.* **2020**, *846*, 156446. [CrossRef]
42. Sudhaik, A.; Raizada, P.; Shandilya, P.; Jeong, D.Y.; Lim, J.H.; Singh, P. Review on Fabrication of Graphitic Carbon Nitride Based Efficient Nanocomposites for Photodegradation of Aqueous Phase Organic Pollutants. *J. Ind. Eng. Chem.* **2018**, *67*, 28–51. [CrossRef]
43. Chen, M.; Jia, Y.; Li, H.; Wu, Z.; Huang, T.; Zhang, H. Enhanced Pyrocatalysis of the Pyroelectric $BiFeO_3$/g-C_3N_4 Heterostructure for Dye Decomposition Driven by Cold-Hot Temperature Alternation. *J. Adv. Ceram.* **2021**, *10*, 338–346. [CrossRef]
44. Armstrong, D.A.; Huie, R.E.; Koppenol, W.H.; Lymar, S.V.; Merenyi, G.; Neta, P.; Ruscic, B.; Stanbury, D.M.; Steenken, S.; Wardman, P. Standard Electrode Potentials Involving Radicals in Aqueous Solution: Inorganic Radicals (IUPAC Technical Report). *Pure Appl. Chem.* **2015**, *87*, 1139–1150. [CrossRef]
45. Song, C.; Zhang, J. Electrocatalytic Oxygen Reduction Reaction. In *PEM Fuel Cell Electrocatalysts and Catalyst Layers: Fundamentals and Applications*; Springer: London, UK, 2008; pp. 89–134. [CrossRef]
46. Blizanac, B.B.; Ross, P.N.; Markovic, N.M. Oxygen Electroreduction on Ag(1 1 1): The PH Effect. *Electrochim. Acta* **2007**, *52*, 2264–2271. [CrossRef]
47. Ge, X.; Sumboja, A.; Wuu, D.; An, T.; Li, B.; Goh, F.W.T.; Hor, T.S.A.; Zong, Y.; Liu, Z. Oxygen Reduction in Alkaline Media: From Mechanisms to Recent Advances of Catalysts. *ACS Catal.* **2015**, *5*, 4643–4667. [CrossRef]
48. Yang, H.-H.; McCreery, R.L. Elucidation of the Mechanism of Dioxygen Reduction on Metal-Free Carbon Electrodes. *J. Electrochem. Soc.* **2000**, *147*, 3420. [CrossRef]
49. Feng, Z.; Georgescu, N.S.; Scherson, D.A. Rotating Ring-Disk Electrode Method for the Detection of Solution Phase Superoxide as a Reaction Intermediate of Oxygen Reduction in Neutral Aqueous Solutions. *Anal. Chem.* **2016**, *88*, 1088–1091. [CrossRef] [PubMed]
50. Bard, A.J.; Faulkner, J.R. *Electrochemical Methods: Fundamental and Applications*, 2nd ed.; Wiley: New York, NY, USA, 2001.
51. Pisarek, M.; Krawczyk, M.; Kosiński, A.; Hołdyński, M.; Andrzejczuk, M.; Krajczewski, J.; Bieńkowski, K.; Solarska, R.; Gurgul, M.; Zaraska, L.; et al. Materials Characterization of TiO_2 nanotubes Decorated by Au Nanoparticles for Photoelectrochemical Applications. *RSC Adv.* **2021**, *11*, 38727–38738. [CrossRef]
52. Kahn, A. Fermi Level, Work Function and Vacuum Level. *Mater. Horiz.* **2016**, *3*, 7–10. [CrossRef]
53. Trasatti, S. The Absolute Electrode Potential: An Explanatory Note (Recommendations 1986). *Pure Appl. Chem.* **1986**, *58*, 955–966. [CrossRef]
54. Kashiwaya, S.; Morasch, J.; Streibel, V.; Toupance, T.; Jaegermann, W.; Klein, A. The Work Function of TiO_2. *Surfaces* **2018**, *1*, 73–89. [CrossRef]
55. Goto, H.; Hanada, Y.; Ohno, T.; Matsumura, M. Quantitative Analysis of Superoxide Ion and Hydrogen Peroxide Produced from Molecular Oxygen on Photoirradiated TiO_2 Particles. *J. Catal.* **2004**, *225*, 223–229. [CrossRef]

Disclaimer/Publisher's Note: The statements, opinions and data contained in all publications are solely those of the individual author(s) and contributor(s) and not of MDPI and/or the editor(s). MDPI and/or the editor(s) disclaim responsibility for any injury to people or property resulting from any ideas, methods, instructions or products referred to in the content.

Article

A Combined Experimental and Computational Study on the Adsorption Sites of Zinc-Based MOFs for Efficient Ammonia Capture

Dongli Zhang [1], Yujun Shen [1,*], Jingtao Ding [1], Haibin Zhou [1], Yuehong Zhang [2], Qikun Feng [3], Xi Zhang [1], Kun Chen [1], Pengxiang Xu [1] and Pengyue Zhang [1]

[1] Academy of Agricultural Planning and Engineering, Key Laboratory of Technologies and Models for Cyclic Utilization from Agricultural Resources, Ministry of Agriculture, Beijing 100125, China
[2] School of Advanced Manufacturing, Guangdong University of Technology, Jieyang 515200, China
[3] State Key Laboratory of Power Systems, Department of Electrical Engineering, Tsinghua University, Beijing 100084, China
* Correspondence: shenyj09b@163.com

Abstract: Ammonia (NH_3) is a common pollutant mostly derived from pig manure composting under humid conditions, and it is absolutely necessary to develop materials for ammonia removal with high stability and efficiency. To this end, metal–organic frameworks (MOFs) have received special attention because of their high selectivity of harmful gases in the air, resulting from their large surface area and high density of active sites, which can be tailored by appropriate modifications. Herein, two synthetic metal–organic frameworks (MOFs), 2-methylimidazole zinc salt (ZIF-8) and zinc-trimesic acid (ZnBTC), were selected for ammonia removal under humid conditions during composting. The two MOFs, with different organic linkers, exhibit fairly distinctive ammonia absorption behaviors under the same conditions. For the ZnBTC framework, the ammonia intake is 11.37 mmol/g at 298 K, nine times higher than that of the ZIF-8 framework (1.26 mmol/g). In combination with theoretical calculations, powder XRD patterns, FTIR, and BET surface area tests were conducted to reveal the absorption mechanisms of ammonia for the two materials. The adsorption of ammonia on the ZnBTC framework can be attributed to both physical and chemical adsorption. A strong coordination interaction exists between the nitrogen atom from the ammonia molecule and the zinc atom in the ZnBTC framework. In contrast, the absorption of ammonia in the ZIF-8 framework is mainly physical. The weak interaction between the ammonia molecule and the ZIF-8 framework mainly results from the inherent severely steric hindrance, which is related to the coordination mode of the imidazole ligands and the zinc atom of this framework. Therefore, this study provides a method for designing promising MOFs with appropriate organic linkers for the selective capture of ammonia during manure composting.

Keywords: metal–organic frameworks (MOFs); organic linkers; ammonia; adsorption capacity; manure composting

1. Introduction

Due to the worldwide intensive industrialized development of livestock and poultry breeding, a large amount of breeding wastes has been generated, posing a serious threat to the ecological environment and human and animal health. Thus, the reduction, detoxification, and resource utilization of livestock and poultry breeding waste have attracted much attention; composting technology has proven to be the most effective treatment method [1–4]. However, the applications of composting technology are limited due to the odor generated during the composting process. Constituted of complex gases, ammonia in particular, the odor not only harms the environment and health of humans and animals but also reduces the fertilizer efficiency and agricultural value of the compost [5–7]. Therefore, the effective control of and reduction in the emissions of odorous gases during the composting process play a vital role in achieving high-quality, efficient, and pollution-free composting given the urgent need for ecological and environmental protection.

In recent years, many practical methods, such as the adjustment of the compost preparation parameters [8,9], change in the aeration rate [10,11], and addition of external additives [12–14], have been developed to promote the composting process and reduce the production and discharge of odor during aerobic composting. Thus far, the external additives method has attracted widespread attention from scholars because of its good performance not only in promoting composting efficiency and reducing odor emissions but also in increasing organic matter conversion and reducing nitrogen losses [15,16]. Despite the fact that biochar is most commonly employed as an adsorbent among all external additives, this material has a low adsorption capacity and cannot be easily recovered, which prevents it from being the premium solution to the breeding waste processing problem.

Metal–organic frameworks (MOFs), on the other hand, given their composition and structure diversity, ultra-high specific surface area, high and adjustable porosity, and open metal sites, have received special attention regarding the selective capture of harmful gases from the air, especially ammonia, as reported by several studies [17–22]. As reported by Li et al. [23], the MOF material $Cu(INA)_2$ has repeatable and remarkable ammonia adsorption and desorption capacities of up to 13 mmol/g due to its reversible structural transformation with the adsorption and desorption of ammonia. In addition, the new IL@MOF composite $[BOHmim][Zn_2Cl_5]@MIL-101(Cr)$ developed by Zhong et al. [24] exhibits a record NH_3 uptake of 24.12 mmol/g, which can be attributed to the multiple adsorption sites and large free volume for NH_3 provided by the IL confined in the framework of the MOF. Moreover, an MOF-74 analog, M-MOF-74 (M = Zn, Co), was prepared and studied by Glover et al. [25], showing a strong ammonia removal ability of 7.60 mol kg^{-1} and 6.70 mol kg^{-1} for Mg-MOF-74 and Co-MOF-74, respectively, in dry conditions of 0% relative humidity (RH).

Regardless of the potential of MOF materials to reduce ammonia emissions, as exemplified above, their application in composting is rather limited due to the following: As part of the agricultural field, the additives and materials employed in composting should not be detrimental to the environment and soil and, therefore, should not introduce any harmful elements to the composting system, which screens out many MOF materials with heavy metal elements [26,27]. Moreover, the complex composting process is usually carried out in a humid and complex atmosphere, which requires the stability of MOF materials in such a harsh environment. Taking the multiple requirements of composting additives into consideration, a reasonable preparation strategy for MOF materials is urgently needed and the factors affecting NH_3 removal must be analyzed in order to make full use of the structural advantages of MOFs and further improve their adsorption performance.

In this study, two MOFs, ZIF-8 and ZnBTC, with different organic linkers were designed and synthesized by a solvothermal method. Ammonia adsorption over ZIF-8 and ZnBTC was systematically measured, revealing the distinct ammonia adsorption capacities of ZIF-8 (1.26 mmol/g) and ZnBTC (11.37 mmol/g). Furthermore, surface area, porosity measurements, and DFT calculations were carried out to illuminate the different ammonia adsorption mechanisms between the two MOFs, confirming the significance of the judicious selection for organic linkers associated with different coordination modes and different exposed active sites.

2. Experimental Methods

2.1. Materials and Reagents

$Zn(NO_3)_2 \cdot 6H_2O$ (AR, 99%) and $ZnAc_2 \cdot 2H_2O$ were obtained from Macklin Co., Ltd. (Shanghai, China). Dimethylglyoxaline dimethylimidazole and benzene-1,3,5-tricarboxylic acid were purchased from Sinopharm Chemical Reagent Co.,Ltd. (Shanghai, China). Sodium hydroxide (NaOH, 95%) was obtained from Beijing Huarong Chemical factory

(Beijing, China). The DMF and H$_2$O were commercially available and used as supplied without further purification.

2.2. Synthesis of ZIF-8 and ZnBTC

ZnBTC was prepared according to the methods reported in the literature [28–30]. The detailed preparation process is described as follows: First, 1.8 g of ZnAc$_2$·6H$_2$O was dissolved in 60 mL of deionized water with constant stirring for 20 min and named solution A. Then, 0.84 g of C$_9$H$_6$O$_6$ and 0.48 g of NaOH were added to the mixture of the water and ethanol (3:1), stirred until evenly dissolved and named solution B. Solution B was slowly added to solution A under stirring conditions. Then, the mixture was transferred onto glass plates and dried in an oven at 120 °C for 12 h. After that, the autoclave was cooled down to room temperature. The resulting white crystal of Zn-BTC precipitate was filtered and washed with ethanol and deionized (DI) water several times until the filtrate presented as neutral. Finally, the Zn-BTC was dried in a vacuum oven at 60 °C for 3 h.

The ZIF-8 was prepared according to the methods reported in the literature [31].

2.3. Adsorption Experiments

The pure NH$_3$ adsorption and desorption curves of the samples were determined by a static capacity sorption analyzer (Bei Shi De, BSD-PSPM, Beijing, China). Before testing, the samples were activated at 50 °C for at least 2 h until the mass no longer changed. For the adsorption experiments in the NH$_3$/H$_2$O system, about 20 mL of NH$_3$/H$_2$O (4:1, v:v) and 0.10 g of adsorbent were placed in a covered container. The adsorption during the composting process was implemented in a similar manner. The adsorbent was added at mesophilic and earlier thermophilic phases during the aerobic composting of manure. After the adsorption reached saturation, the mixture was added to a saturated solution of potassium chloride using a temperature-controlled shaker at room temperature. The mixed- solution was then filtered and analyzed using an Automatic Discrete Analyzer (SmartChem 140, Catania, Italy) to estimate the adsorption intake of NH$_3$.

2.4. Characterization

The morphologies of the prepared MOFs were studied by field-emission scanning (SEM, S-4700, ISS Group Services Ltd., Manchester, England). Fourier transform infrared (FT-IR) spectroscopy was carried out by a Nicolet 6700FT-IR spectrophotometer (Thermo Fisher Scientific Inc., Waltham, MA, USA). The crystal structures of the prepared MOFs were fared, performed by an X-ray diffractometer (Bruker AXS D8-Advance, Bruker, Billerica, MA, USA) in the 2θ range from 5° to 40° at a scan rate of 10° min^{-1}. N$_2$ adsorption/desorption isotherms were recorded by a JWGB Sci & Tech Ltd (Beijing, China). The Brunauer–Emmett–Teller (BET) surface area of the prepared MOFs was tested through a volumetric method. The pure NH$_3$ adsorption and desorption curves of the samples were determined by a static capacity sorption analyzer (BSD-PSPM, Microtrac, Duesseldorf, Germany), and the samples were activated at 50 °C for at least 2 h before testing or until the mass no longer changed. Theoretical calculations were carried out using the Quickstep algorithm of the CP2K package.26 to elucidate the adsorption behavior and inherent mechanism. The Perdew–Burke–Ernzerhof (PBE) function [32] with Grimme's dispersion correction and with Becke–Johnson damping (D3BJ) [33,34] was employed for all the calculations. The Gaussian and plane wave methods were used, and the wave function was expanded in the Gaussian double-ζ valence polarized (DZVP) basis set. A convergence criterion of 3.0×10^{-6} a.u. was used for the optimization of the wave function. The adsorption energy of ammonia ΔE_n was calculated as follows:

$$\Delta E_n = E_{ZIF8(n)} - E_{ZIF8(n-1)} - E_{NH_3} \quad \text{(for ZIF8 system)} \tag{1}$$

$$\Delta E_n = E_{ZnBTC(n)} - E_{ZnBTC(n-1)} - E_{NH_3} + E_{H_2O} \quad \text{(for ZnBTC system)} \tag{2}$$

where the Es with different subscripts are the energies of each species obtained from DFT calculations, and n refers to the nth NH_3 molecule that was adsorbed onto the MOF.

3. Results and Discussion

3.1. Characterization of MOF Materials

As shown in Figure 1, two types of Zn-based MOFs with different organic ligands, ZnBTC and ZIF-8, were synthesized by a solvothermal method. The $ZnAc_2 \cdot 6H_2O$ and H_3BTC were dissolved in the selected solvents, and the resulting solutions were mixed and then placed in the Teflon-lined stainless steel autoclave. The stirring lasted for 9 h at 150 °C. Grayish-white crystals were obtained by washing with deionized water and drying at ambient temperature.

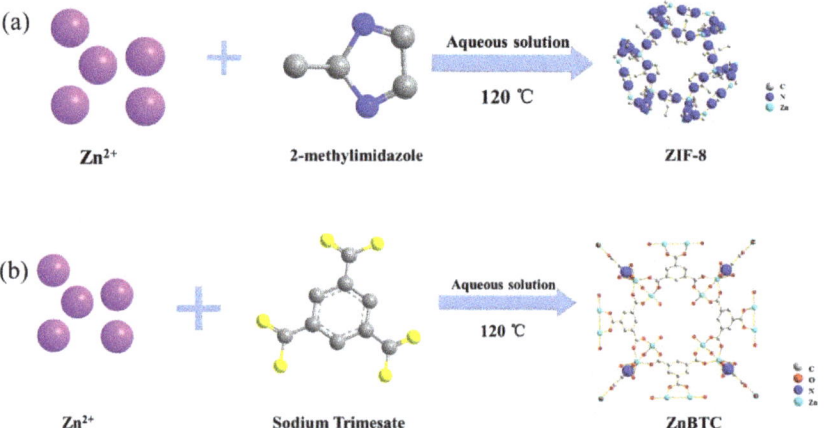

Figure 1. Diagrammatic flow charts of ZIF-8 (**a**) and ZnBTC (**b**).

In the reaction systems, infinite Zn chains are interconnected by the organic ligand linkers into a three-dimensional microporous framework. Due to their porous structures with different functionalized channels, the ZnBTC and ZIF-8 show promise for use as adsorption materials.

The morphologies and microstructures of the as-prepared two MOFs were characterized by scanning electron microscopy (SEM). As shown in Figure 2, both ZIF-8 and ZnBTC possess smooth surfaces. However, the two MOFs exhibit quite different morphologies and structures, which can be attributed to the different coordination modes of the different organic ligands. A relatively homogenous and hexagonal-structured morphology was observed for ZIF-8, while ZnBTC has nonuniform rod-like structures with lengths ranging from 0.5 to 3 µm. As exemplified by the calculations, one zinc ion is coordinated by four imidazole ligands in ZIF-8 and adopts a tetragon coordination mode. ZnBTC employs an octagon coordination mode, which can be easily attacked by polar molecules.

The powder XRD patterns for both ZIF-8 and ZnBTC were obtained before and after the capture of NH_3. As shown in Figure 3, the PXRD patterns of the pristine ZIF-8 and ZnBTC are well-matched with those reported [30,32], revealing the crystalline nature of ZIF-8 and ZnBTC. In detail, the characteristic peaks of ZIF-8 at 2θ = 7.3°, 10.3°, 12.7, 14.8, 16.4, 18.0, 24.6, and 26.7 can be attributed to the crystalline planes (011), (002), (112), (022), (013), (222), (233), and (134), respectively. The as-prepared ZnBTC displayed an intense diffraction peak at 2θ = 10°, consistent with that in the literature [35], confirming the formation of ZnBTC. Nevertheless, the PXRD of ZnBTC significantly changed after the co-adsorption of H_2O/NH_3, and it was speculated that NH_3 and H_2O destroyed the structure of ZnBTC to some extent. From the PXRD of ZIF-8 before and after adsorption of NH_3, it can be seen that the diffraction peak of ZIF-8 does not obviously change, indicating that NH_3 has less of an effect on ZIF-8.

Figure 2. SEM images of the ZIF-8 (**a**) and ZnBTC (**b**); enlarged images of ZIF-8 (**c**) and ZnBTC (**d**).

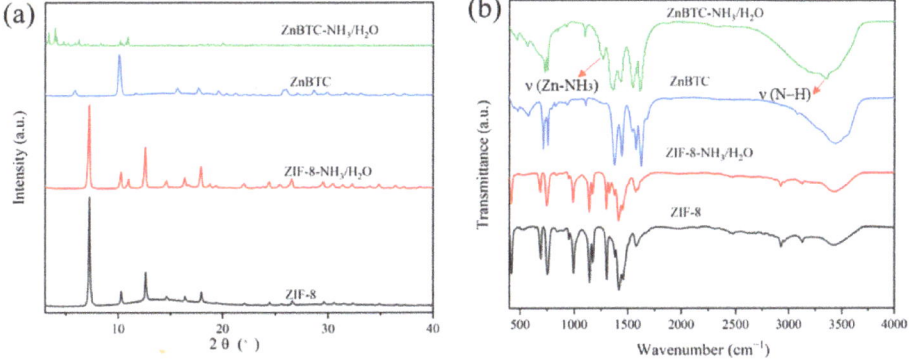

Figure 3. The XRD patterns (**a**) and FTIR (**b**) of ZIF-8 and ZnBTC.

In order to gain insights into the ammonia adsorption mechanisms of the two MOF materials, the FT-IR spectra were measured before and after NH_3 absorption. As shown in Figure 3b, compared with that of pristine ZnBTC material, two additional peaks at 3300 and 1276 cm^{-1}, which correspond to ν (N–H) and δ (Zn-NH_3) [36], respectively, can be observed after the absorption of NH_3. In addition, the large, broad peak in the range of 3200–3600 cm^{-1} was attributed to the absorbed H_2O. The observed results suggested obvious interactions between the zinc ions in ZnBTC and the absorbed ammonia molecules. No apparent change in the absorption signals was observed for ZIF-8 after the ammonia absorption, indicating relatively weak interactions between the zinc ions in ZIF-8 and the ammonia molecules. The stark contrast of the two MOF materials before and after ammonia absorption revealed the effect of different ligands on the interactions between the absorbed ammonia molecules and the MOF materials. For ZIF-8, the absorption peaks of 3134 and 2927 cm^{-1} are C–H stretching vibration peaks of aromatic and aliphatic groups in the imidazole ligand, and the peak at 1573 cm^{-1} is due to the stretching vibration of C=N on the imidazole ring. The signals within the range from 1500 to 1350 cm^{-1} and those from 1500 to 600 cm^{-1} could be attributed to the stretching vibration and bending vibration of the imidazole ring, respectively. The peak at 422 cm^{-1} is the stretching vibration peak of Zn–N [37].

3.2. NH$_3$ Adsorption Studies

The NH$_3$ adsorption and desorption curves of ZnBTC and ZIF-8 were investigated to evaluate the adsorption capacity. As shown in Figure 4, ZnBTC exhibited an adsorption capacity of 113.77 mL/g (5.04 mmol/g) at low pressure. The excellent adsorption capacity of the ZnBTC prepared in this study may be due to the metal vacancy coordination created in the ZnBTC structure, which facilitated the NH$_3$ to enter the pores and be quickly adsorbed in the vacant coordination. With the increase in pressure, NH$_3$ was adsorbed in the void, and the final adsorption capacity was as high as 254.88 mL/g (11.37 mmol/g). The NH$_3$ desorption curve of ZnBTC shows that some of the NH$_3$ was desorbed with the decrease in pressure, and this part of NH$_3$ should be absorbed with physical effects. However, most of the NH$_3$ was still not desorbed, which means that this part of the NH$_3$ gas was chemisorbed and absorbed inside the structure. Compared with ZnBTC, the adsorption capacity of ZIF-8 is low, and all the NH$_3$ would have been removed from the pore structure with the decrease in pressure. This may have occurred mainly because there are no metal vacancies and adsorption functional groups inside the structure of MOFs.

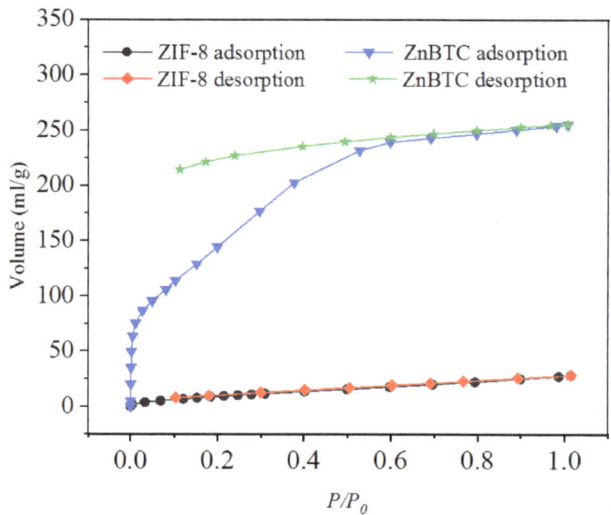

Figure 4. NH$_3$ adsorption and desorption curves obtained for ZnBTC and ZIF-8.

To further demonstrate the application of the MOFs for the absorption of NH$_3$, adsorption experiments were also carried out under ammonia water during the composting process. The adsorption performance during the thermophilic phases during aerobic composting was defined as ZIF-8-1 and ZnBTC-1. ZIF-8-2 and ZnBTC-2 represent the adsorption performance under a 25% NH$_3$ solution steam atmosphere. ZIF-8-3 and ZnBTC-3 were tested with pure ammonia for comparison. It can be seen that the adsorption capacity of ZnBTC is better than that of ZIF-8, as shown in Figure 5, which indicates that the adsorption performance of ZnBTC maintains superiority in contrast with ZIF-8 under the same conditions. In addition, it can be observed that the adsorption capacity during the composting process was lower than that in ammonia water and pure ammonia. The performance parameters of adsorption capacity under pure ammonia were the highest, followed by the 25% NH$_3$ solution steam atmosphere during aerobic composting. As can be seen, ZnBTC-3 and ZnBTC-2 show excellent adsorption capacities of 11.37 and 6.04 mmol/g, respectively. The adsorption capacities of ZIF-8-3 (1.26 mmol/g) and ZIF-8-2 (0.09 mmol/g) were much lower. The adsorption capacities of ZnBTC-1 (2.55) and ZIF-8-1 (undetected) during aerobic composting were the lowest. There may have been two reasons for these findings: On the one hand, the adsorption capacity of ammonia increases with its concentration. On the other hand, competitive adsorption may exist between NH_3 and other gases, such as H_2O under a 25% NH_3 solution steam atmosphere, and the H_2S and N_2O in the composting process occupied the adsorption sites of MOF materials.

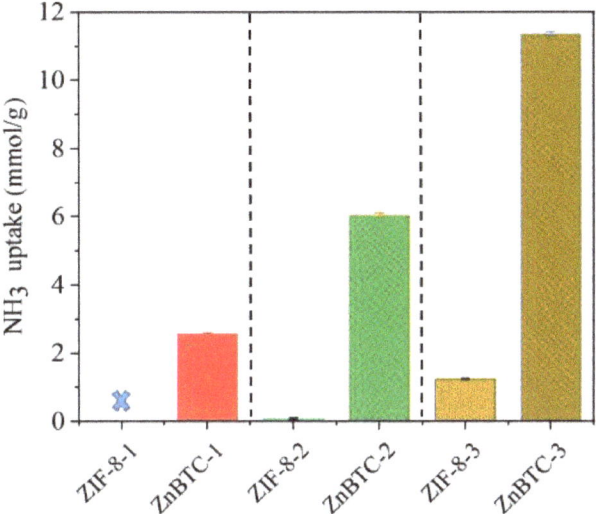

Figure 5. NH_3 adsorption capacity of ZIF-8 and ZnBTC in different settings.

3.3. Adsorption Mechanism

To further characterize the porosity of the two MOFs, the N_2 adsorption–desorption isotherms of samples were measured by the static method at 77 K. The specific surface area and pore structure of ZnBTC and ZIF-8 materials were measured by the automatic rapid specific surface and pore analyzer (Microtrac, Duesseldorf, Germany) (as shown in Figure 6). The built-in software of the instrument was used. The specific surface area of the material was calculated by the BET or Langmuir model. Before the test, about 90 mg of the MOF material was weighed and heated for more than 150 h under vacuum conditions to remove the moisture and impurities in the material.

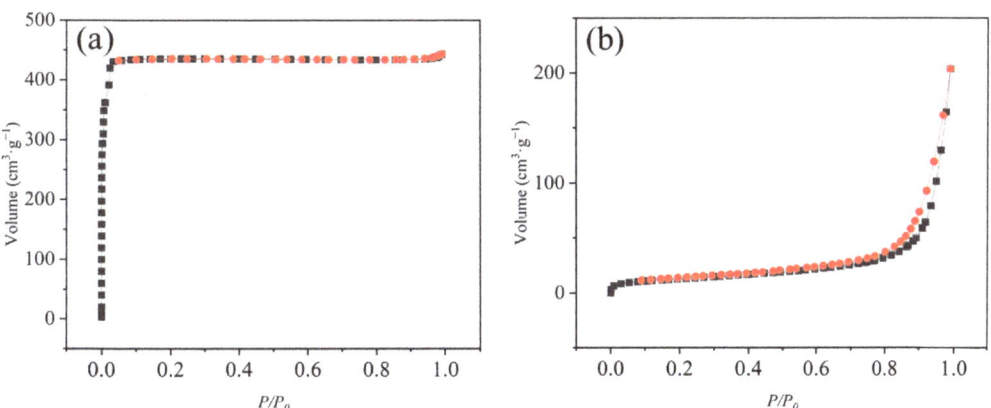

Figure 6. N_2 adsorption−desorption isotherms of (**a**) ZIF-8 (**b**) ZnBTC.

As a result, we found the ZnBTC and ZIF-8 possess a Brunauer–Emmett–Teller (BET) surface area of 1298.50 and 44.47 m^2 g^{-1}, respectively. The surface area of ZIF-8 is apparently higher than that of ZnBTC. Hence, research is needed to determine whether it is the specific surface area or adsorption strength that determines the adsorption capacity of ammonia gas.

The density functional theory calculation was then performed to elucidate the adsorption behavior of the two materials and their inherent mechanism. Rather than simply calculating the adsorption of a single ammonia molecule, we made the two MOF materials gradually adsorb one to four ammonia molecules and calculated the adsorption energy of each adsorbed ammonia molecule. Initially, ZIF-8 was directly employed in the reported crystal structure (CCDC No. 602542), while ZnBTC was modified from the structure (CCDC No. 963916) in which the nitrate ligand was exchanged for chloride for easy computation, and the H$_2$O ligand was maintained. The calculated results showed that each NH$_3$ molecule adsorbed onto ZnBTC replaced the H$_2$O ligand and led to a negative binding energy. Therefore, the NH$_3$ gas adsorbed by ZnBTC is stable (Figure 7b). Comparably, the first two NH$_3$ molecules seemed quite difficult to adsorb onto ZIF-8 due to extremely high binding energy (Figure 7a). The metal–ligand binding could also be revealed by the bond length (Figure 7c). Remarkably, the bond length in ZIF-8 is more than 4 Å, which is far beyond any coordination bond length. In contrast, the bond length in ZnBTC is stably localized at 2.08 Å, a typical coordination bond length. Subsequently, the average binding energy of ZIF-8 is larger than zero, indicating that ZIF-8 is weak for adsorbing NH$_3$. The adsorbability difference between ZIF-8 and ZnBTC results from their crystal structures. The zinc in ZnBTC is hexacoordinated, and the coordination of NH$_3$ instead of H$_2$O was sterically-allowed. In contrast, the zinc in ZIF-8 is tetracoordinated, and there is less room for an extra NH$_3$. As a total consequence, ZnBTC exhibited supreme NH$_3$ adsorbability (Figure 7d).

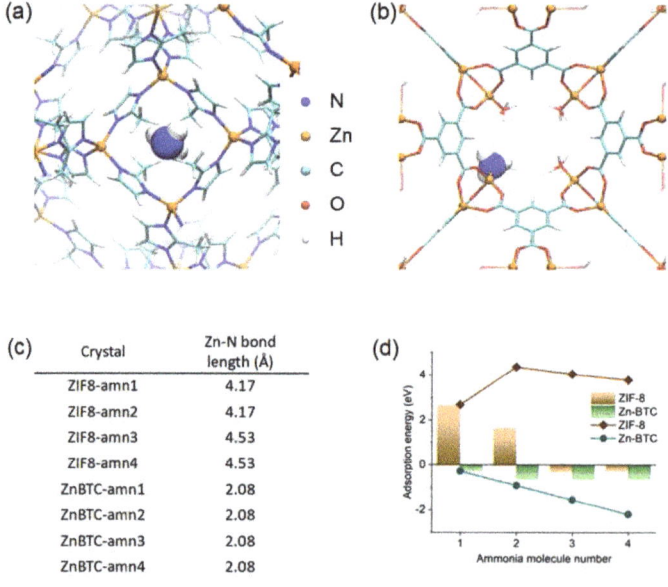

Figure 7. DFT calculation results of ammonia adsorption. (**a**) The structure of ZIF-8 with one adsorbed ammonia molecule (ZIF-8-amn1); (**b**) the structure of ZnBTC with one adsorbed ammonia molecule (ZnBTC-amn1), both of which were drawn by VMD (1.9.3) package [34]; (**c**) the Zn-N bond length in each adsorption structure; (**d**) the adsorption energy (column) and cumulative adsorption energy (line) of each crystal material.

4. Conclusions

In summary, two kinds of MOFs with different organic ligands were designed and analyzed with experimental and theoretical calculations. XRD, FTIR, and other tests proved that the MOFs were successfully prepared. The ZnBTC showed a higher NH_3 uptake of 11.37 mmol/g than ZIF-8 (1.26 mmol/g) due to the multiple adsorption sites, which helps the composting process reduce environmental pollution. The theoretical calculation also showed that ZnBTC has a higher adsorption capacity than ZIF-8. Such experimental adsorption capacities of MOFs were sufficiently consistent with the theoretical calculation. As far as we are aware, we are the first to design structural systems suitable for the materials that are available for the capturing of NH_3 from the composting process. These results not only provide a candidate method for NH_3 abatement but also provide a theoretical basis for material design for specific adsorption targets in the composting process.

Author Contributions: D.Z. synthesized MOFs, analyzed the adsorption data and properties for NH_3, and was the main contributor to the manuscript; Y.S. revised the manuscript and helped design the experiment; J.D. and H.Z. helped solve various problems encountered in the experiment; Y.Z. revised the manuscript several times and helped analyze the data of MOF materials; Q.F., X.Z., K.C., P.X. and P.Z. helped revise the manuscript. All authors have read and agreed to the published version of the manuscript.

Funding: Thanks to the project of Beijing Municipal Bureau of Agriculture "Beijing Innovation Team of Modern Agricultural Industrial Technology System", Independent research and development project of Academy of Agricultural Planning and Engineering (LJRC-2021-02).

Institutional Review Board Statement: Not applicable.

Informed Consent Statement: Not applicable.

Data Availability Statement: The authors do not have permission to share data.

Conflicts of Interest: The authors declare no conflict of interest.

References

1. Yin, Y.; Yang, C.; Li, M.; Zheng, Y.; Ge, C.; Gu, J.; Li, H.; Duan, M.; Wang, X.; Chen, R. Research progress and prospects for using biochar to mitigate greenhouse gas emissions during composting: A review. *Sci. Total Environ.* **2021**, *798*, 149294. [CrossRef] [PubMed]
2. Cao, Y.; Wang, X.; Liu, L.; Velthof, G.L.; Misselbrook, T.; Bai, Z.; Ma, L. Acidification of manure reduces gaseous emissions and nutrient losses from subsequent composting process. *J. Environ. Manag.* **2020**, *264*, 110454. [CrossRef] [PubMed]
3. Wang, S.; Zeng, Y. Ammonia emission mitigation in food waste composting: A review. *Bioresour. Technol.* **2018**, *248*, 13–19. [CrossRef] [PubMed]
4. Wang, R.; Zhao, Y.; Xie, X.; Mohamed, T.A.; Zhu, L.; Tang, Y.; Chen, Y.; Wei, Z. Role of NH_3 recycling on nitrogen fractions during sludge composting. *Bioresour. Technol.* **2020**, *295*, 122175. [CrossRef]
5. Zhang, H.; Li, G.; Gu, J.; Wang, G.; Li, Y.; Zhang, D. Influence of aeration on volatile sulfur compounds (VSCs) and NH_3 emissions during aerobic composting of kitchen waste. *Waste Manag.* **2016**, *58*, 369–375. [CrossRef]
6. Mao, H.; Zhang, H.; Fu, Q.; Zhong, M.; Li, R.; Zhai, B.; Wang, Z.; Zhou, L. Effects of four additives in pig manure composting on greenhouse gas emission reduction and bacterial community change. *Bioresour. Technol.* **2019**, *292*, 121896. [CrossRef]
7. Fukumoto, Y.; Suzuki, K.; Kuroda, K.; Waki, M.; Yasuda, T. Effects of struvite formation and nitratation promotion on nitrogenous emissions such as NH_3, N_2O and NO during swine manure composting. *Bioresour. Technol.* **2011**, *102*, 1468–1474. [CrossRef]
8. Fukumoto, Y.; Osada, T.; Hanajima, D.; Haga, K. Patterns and quantities of NH_3, N_2O and CH_4 emissions during swine manure composting without forced aeration—Effect of compost pile scale. *Bioresour. Technol.* **2003**, *89*, 109–114. [CrossRef]
9. Shen, Y.; Ren, L.; Li, G.; Chen, T.; Guo, R. Influence of aeration on CH_4, N_2O and NH_3 emissions during aerobic composting of a chicken manure and high C/N waste mixture. *Waste Manag.* **2011**, *31*, 33–38. [CrossRef]
10. Szanto, G.L.; Hamelers, H.V.M.; Rulkens, W.H.; Veeken, A.H.M. NH_3, N_2O and CH_4 emissions during passively aerated composting of straw-rich pig manure. *Bioresour. Technol.* **2007**, *98*, 2659–2670. [CrossRef]
11. Guo, R.; Li, G.; Jiang, T.; Schuchardt, F.; Chen, T.; Zhao, Y.; Shen, Y. Effect of aeration rate, C/N ratio and moisture content on the stability and maturity of compost. *Bioresour. Technol.* **2012**, *112*, 171–178. [CrossRef] [PubMed]
12. Luo, Y.; Li, G.; Luo, W.; Schuchardt, F.; Jiang, T.; Xu, D. Effect of phosphogypsum and dicyandiamide as additives on NH_3, N_2O and CH_4 emissions during composting. *J. Environ. Sci.* **2013**, *25*, 1338–1345. [CrossRef]
13. Jiang, T.; Ma, X.; Tang, Q.; Yang, J.; Li, G.; Schuchardt, F. Combined use of nitrification inhibitor and struvite crystallization to reduce the NH_3 and N_2O emissions during composting. *Bioresour. Technol.* **2016**, *217*, 210–218. [CrossRef]

14. Yang, Y.; Awasthi, M.K.; Wu, L.; Yan, Y.; Lv, J. Microbial driving mechanism of biochar and bean dregs on NH_3 and N_2O emissions during composting. *Bioresour. Technol.* **2020**, *315*, 123829. [CrossRef] [PubMed]
15. Wang, Q.; Awasthi, M.K.; Ren, X.; Zhao, J.; Li, R.; Wang, Z.; Wang, M.; Chen, H.; Zhang, Z. Combining biochar, zeolite and wood vinegar for composting of pig manure: The effect on greenhouse gas emission and nitrogen conservation. *Waste Manag.* **2018**, *74*, 221–230. [CrossRef]
16. Wang, Q.; Awasthi, M.K.; Ren, X.; Zhao, J.; Li, R.; Wang, Z.; Chen, H.; Wang, M.; Zhang, Z. Comparison of biochar, zeolite and their mixture amendment for aiding organic matter transformation and nitrogen conservation during pig manure composting. *Bioresour. Technol.* **2017**, *245*, 300–308. [CrossRef]
17. Islamoglu, T.; Chen, Z.; Wasson, M.C.; Buru, C.T.; Kirlikovali, K.O.; Afrin, U.; Mian, M.R.; Farha, O.K. Metal–organic frameworks against toxic chemicals. *Chem. Rev.* **2020**, *120*, 8130–8160. [CrossRef]
18. Zhou, W.; Li, T.; Yuan, M.; Li, B.; Zhong, S.; Li, Z.; Liu, X.; Zhou, J.; Wang, Y.; Cai, H.; et al. Decoupling of inter-particle polarization and intra-particle polarization in core-shell structured nanocomposites towards improved dielectric performance. *Energy Storage Mater.* **2021**, *42*, 1–11. [CrossRef]
19. Martínez-Ahumada, E.; López-Olvera, A.; Jancik, V.; Sánchez-Bautista, J.E.; González-Zamora, E.; Martis, V.; Williams, D.R.; Ibarra, I.A. MOF Materials for the Capture of Highly Toxic H_2S and SO_2. *Organometallics* **2020**, *39*, 883–915. [CrossRef]
20. Kajiwara, T.; Higuchi, M.; Watanabe, D. A systematic study on the stability of porous coordination polymers against ammonia. *Chem. Eur. J.* **2014**, *20*, 15611–15617. [CrossRef]
21. Khan, N.A.; Hasan, Z.; Jhung, S.H. Adsorptive removal of hazardous materials using metal-organic frameworks (MOFs): A review. *J. Hazard. Mater.* **2013**, *244*, 444–456. [CrossRef] [PubMed]
22. Kim, K.C. Design strategies for metal-organic frameworks selectively capturing harmful gases. *J. Organomet. Chem.* **2018**, *854*, 94–105. [CrossRef]
23. Martínez-Ahumada, E.; Díaz-Ramírez, M.L.; Velásquez-Hernández, M.J.; Jancik, V.; Ibarra, I.A. Capture of toxic gases in MOFs: SO_2, H_2S, NH_3 and NO_x. *Chem. Sci.* **2021**, *12*, 6772–6799. [CrossRef] [PubMed]
24. Chen, Y.; Li, L.; Yang, J.; Wang, S.; Li, J. Reversible flexible structural changes in multidimensional MOFs by guest molecules (I_2, NH_3) and thermal stimulation. *J. Solid State Chem.* **2015**, *226*, 114–119. [CrossRef]
25. Han, G.; Liu, Y.; Yang, Q.; Liu, D.; Zhong, C. Construction of stable IL@ MOF composite with multiple adsorption sites for efficient ammonia capture from dry and humid conditions. *Chem. Eng. J.* **2020**, *401*, 126106. [CrossRef]
26. Glover, T.G.; Peterson, G.W.; Schindler, B.J.; Britt, D.; Yaghi, O. MOF-74 building unit has a direct impact on toxic gas adsorption. *Chem. Eng. Sci.* **2011**, *66*, 163–170. [CrossRef]
27. Chen, Y.; Du, Y.; Liu, P.; Yang, J.; Li, L.; Li, J. Removal of Ammonia Emissions via Reversible Structural Transformation in M (BDC)(M= Cu, Zn, Cd) Metal–Organic Frameworks. *Environ. Sci. Technol.* **2020**, *54*, 3636–3642. [CrossRef]
28. Nguyen, T.N.; Harreschou, I.M.; Lee, J.H.; Stylianou, K.C.; Stephan, D.W. A recyclable metal–organic framework for ammonia vapour adsorption. *Chem. Commun.* **2020**, *56*, 9600–9603. [CrossRef]
29. Osman, S.; Senthil, R.A.; Pan, J.; Li, W. Highly activated porous carbon with 3D microspherical structure and hierarchical pores as greatly enhanced cathode material for high-performance supercapacitors. *J. Power Sources* **2018**, *391*, 162–169. [CrossRef]
30. Huang, X.; Chen, Y.; Lin, Z.; Ren, X.; Song, Y.; Xu, Z.; Dong, X.; Li, X.; Hu, C.; Wang, B. Zn-BTC MOFs with active metal sites synthesized via a structure-directing approach for highly efficient carbon conversion. *Chem. Commun.* **2014**, *50*, 2624–2627. [CrossRef]
31. Wu, J.; Li, Z.; Tan, H.; Du, S.; Liu, T.; Yuan, Y.; Liu, X.; Qiu, H. Highly Selective Separation of Rare Earth Elements by Zn-BTC Metal–Organic Framework/Nanoporous Graphene via In Situ Green Synthesis. *Anal. Chem.* **2020**, *93*, 1732–1739. [CrossRef] [PubMed]
32. Perdew, J.P.; Burke, K.; Ernzerhof, M. Generalized Gradient Approximation Made Simple. *Phys. Rev. Lett.* **1996**, *77*, 3865. [CrossRef]
33. Grimme, S.; Ehrlich, S.; Goerigk, L. Effect of the damping function in dispersion corrected density functional theory. *J. Comput. Chem.* **2011**, *32*, 1456–1465. [CrossRef] [PubMed]
34. Grimme, S.; Antony, J.; Ehrlich, S.; Krieg, H. A consistent and accurate ab initio parametrization of density functional dispersion correction (DFT-D) for the 94 elements H-Pu. *J. Chem. Phys.* **2010**, *132*, 154104. [CrossRef] [PubMed]
35. Pei, J.; Chen, Z.; Wang, Y.; Xiao, B.; Zhang, Z.; Cao, X.; Liu, Y. Preparation of phosphorylated iron-doped ZIF-8 and their adsorption application for U (VI). *J. Solid State Chem.* **2022**, *305*, 122650. [CrossRef]
36. Cheng, Y.; Wang, X.; Jia, C.; Wang, Y.; Zhai, L.; Wang, Q.; Zhao, D. Ultrathin mixed matrix membranes containing two-dimensional metal-organic framework nanosheets for efficient CO_2/CH_4 separation. *J. Membr. Sci.* **2017**, *539*, 213–223. [CrossRef]
37. Sacconi, L.; Sabatini, A.; Gans, P. Infrared spectra from 80 to 2000 cm^{-1} of some metal-ammine complexes. *Inorg. Chem.* **1964**, *3*, 1772–1774. [CrossRef]

Article

Tunable Ammonia Adsorption within Metal–Organic Frameworks with Different Unsaturated Metal Sites

Dongli Zhang [1], Yujun Shen [1,*], Jingtao Ding [1], Haibin Zhou [1], Yuehong Zhang [2], Qikun Feng [3], Xi Zhang [1], Kun Chen [1], Jian Wang [1], Qiongyi Chen [1], Yang Zhang [1] and Chaoqun Li [4]

[1] Academy of Agricultural Planning and Engineering, Key Laboratory of Technologies and Models for Cyclic Utilization from Agricultural Resources, Ministry of Agriculture, Beijing 100125, China
[2] School of Advanced Manufacturing, Guangdong University of Technology, Jieyang 515200, China
[3] State Key Laboratory of Power Systems, Department of Electrical Engineering, Tsinghua University, Beijing 100084, China
[4] College of Materials Science and Technology, Hainan University, Haikou 570228, China
* Correspondence: shenyj09b@163.com

Abstract: Ammonia (NH_3) emissions during agricultural production can cause serious consequences on animal and human health, and it is quite vital to develop high-efficiency adsorbents for NH_3 removal from emission sources or air. Porous metal–organic frameworks (MOFs), as the most promising candidates for the capture of NH_3, offer a unique solid adsorbent design platform. In this work, a series of MOFs with different metal centers, ZnBTC, FeBTC and CuBTC, were proposed for NH_3 adsorption. The metal centers of the three MOFs are coordinated in a different manner and can be attacked by NH_3 with different strengths, resulting in different adsorption capacities of 11.33, 9.5, and 23.88 mmol/g, respectively. In addition, theoretical calculations, powder XRD patterns, FTIR, and BET for the three materials before and after absorption of ammonia were investigated to elucidate their distinctively different ammonia absorption mechanisms. Overall, the study will absolutely provide an important step in designing promising MOFs with appropriate central metals for the capture of NH_3.

Keywords: metal–organic frameworks (MOFs); unsaturated metal sites; ammonia; adsorption capacity; absorption mechanisms

1. Introduction

Ammonia (NH_3), as a highly toxic and corrosive gas, is a rather common pollutant in agricultural production. Additionally, ammonia has a stimulating effect on human upper airways and eyes, even in small concentrations. A large amount of NH_3 emissions have brought huge hidden dangers to the environment and economy [1–3]. Hence, ammonia pollution is such a serious problem that it has aroused concerns from the whole society. Furthermore, the research interests in the treatment of NH_3 pollution have increased among relevant departments of academia and government [4–6]. It is well known that poultry and animal husbandry, fossil fuel combustion, refrigeration industry, fertilizer manufacturing and coke manufacturing could generate a large quantity of NH_3 [7,8]. The reduction in NH_3 emissions has been placed in an important position. In addition, due to the dispersion and uncertainty of NH_3 pollution, the effective mechanism for resolving NH_3 pollution has not been systematically established. There are two primary types of ammonia adsorption: physisorption and chemisorption. The physisorption is related to the surface area and pore volume, while chemisorption depends on coordinate bonding, including the unsaturated metal sites, functional groups, acidic binding sites, and defective sites. The adsorption mechanism for ammonia adsorption still remains an important issue to be solved both theoretically and practically.

Activated carbon, zeolite molecular sieves and metal oxides, belonging to traditional ammonia adsorption materials, are believed to be a practical solution for ammonia pollution. However, they suffer from low selectivity and low capacity for ammonia [9–13]. Nevertheless, MOFs (metal–organic frameworks), as a typical porous material, were formed by the self-assembly of organic ligands with metal ions or metal clusters and have the advantages of large specific surface area, high porosity, ordered pore structure and modifiable structure [14–19]. MOFs are widely used in catalysis, gas storage and separation and molecular recognition because of their merits. In addition, MOFs have been postulated as a promising option for the removal of NH_3 [20–23].

Among the widely studied MOFs, ZIF-8, Cu-BTC, UiO and MIL series MOFs materials have been widely employed for ammonia adsorption in previous research works. ZIF-8, as a well-known MOF, can even be mass-produced and put into application properly, but its ammonia adsorption capacity is low [24]. CuBTC, as reported by Peterson et al., has very high ammonia uptake [25] because it could interact strongly with ammonia due to the presence of open metal sites. Moreover, MOFs with the incorporation of different functional groups have also been explored for NH_3 removal. As reported by Wu et al. [26], the incorporation of biphenyl and bipyridine groups into MOFs UiO-67 and UiO-bpydc could lead to drastically different NH_3 adsorption properties because the bipyridine moieties can induce flexibility to the framework without significant pore volume alteration. In addition, Chen's group reported that MIL-100 and MIL-101 have a large NH_3 uptake of 8 and 10 mmol/g, respectively, and the NH_3 adsorption capacity could be improved by modified amino functional groups. More importantly, all MOF materials exhibit excellent stability and recycled NH_3 removal [27]. Although some common MOFs were reported in the field of NH_3 adsorption [28,29], many other MOFs, such as ZnBTC [30], are rarely involved. Specifically, the effectiveness of MOFs with different open metal sites for ammonia capture was not systematically studied.

In this study, three monometallic adsorbents, ZnBTC, FeBTC and CuBTC, were prepared by hydrothermal methods and their NH_3 adsorption property has been investigated. The phase structures and adsorption performances of synthesized MBTC materials with different metal central ions were characterized by SEM, BET, XRD and FTIR, respectively. Their main adsorption paths and mechanisms were further explored by DFT calculations, which could provide theoretical support for the preparation of the adsorbents for ammonia in the later stage.

2. Experimental Section

2.1. Materials and Instruments

All reagents were used as supplied without further purification. The benzene-1,3,5-tricarboxylic acid was obtained from the Sinopharm Chemical Reagent Co., Ltd. (Beijing, China). $Zn(NO_3)_2·6H_2O$ (AR, 99%), $Cu(NO_3)_2·3H_2O$ (AR, 98%) and $Fe(NO_3)_2·9H_2O$ and $ZnAc_2·2H_2O$ were purchased from Macklin Co. Ltd. (Shanghai, China). Ethanol and H_2O were commercially available.

2.2. Synthesis of ZnBTC, FeBTC and CuBTC

The synthetic methods of ZnBTC, FeBTC and CuBTC were determined according to the reported literature [28–30]. The preparation process comprises the following steps. First, $Zn(NO_3)_2·6H_2O$ (1.8 g), H_3BTC (0.6 g), and ethyl alcohol-H_2O (1:1, 60 mL) were well-mixed in a 100 mL vial by constantly stirring for 20 min. Then, the mixture was heated at 120 °C for 16 h. After cooling down to room temperature, the white crystals of Zn-BTC precipitate were filtered and then washed with ethanol and deionized (DI) water several times until the filtrate was neutral. Finally, the Zn-BTC was dried in a vacuum oven at 60 °C for 12 h.

FeBTC and CuBTC were prepared in similar methods [27,31].

2.3. Adsorption Experiments

The static capacity sorption analyzer of Bei Shi De (BSD-PSPM) was used to estimate the adsorption uptake of the pure NH_3. Before testing, the samples were activated at 50 °C for at least 2 h until the mass no longer changed. For the co-adsorption of NH_3/H_2O, 0.10 g of adsorbent was in an enclosed area with 20 mL of NH_3/H_2O (4:1, *v:v*). After the saturated adsorption was reached, the mixture was added to a saturated solution of potassium chloride using a temperature-controlled shaker to extract ammonia nitrogen. The filter liquid was achieved by membrane filtration methods and analyzed by Automatic Discrete Analyzer (SmartChem 140) to evaluate the adsorption uptake of NH_3.

2.4. Characterization

Scanning electron microscopy (SEM) images of the prepared MOFs were obtained by a field-emission scanning (SEM, S-4700). Fourier transform infrared (FT-IR) spectroscopy was carried out by Nicolet 6700FT-IR spectrophotometer. The crystal structure and phase purity of materials were collected on an X-ray diffractometer (Bruker AXS D8-Advance) in 2θ range from 5° to 40° at a scan rate of 10° min^{-1}. The specific surface area of the materials in this study was obtained by using a surface and micropore analyzer of BEST Instrument Technology Co., Ltd (Beijing, China). The Brunauer–Emmett–Teller (BET) surface area was obtained using a volumetric method. Using the Gaussian09 (Revision D.01) package [32], the calculations of ZnBTC, FeBTC and CuBTC in this study were carried out by the density functional theory (DFT) method, which employs the B3LYP [33] functional. The basis set was employed, LANL2DZ, for metal elements and 6–31 G(d) [34] for others. The values 2.7×10^{-4} eV (energy), 0.05 eV/Å (gradi-ent), and 0.005 Å (displacement) are defined as the parameters of the convergence threshold [35]. The binding energy (Δ*E*) was calculated as follows:

For MOFs + NH_3 MOFs$_{(NH_3)}$

$$\Delta E = E_{MOFs(NH_3)} - E_{MOFs} - E_{NH_3} \quad (1)$$

where E_{MOF}, E_{NH_3} and $E_{MOFs(NH_3)}$ represent the energy of the MOF unit, NH_3, and the MOFs after adsorption of ammonia, respectively.

3. Results and Discussion

The crystal morphologies of the as-synthesized materials, ZnBTC, FeBTC and CuBTC, were characterized by SEM. Before observation, the samples were sprayed with Au particles. As illustrated in Figure 1, the particles of the three crystalline materials are not uniform in size, with a particle size distribution of 1–5 μm. The surface morphology of the material is relatively rough. Taking the CuBTC, for example, there are some fixtures on the surface (Figure 1e,f). It is worth noting that the three MOFs bearing different metal centers take on quite different microstructures. In detail, ZnBTC is a rod-like structure with uneven length, while FeBTC and CuBTC exhibit an irregular blocky structure and a sheet-like structure, respectively. These rather distinctive morphology images of the three different materials may be due to their different metal centers involving different coordination patterns between the metal centers and the ligands.

The TGA analysis was adopted to evaluate the thermal stability of the three MOFs. As shown in Figure 2a, the pyrolysis processes of ZnBTC and CuBTC are similar and can be mainly divided into three stages. It can be found that ZnBTC and CuBTC exhibit visible weight loss from 30 to 200 °C (calculated: 6.42% for ZnBTC, 10.02% for CuBTC), which can be ascribed to the desorption of coordinated guest water molecules formed between unsaturated metal sites and oxygen molecule through the high affinity. The weight loss between 300 and 550 °C may be due to the partial decomposition of the ligand, and the final weight loss can be attributed to the carbonization of the sample and the formation of oxides. The weight loss process of FeBTC during pyrolysis is evidently different from those of ZnBTC and CuBTC. The mass loss of FeBTC with the temperature range of 30 to 200 °C originates from the evaporation of adsorbed water (calculated 4.82%), while the weight loss at 200–400 °C may be caused by the partial decomposition of the ligands. The decrease in weight for FeBTC during 400–550 °C should be attributed to the decomposition of some carboxyl groups into graphitized carbon. The final weight loss stage can be attributed to further carbonization of the sample and reduction in iron species. The residual weights of ZnBTC, CuBTC and FeBTC at 700 °C are 40.29%, 38.03% and 30.04%, respectively, indicating that the thermal stability of ZnBTC and CuBTC is superior to that of FeBTC. It is worth noting that the three compounds behave differently in the first weight loss stage within the temperature range from 30 to 200 °C, with a respective weight loss of 6.42% for ZnBTC, 10.02% for CuBTC, 4.82% for FeBTC, suggesting the increasing adsorption capacity of polar water molecules from FeBTC, ZnBTC, and CuBTC. In other words, the interaction between the metal center Cu and polar molecules is the strongest of the three compounds.

Figure 1. The SEM images of ZnBTC (**a**), the enlarged images of ZnBTC (**b**), FeBTC (**c**), FeBTC (**d**), CuBTC (**e**), and CuBTC (**f**).

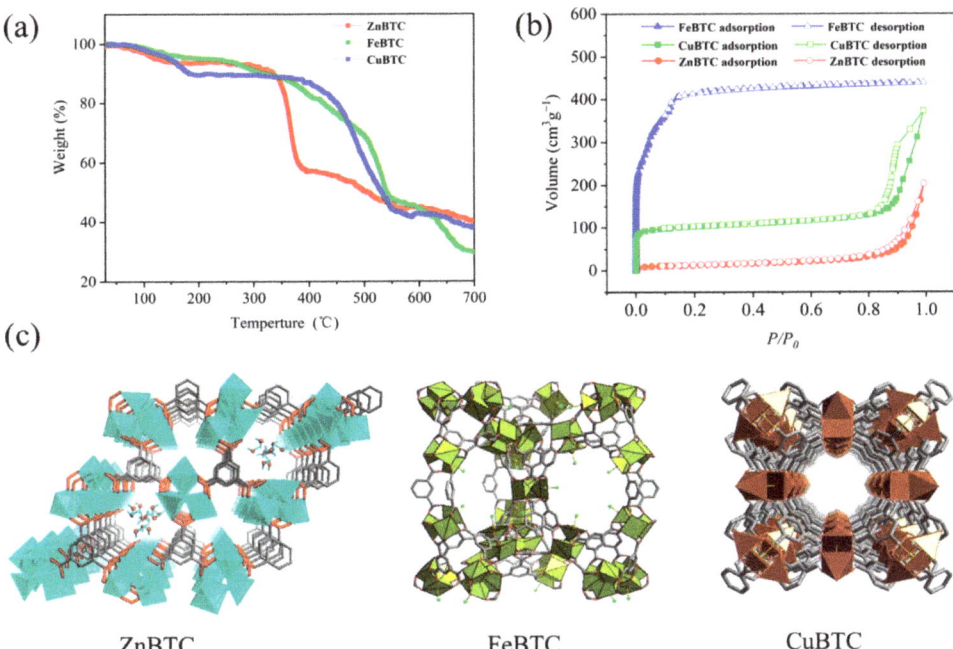

Figure 2. The TGA curve (**a**), N_2 adsorption−desorption isotherms (**b**) and the crystal character of ZnBTC, FeBTC and CuBTC (**c**).

Isothermal adsorption and desorption measurements were conducted to characterize parameters such as specific surface area. Before testing, a degassing pretreatment was performed under vacuum. The specific surface area of the three MOFs was determined by the BET and Langmuir methods, respectively. The nitrogen adsorption and desorption isotherms of ZnBTC, FeBTC and CuBTC all show Type I adsorption isotherms (Figure 2b), indicating that the three materials have microporous structures. As can be observed, the adsorption amount of the material rises sharply when the relative pressure is low. This is mainly due to the filling effect of the micropores and the enhancement of the adsorption potential in the micropores, which lead to adsorbate molecules having a strong capture capacity at lower relative pressures. Further, with the increase in relative pressure, the adsorption amount increased slowly and the growth rate was small, indicating that the adsorption almost reached saturation. The adsorption amount increased slowly with relative pressure and may mean it has no potential to rise, as is made evident by the deformed saturation point of FeBTC at 1 bar. When the pressure P/P_0 of CuBTC and ZnBTC was close to 1, the adsorption curve was upturned and the adsorption capacity increased. The main reason is that the two MOF samples are both nanoparticles, and nitrogen molecules condense on the surfaces of the pore structure formed by the accumulation between particles. It can be calculated from the nitrogen adsorption and desorption curves that the specific surface areas of FeBTC, CuBTC and ZnBTC are 1407.04, 388.68 and 44.47 m^2/g, respectively.

To elucidate the NH_3 adsorption properties of the three MOFs, NH_3 adsorption and desorption isotherms of ZnBTC, FeBTC and CuBTC were performed. As shown in Figure 3a, ZnBTC, FeBTC and CuBTC exhibited distinct NH_3 uptake amounts of 5.04, 4.33 and 15.35 mmol/g, respectively, at low adsorption pressure (25 mbar). With the increase in pressure, the adsorption capacity of all three materials slowed down in growth. For FeBTC and CuBTC, their adsorption capacity still slightly improved, while that for ZnBTC plateaued in the high-pressure region. The distinctively different performances of the three MOFs may result from the metal sites in MOFs. In addition, the results indicate that adsorption capacity has no apparent relevance to specific surface area and pore size. Finally, the adsorption amount of NH_3 for ZnBTC, FeBTC and CuBTC reached 11.33, 9.5, and 23.88 mmol/g, respectively, at 1 bar.

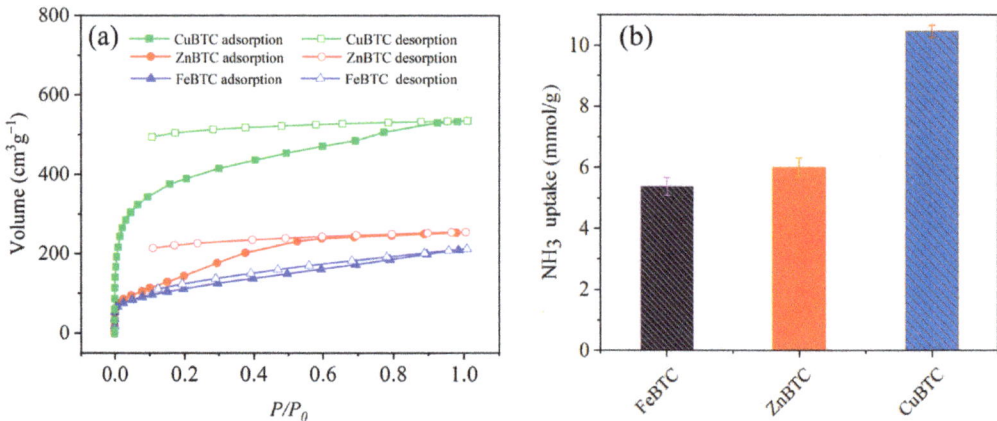

Figure 3. NH_3 adsorption capacity of ZnBTC, FeBTC and CuBTC with different metal centers (**a**), and NH_3 adsorption capacity of ZnBTC, FeBTC and CuBTC under an ammonia atmosphere (**b**).

During desorption at different pressures, all three materials exhibit apparent lag, suggesting a strong interaction between the open metal sites and NH_3. As shown in the desorption curves, only a small amount of NH_3 is desorbed initially, and there still is a large amount of NH_3 that cannot be completely removed when the pressure drops as low as 0.1 bar. Based on the high absorption amount from the low-pressure region and the fact that a large amount of NH_3 cannot be desorbed, it can be assumed that NH_3 is chemically adsorbed in the three MOFs. The distinctive ammonia adsorption performance of CuBTC should originate from the strong interaction between its metal center copper and the polar molecule ammonia, as exemplified by the strong interaction between copper and water detailed above. In particular, the role of unsaturated metal sites dominates, resulting in a rapid increase in the adsorption capacity. Finally, it calls for putting the measures mentioned above into practice, such as ammonia water, and further demonstrates the application of MOFs to absorb NH_3 in moist conditions. The adsorption amount of NH_3 was calculated by the desorption of ammonium ions. The adsorption capacity for CuBTC, ZnBTC and FeBTC under the atmosphere of ammonia water was 10.47, 6.03, and 5.39 mmol/g, respectively.

To confirm the crystallinity of ZnBTC, FeBTC and CuBTC before and after NH_3 adsorption, XRD patterns of the three MOFs are inspected in Figure 4a. The peak positions of different materials are in complete agreement with the simulated data reported previously, confirming that the material has been successfully synthesized [25,30,36]. After the co-adsorption of H_2O/NH_3, the PXRD of CuBTC changed significantly, indicating it was transformed into $Cu_3(BTC)_2(NH_3)_6(H_2O)_3$ [37]. The adsorption peaks of ZnBTC and FeBTC also changed significantly, indicating that the metal centers of ZnBTC and FeBTC may have been coordinated with NH_3, respectively.

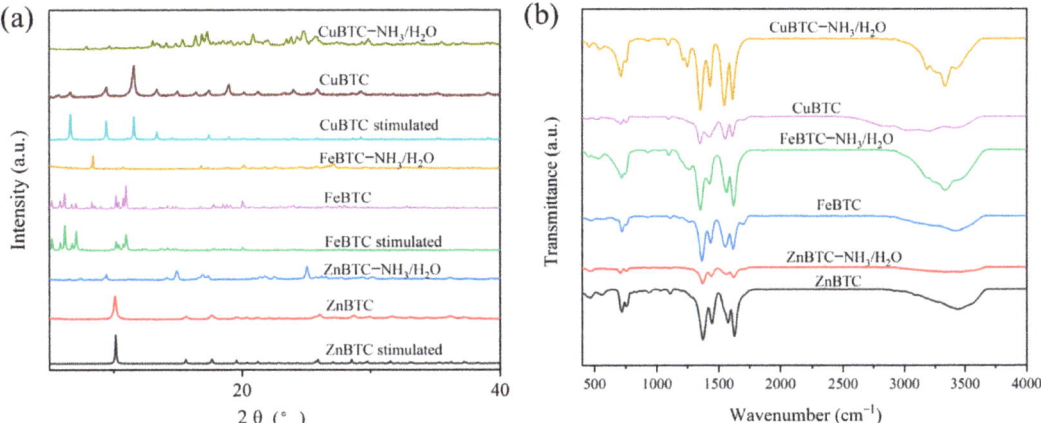

Figure 4. The XRD patterns (**a**) and FTIR (**b**) of ZnBTC, FeBTC and CuBTC before and after NH_3 adsorption.

To explore the types of functional groups in these three synthetic materials, FTIR were performed with the range 4000–4500 cm^{-1}, and the result is consistent with that reported previously [38]. As shown in Figure 4b, all the three MOFs exhibit similar absorption peaks, indicating the presence of similar functional groups in the three materials. Taking FeBTC, for example, the signal at 3345 cm^{-1} corresponds to the -OH peak on the carboxylic acid ligand, and the absorption peak at 1625 cm^{-1} represents the C=O on the carboxylic acid ligand and water H-O-H vibration peaks. The absorption signals at 1384 and 1618 cm^{-1} are the stretching vibration signals of C-O on the carboxyl group, and the vibrational peaks of C-C on the benzene ring were at 1438 cm^{-1}, respectively. The signals at 712 and 761 cm^{-1} are the C-H vibrations on the benzene ring. After NH_3/H_2O co-adsorption, a broad absorption peak of NH_3 at ca. 3300 cm^{-1} appears in the spectrum, indicating that NH_3 molecules are incorporated into the structure of MBTC. Moreover, free organic ligands in each MOF were observed, suggesting the destruction of the frameworks.

To further understand the adsorption mechanism of the NH_3 over ZnBTC, FeBTC and CuBTC, this study simulates the microstructure of MOFs adsorption materials. The crystal structures of Zn-BTC, FeBTC and CuBTC were referenced from the Cambridge Crystallographic Data Centre (CCDC) [27–30,35]. In order to simulate the local environment and reduce the calculation amount, the optimization of MOF structures was essential. The all-atom frameworks were adopted to analyze the canonical clusters of the MOFs. For example, to retain the correct hybridization, the methyl ($-CH_3$) groups were used to cut off the dangling bonds from these fragmented clusters.

The adsorption sites, configurations, and adsorption energies of NH_3 over Zn-BTC, FeBTC and CuBTC are obtained by the calculation. As illustrated in Figure 5, the NH_3 molecules attacked the metal center of MBTC (M=Cu, Zn, Fe) in different ways. The CuBTC and ZnBTC are attacked relatively easily as they have high polarity, strong coordination ability and simple configurations. Hence, the NH_3 could attack CuBTC and ZnBTC in rectilinear form. Conversely, FeBTC is not vulnerable to the attack due to the cage structure, which forms a large steric hindrance effect. Overall, the connection between the metals (Cu, Zn, Fe) and BTC ligands was damaged, following the combination of NH_3 molecules.

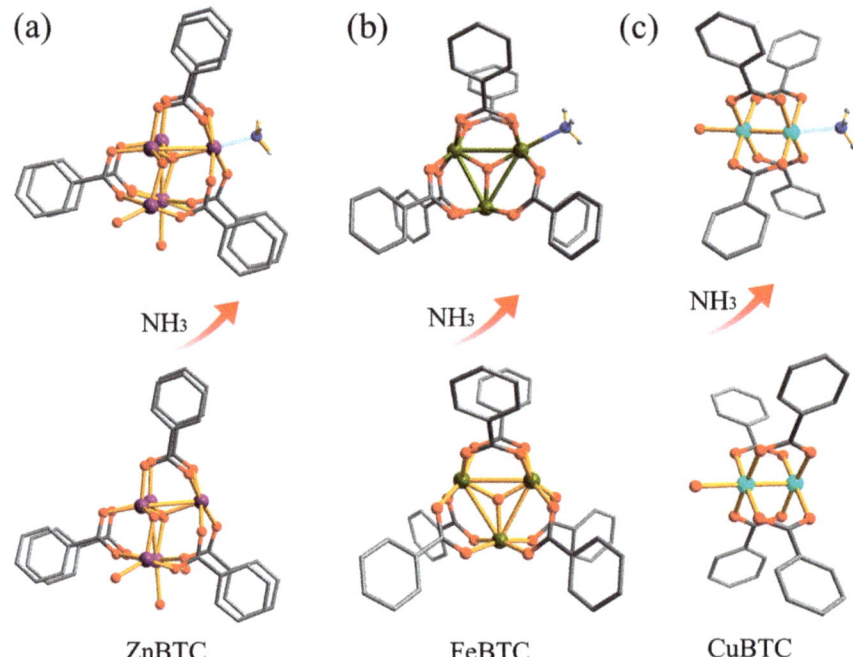

Figure 5. Adsorption configuration and distance of NH_3 on metal site: (**a**) ZnBTC, (**b**) FeBTC and (**c**) CuBTC.

Moreover, the binding energy on M sites over ZnBTC, FeBTC and CuBTC are −32.9, −18.7 and −14.7 kcal/mol, respectively, indicating that the ZnBTC owns higher adsorption strength than CuBTC and FeBTC (Table 1). The different trend conforms with the distance from the M sites to NH_3, which are 2.146, 2.298 and 2.282 Å for ZnBTC, FeBTC and CuBTC, respectively. Our preliminary results suggest that the BET and the distance from the M sites to NH_3 have a great correlation with adsorption capacity but are not the only factors that decide the adsorption capacity.

Table 1. The results of the theoretical calculation of Zn-BTC, FeBTC and CuBTC.

MOFs	Adsorbate	$E_{MOF(NH_3)}$ (kcal/mol)	E_{MOF} (kcal/mol)	E_{NH_3} (kcal/mol)	ΔE (kcal/mol)
CuBTC	NH_3	−1,384,454.491	−1,348,955.985	−35,483.834	−14.7
FeBTC	NH_3	−1,960,018.095	−1,924,515.523	−35,483.834	−18.7
ZnBTC	NH_3	−2,158,051.865	−2,122,535.112	−35,483.834	−32.9

4. Conclusions

In summary, three metal–organic frameworks (MOFs) with different metal centers, ZnBTC, FeBTC and CuBTC, were designed, synthesized, and characterized by a series of techniques such as XRD, FT-IR. The results illustrate that CuBTC exhibits a rather large ammonia adsorption capacity of 23.88 mmol/g compared with the other two (ZnBTC (11.33 mmol/g) and FeBTC (9.5 mmol/g)). As disclosed by theoretical calculations, this is mainly due to the different metal coordination in the different MOFs, which may lead to different attack modes and power of NH_3 to the metal center. In general, this study not only provides an important first step in designing suitable MOFs for the removal of NH_3 from the composting process but also establishes a theoretical system for specific adsorption targets in the composting process.

Author Contributions: D.Z.: Conceptualization, Data curation, Formal analysis, Writing—original draft, Methodology; Y.S.: Writing—review & editing; J.D. and H.Z.: Project administration, Resources; Y.Z. (Yang Zhang) Supervision, Validation, Visualization, Methodology; Q.F., X.Z. and K.C.: Resources, Validation, Software; Q.C., J.W., Y.Z. (Yuehong Zhang) and C.L.: Investigation, review. All authors have read and agreed to the published version of the manuscript.

Funding: This work was supported by the Independent Research and Development Project of Academy of Agricultural Planning and Engineering (LJRC-2021-02) and the project of the Beijing Municipal Bureau of Agriculture "Beijing Innovation Team of Modern Agricultural Industrial Technology System".

Institutional Review Board Statement: Not applicable.

Informed Consent Statement: Not applicable.

Data Availability Statement: All data generated or analyzed during this study are included in this published article. All data, models, and code generated or used during the study appear in the submitted article.

Acknowledgments: The authors thank the project of the Beijing Municipal Bureau of Agriculture, "Beijing Innovation Team of Modern Agricultural Industrial Technology System" and the Independent research and development project of Academy of Agricultural Planning and Engineering (LJRC-2021-02). The authors are grateful to Dongdong Qi at the Department of Chemistry, University of Science and Technology Beijing for a grant of computer time.

Conflicts of Interest: The authors declare no conflict of interest.

Sample Availability: Samples of the compounds ZnBTC, FeBTC and CuBTC are available from the authors.

References

1. Islamoglu, T.; Chen, Z.; Wasson, M.C.; Buru, C.T.; Kirlikovali, K.O.; Afrin, U.; Mian, M.; Farha, O.K. Metal–organic frameworks against toxic chemicals. *Chem. Rev.* **2020**, *120*, 8130–8160. [CrossRef]
2. Jiang, C.; Wang, X.; Ouyang, Y.; Lu, K.; Jiang, W.; Xu, H.; Wei, X.; Wang, Z.; Dai, F.; Sun, D. Recent advances in metal-organic frameworks for gas adsorption/separation. *Nanoscale Adv.* **2022**, *4*, 2077–2089. [CrossRef] [PubMed]
3. Khan, N.A.; Hasan, Z.; Jhung, S.H. Adsorptive removal of hazardous materials using metal-organic frameworks (MOFs): A review. *J. Hazard. Mater.* **2013**, *244*, 444–456. [CrossRef]
4. Kim, C.K. Design strategies for metal-organic frameworks selectively capturing harmful gases. *J. Organomet. Chem.* **2018**, *854*, 94–105. [CrossRef]
5. Kang, D.W.; Ju, S.E.; Kim, D.W.; Kang, M.; Kim, H.; Hong, C.S. Emerging porous materials and their composites for NH_3 gas removal. *Adv. Sci.* **2020**, *7*, 2002142. [CrossRef]
6. Martínez-Ahumada, E.; Díaz-Ramírez, M.L.; Velásquez-Hernández, M.D.J.; Jancik, V.; Ibarra, I.A. Capture of toxic gases in MOFs: SO_2, H_2S, NH_3 and NO_x. *Chem. Sci.* **2021**, *12*, 772–6799. [CrossRef]
7. Zhai, Z.; Zhang, X.; Hao, X.; Niu, B.; Li, C. Metal–Organic Frameworks Materials for Capacitive Gas Sensors. *Adv. Mater. Technol.* **2021**, *6*, 2100127. [CrossRef]
8. Wang, H.; Lustig, W.P.; Li, J. Sensing and capture of toxic and hazardous gases and vapors by metal–organic frameworks. *Chem. Soc. Rev.* **2018**, *47*, 4729–4756. [CrossRef]
9. Yin, Y.; Yang, C.; Li, M.; Zheng, Y.; Ge, C.; Gu, J.; Li, H.; Duan, M.; Wang, X.; Chen, R. Research progress and prospects for using biochar to mitigate greenhouse gas emissions during composting: A review. *Sci. Total Environ.* **2021**, *798*, 149294. [CrossRef]
10. Wang, Q.; Awasthi, M.K.; Ren, X.; Zhao, J.; Li, R.; Wang, Z.; Wang, M.; Chen, H.; Zhang, Z. Combining biochar, zeolite and wood vinegar for composting of pig manure: The effect on greenhouse gas emission and nitrogen conservation. *Waste Manag.* **2018**, *74*, 221–230. [CrossRef]
11. Wang, Q.; Awasthi, M.K.; Ren, X.; Zhao, J.; Li, R.; Wang, Z.; Chen, H.; Wang, M.; Zhang, Z. Comparison of biochar, zeolite and their mixture amendment for aiding organic matter transformation and nitrogen conservation during pig manure composting. *Bioresour. Technol.* **2017**, *245*, 300–308. [CrossRef]
12. Maitlo, H.A.; Maitlo, G.; Song, X.; Zhou, M.; Kim, K.H. A figure of merits-based performance comparison of various advanced functional nanomaterials for adsorptive removal of gaseous ammonia. *Sci. Total Environ.* **2022**, *822*, 153428. [CrossRef] [PubMed]
13. Melati, A.; Padmasari, G.; Oktavian, R.; Rakhmadi, F.A. A comparative study of carbon nanofiber (CNF) and activated carbon based on coconut shell for ammonia (NH_3) adsorption performance. *Appl. Phys. A* **2022**, *128*, 211. [CrossRef]
14. Glover, T.G.; Peterson, G.W.; Schindler, B.J.; Britt, D.; Yaghi, O. MOF-74 building unit has a direct impact on toxic gas adsorption. *Chem. Eng. Sci.* **2011**, *66*, 163–170. [CrossRef]

15. Ma, K.; Islamoglu, T.; Chen, Z.; Li, P.; Wasson, M.C.; Chen, Y.; Wang, Y.; Peterson, G.; Xin, J.; Farha, O.K. Scalable and template-free aqueous synthesis of zirconium-based metal–organic framework coating on textile fiber. *J. Am. Chem. Soc.* **2019**, *141*, 15626–15633. [CrossRef]
16. Han, G.; Liu, C.; Yang, Q.; Liu, D.; Zhong, C. Construction of stable IL@ MOF composite with multiple adsorption sites for efficient ammonia capture from dry and humid conditions. *Chem. Eng. J.* **2020**, *401*, 126106. [CrossRef]
17. Zhou, W.; Li, T.; Yuan, M.; Li, B.; Zhong, S.; Li, Z.; Liu, X.; Zhou, J.; Dang, Z.M. Decoupling of inter-particle polarization and intra-particle polarization in core-shell structured nanocomposites towards improved dielectric performance. *Energy Storage Mater.* **2021**, *42*, 1–11. [CrossRef]
18. Chen, Y.; Du, Y.; Liu, P.; Yang, J.; Li, L.; Li, J. Removal of Ammonia Emissions via Reversible Structural Transformation in M (BDC)(M = Cu, Zn, Cd) Metal–Organic Frameworks. *Environ. Sci. Technol.* **2020**, *54*, 3636–3642. [CrossRef]
19. Wilcox, O.T.; Fateeva, A.; Katsoulidis, A.P.; Smith, M.W.; Stone, C.A.; Rosseinsky, M.J. Acid loaded porphyrin-based metal–organic framework for ammonia uptake. *Chem. Commun.* **2015**, *51*, 14989–14991. [CrossRef] [PubMed]
20. Wang, Z.; Li, Z.; Zhang, X.G.; Xia, Q.; Wang, H.; Wang, C.; Wang, Y.; He, H.; Zhao, Y.; Wang, J. Tailoring Multiple Sites of Metal–Organic Frameworks for Highly Efficient and Reversible Ammonia Adsorption. *ACS Appl. Mater. Interfaces* **2021**, *13*, 56025–56034. [CrossRef]
21. Moribe, S.; Chen, Z.; Alayoglu, S.; Syed, Z.H.; Islamoglu, T.; Farha, O.K. Ammonia capture within isoreticular metal–organic frameworks with rod secondary building units. *ACS Mater. Lett.* **2019**, *1*, 476–480. [CrossRef]
22. Aulakh, D.; Nicoletta, A.P.; Varghese, J.R.; Wriedt, M. The structural diversity and properties of nine new viologen based zwitterionic metal–organic frameworks. *CrystEngComm* **2016**, *18*, 2189–2202. [CrossRef]
23. Guo, L.; Han, X.; Ma, Y.; Li, J.; Lu, W.; Li, W.; Lee, D.; Silva, I.; Cheng, Y.; Yang, S. High capacity ammonia adsorption in a robust metal–organic framework mediated by reversible host–guest interactions. *Chem. Commun.* **2022**, *58*, 5753–5756. [CrossRef]
24. Kajiwara, T.; Higuchi, M.; Watanabe, D.; Higashimura, H.; Yamada, T.; Kitagawa, H. A systematic study on the stability of porous coordination polymers against ammonia. *Chem. A Eur. J.* **2014**, *20*, 15611–15617. [CrossRef] [PubMed]
25. Chen, Y.; Yang, C.; Wang, X.; Yang, J.; Li, J. Vapor phase solvents loaded in zeolite as the sustainable medium for the preparation of Cu-BTC and ZIF-8. *Chem. Eng. J.* **2017**, *313*, 179–186. [CrossRef]
26. Yoskamtorn, T.; Zhao, P.; Wu, X.P.; Purchase, K.; Orlandi, F.; Manuel, P.; Taylor, J.; Li, Y.; Day, S.; Tsang, S.E. Responses of Defect-Rich Zr-Based Metal–Organic Frameworks toward NH_3 Adsorption. *J. Am. Chem. Soc.* **2021**, *143*, 3205–3218. [CrossRef] [PubMed]
27. Chen, Y.; Zhang, F.; Wang, Y.; Yang, C.; Yang, J.; Li, J. Recyclable ammonia uptake of a MIL series of metal-organic frameworks with high structural stability. *Microporous Mesoporous Mater.* **2018**, *258*, 170–177. [CrossRef]
28. Swaroopa Datta Devulapalli, V.; McDonnell, R.P.; Ruffley, J.P.; Shukla, P.B.; Luo, T.Y.; De Souza, M.L.; Das, P.; Rosi, N.; Johnson, J.; Borguet, E. Identifying UiO-67 Metal-Organic Framework Defects and Binding Sites through Ammonia Adsorption. *ChemSusChem* **2022**, *15*, e202102217. [CrossRef] [PubMed]
29. Chen, Y.; Li, L.; Yang, J.; Wang, S.; Li, J. Reversible flexible structural changes in multidimensional MOFs by guest molecules (I_2, NH_3) and thermal stimulation. *J. Solid State Chem.* **2015**, *226*, 114–119. [CrossRef]
30. Huang, X.; Chen, Y.; Lin, Z.; Ren, X.; Song, Y.; Xu, Z.; Dong, X.; Li, X.; Hu, C.; Wang, B. Zn-BTC MOFs with active metal sites synthesized via a structure-directing approach for highly efficient carbon conversion. *Chem. Commun.* **2014**, *50*, 2624–2627. [CrossRef]
31. Chen, R.; Liu, J. Competitive coadsorption of ammonia with water and sulfur dioxide on metal-organic frameworks at low pressure. *Build. Environ.* **2022**, *207*, 108421. [CrossRef]
32. Frisch, M.; Trucks, G.; Schlegel, H.; Scuseria, G.; Robb, M.; Cheeseman, J.; Scalmani, G.; Barone, V.; Mennucci, B.; Petersson, G. *Gaussian 09 (Revision D.01)*; Gaussian Inc.: Wallingford, CT, USA, 2009.
33. Becke, A.D. Density-functional thermochemistry. III. The role of exact exchange. *J. Chem. Phys.* **1993**, *98*, 5648–5652. [CrossRef]
34. Petersson, G.A.; Bennett, A.; Tensfeldt, T.G. MA A1-Laham, WA Shirlen, J. Mantzaris. *J. Chem. Phys.* **1988**, *89*, 2193. [CrossRef]
35. Wu, L.; Xiao, J.; Wu, Y.; Xian, S.; Miao, G.; Wang, H.; Li, Z. A combined experimental/computational study on the adsorption of organosulfur compounds over metal-organic frameworks from fuels. *Langmuir Acs J. Surf. Colloids* **2014**, *30*, 1080. [CrossRef] [PubMed]
36. Ying, F.; Li, G.; Liao, F.; Ming, X.; Lin, J. Two novel transition metal–organic frameworks based on 1,3,5-benzenetricarboxylate ligand: Syntheses, structures and thermal properties. *J. Mol. Struct.* **2011**, *1004*, 252–256. [CrossRef]
37. Ene, C.D.; Tuna, F.; Fabelo, O.; Ruiz-Perez, C.; Madalan, A.M.; Roesky, H.W.; Andruh, M. One-dimensional and two-dimensional coordination polymers constructed from copper (II) nodes and polycarboxylato spacers: Synthesis, crystal structures and magnetic properties. *Polyhedron* **2008**, *27*, 574–582. [CrossRef]
38. Liu, Z.; Chen, J.; Wu, Y.; Li, Y.; Zhao, J.; Na, P. Synthesis of magnetic orderly mesoporous α-Fe_2O_3 nanocluster derived from MIL-100(Fe) for rapid and efficient arsenic(III, V) removal. *J. Hazard. Mater.* **2018**, *343*, 304–314. [CrossRef]

Review

Two-Dimensional Black Phosphorus: Preparation, Passivation and Lithium-Ion Battery Applications

Hongda Li *, Chenpu Li, Hao Zhao, Boran Tao * and Guofu Wang *

Liuzhou Key Laboratory for New Energy Vehicle Power Lithium Battery, School of Electronic Engineering, Guangxi University of Science and Technology, Liuzhou 545006, China
* Correspondence: hdli@gxust.edu.cn (H.L.); brtao@gxust.edu.cn (B.T.); gfwang@guet.edu.cn (G.W.)

Abstract: As a new type of single element direct-bandgap semiconductor, black phosphorus (BP) shows many excellent characteristics due to its unique two-dimensional (2D) structure, which has great potential in the fields of optoelectronics, biology, sensing, information, and so on. In recent years, a series of physical and chemical methods have been developed to modify the surface of 2D BP to inhibit its contact with water and oxygen and improve the stability and physical properties of 2D BP. By doping and coating other materials, the stability of BP applied in the anode of a lithium-ion battery was improved. In this work, the preparation, passivation, and lithium-ion battery applications of two-dimensional black phosphorus are summarized and reviewed. Firstly, a variety of BP preparation methods are summarized. Secondly, starting from the environmental instability of BP, different passivation technologies are compared. Thirdly, the applications of BP in energy storage are introduced, especially the application of BP-based materials in lithium-ion batteries. Finally, based on preparation, surface functionalization, and lithium-ion battery of 2D BP, the current research status and possible future development direction are put forward.

Keywords: two-dimensional black phosphorus; preparation; surface modification; lithium-ion battery applications

1. Introduction

With the development of a global economy and society, people's demands for and use of energy are increasing day by day. The problems of environmental pollution and fossil energy shortage are becoming more and more obvious. For the sustainable development of human beings, it is urgent to seek the development of new energy and renewable resources. As a chemical energy storage device, lithium-ion batteries (LIBs) are widely used in portable electronic equipment [1], aerospace [2], military equipment [3], and electric vehicles [4] due to their advantages of high specific power, high energy density, long life, low self-discharge rate, and long storage time [5–9]. At present, LIBs have gradually replaced other batteries as the main power battery [10]. The development of smart grid and large-scale energy storage in recent years has put forward higher requirements on the energy density and power density of LIBs, which makes it particularly important to develop new LIBs with high energy density and power density.

Black phosphorus (BP) is the most stable of the three common allotropes of phosphorus (red phosphorus, white phosphorus, and BP) [11]. BP has four kinds of crystal structures: orthogonal, rhombus, simple cubic, and amorphous, and is an orthogonal crystal structure at room temperature and under pressure. BP has a layered structure like graphite, but the phosphorus atoms within the same layer are not in the same plane; it is a kind of honeycombed fold structure. There is a strong covalent bond in the layer, and a single electron pair remains, so each atom is saturated, and the atoms between layers act by van der Waals forces. Similar to graphite, the structure of BP allows the preparation of two-dimensional (2D) BP crystals. Some properties of 2D materials are not found in bulk materials. At

present, graphene and 2D transition metal sulfide are pioneers in this field [12,13]. The performance of monolayer BP is superior to graphene and 2D transition metal sulfide mainly in semiconductor and photoelectric properties. Graphene has very high carrier mobility, but its zero-band gap effect makes it unable to realize semiconductor logic switch. Two-dimensional transition metal sulfide (e.g., $MoTe_2$) has good semiconductor performance, and the transistor prepared by $MoTe_2$ has good electrical regulation performance, but its low carrier mobility limits its application in the field of electronics [14,15]. As a new 2D material, 2D BP has become the star of the anode material for LIBs with adjustable band structure, excellent electrical properties, anisotropic mechanical, thermodynamic, and photoelectrical properties. Since SONY LIBs were commercialized in 1990, LIBs have become one of the research hotspots in the field of new energy due to its characteristics of high specific capacity, long cycle life, no pollution, and good safety.

Two-dimensional BP shows excellent mechanical, electrochemical, and thermodynamic properties, and has great research value in ultra-light materials, energy storage devices, flexible electronics, and other aspects [16–20]. BP was first synthesized by Bridgman [21] under high temperature and high pressure, but it was seldom studied due to the technical conditions at that time. In 2014, Zhang and Ye et al. [22] successfully obtained monolayer BP through the method of sellot-tape stripping, which promoted the further study of BP performance [23].

The emergence of 2D BP also promotes the development of a new energy field. With the progress of society and technology, the existing energy materials available to meet people's needs are becoming less and less, which stimulates people's continuous exploration in the field of new energy. LIBs, sodium ion batteries (SIBs), lithium sulfur batteries (LBS), magnesium ion batteries (MIBs), super capacitors, and other electrochemical energy storage devices are developing rapidly. In these energy storage devices, LIBs have been widely used. The 3C market (mobile phones, computers, cameras, etc.) has its most mature systems. However, LIBs are mostly confined to these small devices and have limited application in the power battery market. Therefore, researchers have been committed to exploring high-performance LIBs with high capacity, high speed, and long life. The electrode material determines the overall performance of LIBs. Most importantly, with a high theoretical specific capacity of 2596 mAh g^{-1}, BP will shine in the field of energy storage [24]. Therefore, a large number of theoretical and experimental studies have been carried out on BP and BP-based electrode materials in order to seek new breakthroughs in the field of LIBs.

Given the advantages of BP in the field of energy storage, this paper summarizes and reviews the preparation, passivation, and lithium-ion battery applications of two-dimensional black phosphorus. Firstly, a variety of BP preparation methods are summarized. Secondly, starting from the environmental instability of BP, different passivation technologies are compared. Thirdly, the applications of BP in energy storage are introduced, especially the application of BP-based materials in lithium-ion batteries. Finally, based on preparation, surface functionalization, and lithium-ion battery of 2D BP, the current research status and possible future development directions are put forward.

2. Preparation of 2D BP

In 1914, Bridgman heated white phosphorus to 200 °C in an experiment and obtained BP crystals under a pressure of 1.2 GPa [25], but it did not get people's attention for a long time. Not until 1981, when Maruyama et al. [26] dissolved white phosphorus in bismuth solution in a reaction kettle and kept it for 20 h at 400 °C. After slow cooling, BP with needle-like or rod-like structure was obtained, with a grain size of $5 \times 0.1 \times 0.07$ mm^3. The use of BP as an anode in LIB only gained attention in 2005. Due to its demanding synthesis method, a simple, efficient, and non-toxic preparation method that can be used in industrial production is extremely important for the development of BP.

The traditional preparation method of 2D BP (single layer or less layer) adopts scotch tape mechanical cleavage method from the bulk BP crystal, which has low preparation efficiency and low purity of 2D BP. Therefore, researchers conducted a large number of experiments to explore a variety of basic research preparation methods, including chemical vapor deposition (CVD) [27], exfoliation method [28], phase transition and solvothermal reaction method, and other methods.

2.1. Synthesis of Bulk BP

Mercury reflux is a method to reduce the pressure required for the preparation of BP. In 1955, Krebs et al. [29] first reported a method to prepare BP using metal mercury, which can be synthesized at the pressure of $3.5 \times 10^7 \sim 4.5 \times 10^7$ Pa. BP can be prepared by mixing white phosphorus with metallic mercury and putting it into a pressure vessel and holding it at a certain temperature for several days. Due to the catalytic action of mercury, the activation energy required for the conversion of white phosphorus to BP is reduced, so the preparation pressure is also reduced. Although the mercury reflux method can produce BP under relatively mild conditions, metallic mercury is greatly harmful to the human body and the environment, it takes a long time to prepare, and it is necessary to remove metallic mercury from BP products in the later stage. Therefore, few researchers carry out relevant studies on it.

The bismuth melting method [26] is to prepare BP by dissolving white phosphorus in liquid bismuth and holding it at 400 °C for a long time (20–48 h). Since red phosphorus is insoluble in liquid bismuth and there is a certain risk in direct contact with white phosphorus, Mamoru Baba et al. [30] first heated red phosphorus as the precursor at a specific temperature to vaporize red phosphorus and condense it to produce white phosphorus. Meanwhile, they heated solid bismuth at 300 °C (melting point is 271.3 °C) to make it liquid. BP can be prepared after white phosphorus is dissolved in liquid bismuth at a high temperature and held for a period of time. The advantage of this preparation is the reduced risk of direct contact with white phosphorus. However, in general, the preparation process is relatively complex, the experimental process consumes a large amount of bismuth, the heat preservation time is longer, but also, the need to use strong acid bismuth from BP removal in the product will produce waste liquid in the process of environmental pollution. Thus, the bismuth melting method also was not developed well.

2.2. Exfoliation Methods

At present, the synthesis of single crystal BP usually uses the mineralizer method [27]. A certain proportion of Sn, SnI4, and RP are sealed in a vacuum quartz tube, and the single crystal is synthesized by setting a temperature gradient. In recent years, researchers have used different kinds of stripping methods to strip single crystal BP.

2.2.1. Mechanical Exfoliation

As BP has a similar layered structure to graphite, it can also be used to prepare BP film by referring to the method of mechanically peeling graphite to prepare graphene. The traditional mechanical stripping method is to peel BP repeatedly with transparent tape to achieve a few layers of BP film, as shown in Figure 1A [31]. Finally, with the clear tape residue, the final product is obtained [22,32,33]. Liu et al. [22] used this method to prepare phosphorus with different layers. As can be seen from the AFM image in Figure 1B, the thickness of single-layer phosphating film is 0.85 nm, larger than the theoretical calculation of 0.6 nm. The photoluminescence (PL) spectra (Figure 1C) shows that the energy gap of the monolayer phosphating film is 1.45 eV. In addition, the prepared BP chip with the thickness of 46 nm has a hole carrier mobility of 286 cm^2 V^{-1} s^{-1}. Guan et al. [34] improved the traditional mechanical stripping method of scotch tape. A layer of gold or silver about 10 nm was deposited on the Si/SiO$_2$ substrate to enhance the adhesion between the BP body and the substrate, and then it was stripped. The phosphorescence prepared by the metal-assisted mechanical stripping method has a thickness of ~4 nm, a transverse size of ~50 μm, and a hole carrier mobility of 68.6 cm^2 V^{-1} s^{-1}. It can be seen that phosphorus obtained by the mechanical stripping method has an irregular size, uncontrollable layers, general electrical performance, and low efficiency, which can only be used for the study of infrastructure and performance [35–38].

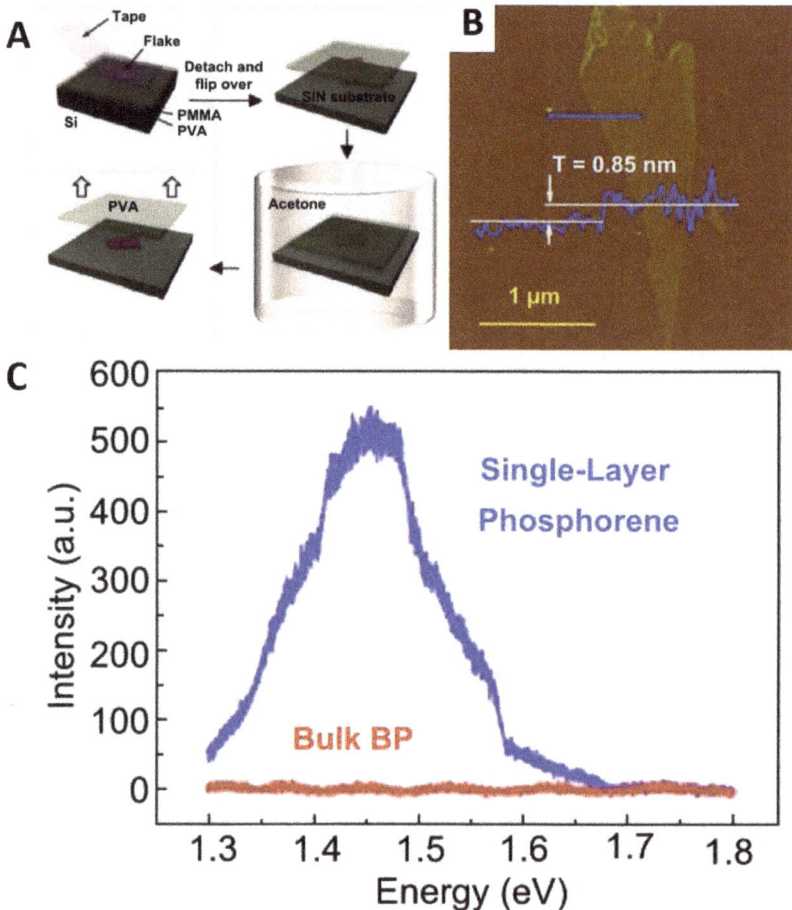

Figure 1. (**A**) Process of preparation of less layer BP by mechanical stripping [31]. (**B**) AFM image of BP prepared by mechanical stripping; (**C**) Photoluminescence spectrum. Reproduced with permission [22]. Copyright 2014, ACS Publications.

2.2.2. Sonication Liquid-Phase Exfoliation

Although the mechanical stripping method required relatively simple experimental conditions, it is labor intensive, time-consuming, has a low yield, and only a single form of phosphorusene can be prepared, so it is only suitable for basic laboratory characterization and research. In contrast, the ultrasonic stripping method can control the ultrasonic power to prepare different forms of BP nanoparticles, such as phosphorusene and BP quantum dots. Due to the advantages of low cost and easy operation, the method is often used for the preparation of nano-BP matrix composites. Brent et al. [39] first reported the study on the preparation of nanometer BP by ultrasonic stripping. They placed BP in N-methylpyrrolidone (NMP), controlled the bath temperature below 30 °C, and obtained BP nanosheets with a size of 200 × 200 nm and a thickness of 3.5–5 nm by continuous ultrasonic stripping for 24 h. Although the prepared nanosheets have high crystallinity, they have poor stability, a time-consuming preparation process, and low yield (less than 10%). In order to improve the stability of the nano-BP and stripping time, Halon et al. [40]

used the solvent for N-cyclohexyl-2-pyrrolidone (CHP). After ultrasonic stripping for 5 h, they achieved a high quality as the multiple centrifugal supernatant fluid layer black phosphorus nanometer film, based on the solvation shell protection principle, and the stability of the preparation of nanometer BP were improved to a certain extent. In order to study the effect of centrifugal rate on the size and morphology of nano-BP, Late et al. [41] centrifuged BP at 3000, 5000, and 10,000 r/min, respectively, with NMP as solvent, and found that the increase of rotational speed was conducive to obtaining small and thin BP nanosheets.

In consideration of environmental friendliness, scale, and simplicity of preparation, Zhao et al. [42] prepared atomically thin BP nanosheets using ionic liquids instead of organic solvents for the first time, and also considered the stripping effect of nine ionic liquids. It was found that 1-hydroxyethyl-3-methyl imidazolium trifluoromethane sulfonate could be used as the stripping solvent to obtain dispersions of BP nanosheets with a concentration of 0.95 mg mL^{-1} (0.4 mg mL^{-1} in NMP), and the dispersions could stably exist in air for one month without obvious polymerization. The introduction of ionic liquid improves the stability and concentration of the dispersion, and is an ideal green stripping agent [43], but the yield of the preparation of BP nanoparticles is still not high, and the price of ionic liquid is more expensive. In order to improve the production rate of BP nanosheets and reduce the production cost of BP nanosheets, Su et al. [44] added an Li2SiF6 assisted intercalation water bath ultrasound for 5 h, and confirmed through comparative experiments that the production rate of BP nanosheets of Li2SiF6 assisted intercalation in DMSO was up to 75%. The BP nanosheets obtained by exfoliation have high purity and high crystallinity. The apparent monolayer thickness is (2.04 ± 0.18) nm, the average number of layers, and the transverse size are about 4 layers and 3.74 mm, respectively. In addition, ultrasonic stripping can prepare BP quantum dots as well as phosphorusene [45].

Although the sonication liquid-phase exfoliation method is simple to operate, not very time consuming, and has a high yield, a variety of organic solvents used will pollute the environment, and organic solvent molecules are easily adsorbed on the final products, affecting their inherent properties.

2.2.3. Electrochemical Exfoliation

The electrochemical stripping method can regulate the voltage size in order to control the morphology of nano-BP, and has advantages of being a simple operation and having a low cost, and it can prepare BP, black phosphate quantum dots, and even the structure of the new nano-BP-perforated BP, 3D BP, a different morphology of nanometer BP. Thus, at present, it is a commonly used method for the preparation of BP.

Depending on the lamellar materials (BP) and the position (cathode, anode, and electrolyte), the electrochemical stripping method can be summarized as the anode stripping method [28], cathodic stripping method [46–49], or electrolytic stripping method [47]. These three kinds of stripping methods have a similar principle: in the selected electrolyte, the applied electric field of the generated gas or ions are inserted into the layer interface between the layers, thus it requires less stripping than even a single layer. Among them, the anode stripping method and electrolyte stripping method are more green and environmentally friendly (inorganic solution is mostly used as the stripping solvent), but the variety of nano-BP prepared is smaller and is easily oxidized, which is not conducive to the application of nano-BP. In contrast, the BP nanoparticles prepared by the cathode stripping method are rich in variety, and oxygen free radicals are not generated during the stripping process, so the BP nanoparticles prepared are relatively stable.

In the process of electrochemical stripping, the bulk BP acts as the working electrode and generates current by applying voltage. Due to the joint action of current and electrolyte on the layered structure, the bulk BP will be stripped into phosphorusene. The electrochemical stripping efficiency of BP depends on the selection of voltage and electrolyte. Erande et al. [28] used a massive BP crystal as anode, platinum wire as the counter electrode, 0.05 M Na_2SO_4 as electrolyte, and applied a voltage of +7 V, equivalent to 1 mA current. After 25 min of reaction, the solution turned light yellow, and after 90 min, the power was turned off. Finally, the solution was centrifuged to obtain atomic thin layer of phosphorusene, and the yield could reach more than 80%. However, the BP nanosheets after electrochemical stripping showed a wide thickness range (1.4–10 nm, corresponding to 3–15 layers). Due to the uneven size of the nanosheets, the BP mobility was only 7.3 cm^2 V^{-1} s^{-1}. Li et al. [47] use different electrolytes (0.001 M Tetraalkylammonium (TAA) salts and DMSO) and voltage (−5 V low voltage) to achieve rapid expansion and stripping of large BPS in a few minutes. The prepared BP has good size uniformity and electrical properties, with an average thickness of five layers, an average transverse area of 10 μm^2, and a hole carrier mobility of 100 cm^2 V^{-1} s^{-1}. In addition, H_2SO_4 [50] and tetrabutylammonium hexafluorophosphate (TBA) [46] have also been reported as electrolytes for BP electrochemical stripping.

Therefore, high quality and uniform phosphorusene can be prepared by selecting suitable electrolyte and voltage. Electrochemical stripping opens up new possibilities for industrial production of phosphating materials.

2.3. Chemical Vapor Deposition (CVD) Method

The CVD method is widely used in the preparation of 2D materials. It has a bottom-up method to manufacture large size and high-quality 2D materials, including graphene, hexagonal boron nitride, transition metal dichalcogenides (TMDs), etc. [14,51,52]. Because of this, the use of the CVD method to prepare 2D materials has attracted the attention of scholars.

In 2016, the direct synthesis of 2D BP via CVD was reported by Smith et al. [53]. The experimental device is shown in Figure 2A,B. The temperature and pressure are monitored during the experiment, and a typical ladder diagram is shown in Figure 2C. Through this method, amorphous red phosphorus was directly grown on the silicon substrate, and the thickness of the sample was about 600 nm. This work finally obtained four layers of BP film through the experiment, but also improved the safety of the experiment, and played a role in promoting the mass production of BP film. In the same year, Xu et al. [54] used the gas phase growth strategy of epitaxial nucleation design and further transverse growth control to directly grow large-size BP films on insulating silicon substrates. The maximum transverse size can reach the level of millimeters, and the thickness can be regulated from several nanometers to several hundred nanometers. The experimental temperature gradient is shown in Figure 2F. After testing and analysis, the BP films obtained have excellent electrical properties. The field-effect mobility and Hall mobility at room temperature are over 1200 cm^2 V^{-1} s^{-1} and 1400 cm^2 V^{-1} s^{-1}, respectively, which are similar to the films stripped from BP bulk crystal materials [55]. As shown in Figure 2D,E, by comparing the microstructure of the layered BP film grown on Si/SiO$_2$ substrate with that of the conventional layered BP crystal, it is found that the BP film synthesized by the substrate has a unique layered structure, which is quite different from that of the conventional BP crystal film. At the same time, the prepared BP also shows excellent optical properties in the infrared band, with enhanced infrared absorption and photoluminescence characteristics.

The gas phase growth method provides a new way for the controllable preparation of large-area and high-quality BP thin films, which is expected to be widely used in general optoelectronic devices and integrated circuits [56,57]. Liu et al. [58] achieved high quality BP growth using an efficient short distance transport (SDT) growth method with a yield of 98%. This method does not need to set temperature gradient and is a short distance transportation growth method at uniform temperature. Lange et al. [59] proposed to use the polyphosphate Au3SnP7 as the nucleation site, and realized the large-scale growth of BP film with high crystallinity on silicon and other insulating substrates for the first time.

However, there are still some problems in the process of the experiment, such as the difficult sealing of the ampoule, the formation of white phosphorus in the reaction process, and the harsh reaction conditions (not reaching the appropriate temperature and pressure will lead to the rupture of the ampoule).

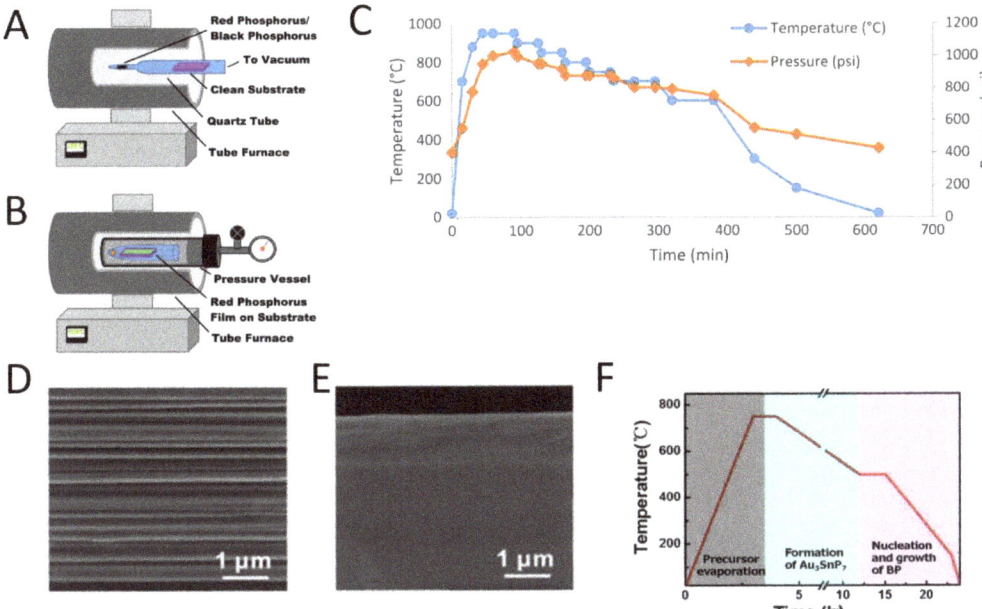

Figure 2. Synthesis of 2D BP nanosheets by the CVD method: (**A**) Schematic of amorphous red phosphorus thin film growth from vapor deposition of red phosphorus powder/BP piece; (**B**) Standard temperature/pressure ramp for growth of substrate BP (SBP) from the red phosphorus thin films. Temperature recorded was that of the tube furnace; (**C**) Schematic for growth of SBP on substrate from amorphous red phosphorus thin film in pressure vessel reactor [53]. (**D**) Comparison of microstructure between layered BP films grown on Si/SiO$_2$ substrates and conventional compact extruded layered BP bulk crystals; (**E**) Conventional bulk BP crystals; (**F**) Diagram of experimental reaction temperature gradient [54].

2.4. Phase Transition and Solvothermal Reaction Method

In 2018, Tian et al. [60] successfully prepared small BP nanosheets using a simple, scalable, and low-cost method, the key factors of which are the choice of solvent and temperature, as shown in Figure 3A. Using white phosphorus as raw material and ethylenediamine as solvent, small layer BP was successfully prepared at 60–140 °C. As shown in Figure 3B, typical Raman peaks at 360.2, 437.5 and 464.6 cm−1 can be observed by Raman spectroscopy, corresponding to the characteristic peaks A1g, B2g, and A2g [61] of BP, which further proves the formation of BP. The solvothermal method can improve the stability of BP nanosheets, which provides conditions for rapid development and application of BP. Zhao et al. [62] used ammonium fluoride to reduce the surface activation energy of red phosphorus. Assisted by NH4F and based on the phase transformation of phosphorus and Gibbs free energy theory, a new type of 2D polycrystalline BP nanosheet was prepared for the first time by mild method, as shown in Figure 3C. NH4F can make the surface of phosphorus smooth, and in the reaction process, phosphorus is cut into small pieces by water, which plays an important role in the structure and morphology characteristics of BP cambium.

In 2016, Zhang et al. [63] demonstrated a sublimation-induced approach to prepare few-layer 2D holey phosphorus-based nanosheets from bulk RP under a wet-chemical solvothermal reaction. The mechanism of this approach includes solid–vapor–solid transformation driven by continuous vaporization condensation process, as well as sub-sequent bottom-up assembly growth.

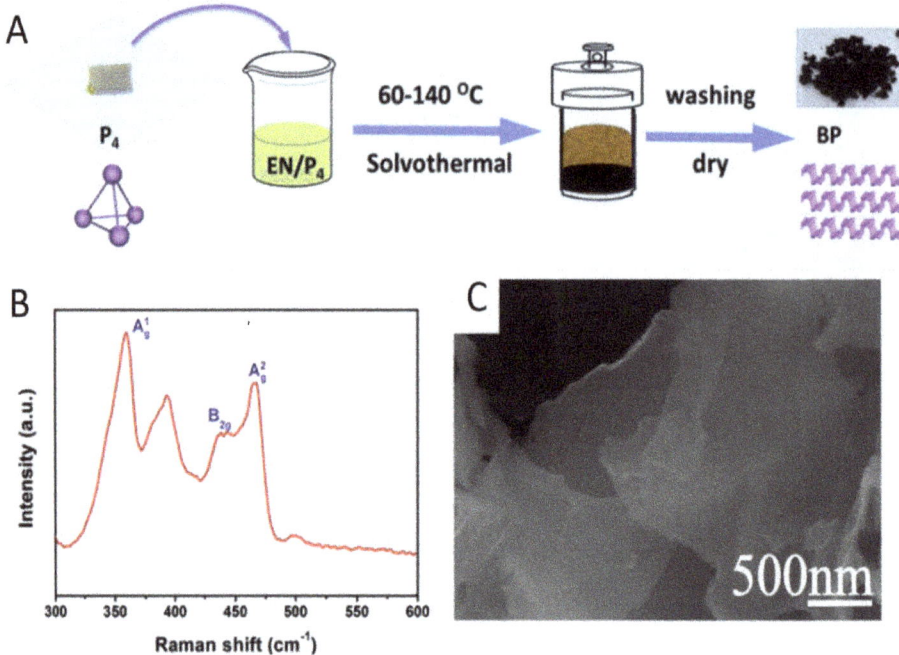

Figure 3. (**A**) Layer BP was synthesized from white phosphorus and ethylenediamine. (**B**) Raman spectrum. The excitation wavelength is 633 nm [60]. (**C**) SEM image of BP nanosheets synthesized by ammonium fluoride. Reproduced with permission [62]. Copyright 2016, Wiley.

Currently, the bottom-up method development is still at the initial stage with quite a lot of challenges to be solved. All reports related to the direct growth of thin-layer BP must start from RP, followed by phase transition.

This section summarized some classical methods of preparing 2D BP and compared their characteristics, as shown in Table 1.

Table 1. Comparison of some classical preparation methods and characteristics of 2D BP.

Method	Experiment	Thickness	Characteristic	Ref.
Chemical vapor deposition	—	~4 layers	Flexible size, a variety of lateral size samples	[53,54,58]
Pulsed laser deposition	—	2–8 nm	Flexible size with limited carrier mobility due to disordered structure	[64]
Mechanical exfoliation	Sticky-tape	<7.5 nm	High carrier mobility with low production yield limit	[65,66]
	Metal-assisted	~4 nm	Long time consuming, low yield, uncontrolled size	[34]
Sonication liquid-phase exfoliation	Sonic exfoliated in NMP	3–5 layers	A 200 × 200 nm^2 lateral dimension	[39]
	Sonic exfoliated in formamide	~3 layers	A 50–300 nm lateral dimension, 38% yield	[67]
Electrochemical exfoliation	Two electrode system (Pt and bulk BP)	3–15 layers	A 0.5 to 30 μm lateral dimension, yield excess 80%, non-uniform size	[28,68]
	Electrolyte TAA and −5 V low voltage	~5 layers	Fine dimensional uniformity and electrical properties, average cross area of 10 μm^2	[47]
Solvent hot method	White phosphorus as raw material, ethylenediamine as solvent	—	The synthesis is stable, and the transverse size is 0.8–1.0 μm	[60]
	Red phosphorus as raw material, ammonium fluoride as solvent	~2 layers	Large scale preparation, low cost	[62]

3. Passivation of 2D BP

Compared with graphene and other traditional 2D nanomaterials, BP shows great potential in chip manufacturing, photoelectric sensing, biomedicine, and other fields because of its lamellar tunable direct band gap [63], high carrier mobility, and anisotropic thermoelectric properties [69]. In addition, BP with few layers also has excellent mechanical strength, excellent thermal stability, and high specific surface area, which makes it a potential nanofiller for preparing ideal composite materials [70].

However, in the practical application of BP, in addition to the lack of large area of high-quality sample synthesis method, its easy degradation in air is also one of the main shortcomings hindering the development of BP. Related studies [71,72] showed that the reasons for the unsteady layer BP structure are that each phosphorous has five valence electrons ($3s^2 3p^3$), three electronic distribution on three 3p orbital, one lone pair electron distribution in 3p on a 3s orbital orbit of electron, and three adjacent covalent bindings of phosphorus atoms, and on the 3s orbital electrons are retained these 3s orbital lone pair electrons that have high reaction activity with oxygen and form the P_xO_y group, leading to rapid degradation of BP under environmental conditions. Therefore, how to passivate the 3s orbital lone pair electrons that prevent phosphorus atom from reacting with oxygen should be the key to improve the stability of BP.

3.1. Ion Modification

The principle of metal ion modification is to start from the direction of lone pair electrons occupying BP, so that BP lone pair electrons can be combined with metal cations, thus preventing BP from reacting with oxygen. In 2017, Guo et al. [38] calculated by density functional theory (DFT) that the binding energy of silver ion and BP was 41.8 cal, indicating that free Ag^+ could adsorb on the surface of BP and the interaction between them was strong enough. Therefore, Ag^+ was selected for surface modification of BP. The modification mechanism is shown in Figure 4A. The relationship between Ag+ and phosphorus atoms is one-to-many rather than one-to-one [73]. Compared with the oxidation degree of the modified BP nanosheet exposed to air (Figure 4B), it was found that the electrical performance of the BP transistor modified based on Ag^+ was significantly improved, with a hole mobility of 1666 cm^2 V^{-1} s^{-1} and a switching ratio of 2.6×10^6, which were two and 44 times higher than that of the bare BP transistor, respectively. In addition to Ag^+, the effects of Mg^{2+}, Fe^{3+}, and Hg^{2+} on crystal stability and transistor performance were also studied. The results show that different ions have different extranuclear electrons, which improve crystal stability and transistor performance in different degrees [74]. In 2018, Feng et al. [75] used Poly Dimethyl Diallyl Ammonium to carry out surface modification of BP (Figure 4C). The PDDA is selected to spontaneously and uniformly adsorb on the surface of few-layer BP via electrostatic interaction (Figure 4D). The positive charge-center of N atom of the PDDA, which passivates the lone-pair electrons of P, plays a critical role in stabilizing the BP. Meanwhile, the PDDA could serve as hydrophilic ligands to improve the dispersity of exfoliated BP in water [76]. The thinner PDDA-BP nanosheets can stabilize in both air and water even after 15 days exposure. Finally, the uniform PDDA-BP-polymer film was used as saturable absorber to realize passive mode-locking operations in a fiber laser, delivering a train of ultrafast pulses with the duration of 1.2 ps at 1557.8 nm. This work provides a new way to obtain highly stable few-layer BP, which shows great promise in ultrafast optics application [77].

In 2019, Zhang et al. [78] first tried to stabilize BP with cationic cisplatin drugs. They reacted cisplatin, oxaliplatin, and cisplatin oxidized by H_2O_2 with BP, respectively, and observed no significant change in binding energy. Subsequently, the two chloride ions of cisplatin were replaced by nitrate to improve the positive of cisplatin, and the binding energy of Cisplatin and BP reached 133.0 eV. After modification, the surface morphology of PT@BP did not change significantly after 24 h exposure in water and PBS. Finally, the stability of exposed BP and PT@BP in a longer period was further detected. As shown in Figure 5A, after 10 days in environmental conditions, the surface morphology of exposed

BP completely disappeared, while PT@BP did not change significantly, which proved that cisplatin modification played a good role in the protection of BP. In 2021, Tofan et al. [79] developed a low-cost, high-efficiency liquid phase experimental method for surface modification of BP using 13 Group Lewis acid and demonstrated its effectiveness in environmental stability and regulation of electronic properties. Because of the electrophilic property of the reagent, Lewis acid completely inhibits the N-type conductivity [80]. The reagent can not only form Lewis adduct with BP surface, but also capture electrons at high gate voltage state. Three factors determine the effectiveness of Lewis acid in protecting BP from environmental oxidation: the electrophilicity of the Lewis acid, the Pearson hard-soft match between the acid and the phosphine, and the steric bulk of the ancillary ligands. Therefore, the use of 13 group halides (such as $AlBr_3$ or $GaCl_3$) can be the strongest adduct, as shown in Figure 5B. Subsequently, the samples were exposed to air, and when exposed to the environment for 3 h, BP showed visible degradation, while Lewis acid-treated BP was significantly more stable, as shown in Figure 5C.

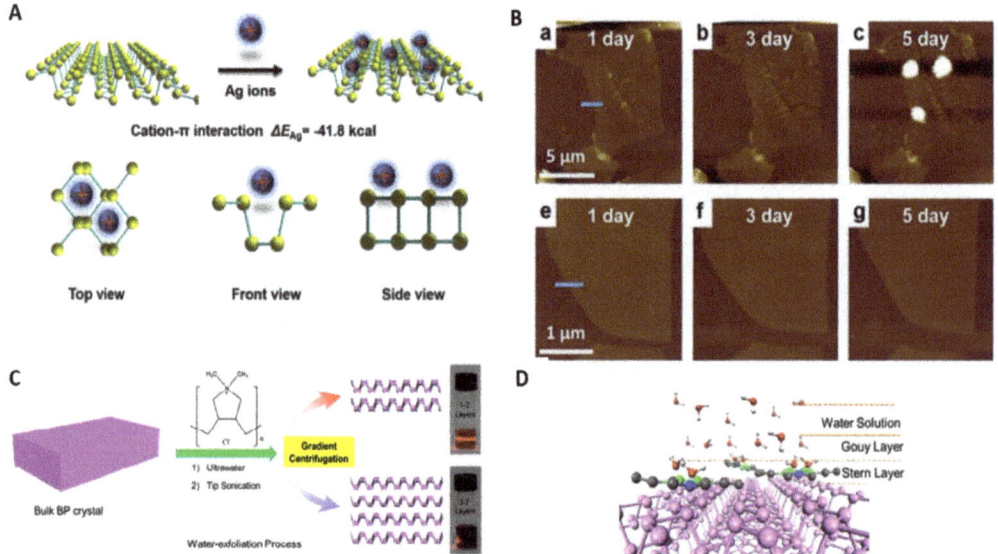

Figure 4. (**A**) Schematic for Ag^+ adsorbing on BP and their three different views; (**B**) AFM images of an unmodified BP sheet (top) and Ag^+—modified BP sheet (bottom) exposed to air for one day, three days, and five days. Reproduced with permission [38]. Copyright 2017, Wiley. (**C**) Schematic of the fabrication process of exfoliated few-layer BP in water by PDDA, Tyndall effect of BP colloid solution via centrifugation at 15,000 rpm (top) and 11,000 rpm (down) with bulk BP crystal (10 mg). (**D**) Schematic of adsorption process for PDDA-BP nanosheets in water. Reproduced with permission [74]. Copyright 2018, ACS Publications.

3.2. Coating

The current modification of BP mainly involves coating its surface to prevent contact with oxygen and water in the air. In recent years, there are many kinds of coatings, such as ionic liquid [81], polymer [82], organic matter [83–85], and layered inorganic oxide [86–89]. Two-dimensional BP can be modified by organic matter, ionic liquid, inorganic oxide, and others, coated on the surface of BP, so that BP does not contact with oxygen and water in the air, so as to achieve the effect of BP passivation. Li et al. [87] put forward a kind of convenient, environmental protection, widely used to inactivate the BP method of nanometer sheet, 3-amino propyl-triethoxy silane, and methyl triethyl silane hydrolysis condensation on the surface of BP form hydrophobic shell and on the physical layer of BP nanosheets to prevent contact with water and oxygen, and then introduced hydrophilic silica shell. Thus, BP is hydrophobic. As shown in Figure 6A, within 15 days, the degradation degree of exposed BP reached 62.6%, while the degradation degree of BP suspension modified by SiO$_2$ was only 33.7%. It can be proved that the BP modified by SiO$_2$ has hydrophobicity, which can effectively shield the water on the surface of BP and slow down its degradation. Liang et al. [85] proposed passivating BP in chloroform using a self-assembled monomolecular layer of h6-methylenedi amine (HMA) to eliminate its instability in oxygen and moisture conditions, as shown in Figure 6B. HMA coating not only keeps the original honeycomb structure of BP, but also has good electrical conductivity, and enhances the stability and dispersion of BP in aqueous solutions. Its stability allows BP to remain in its original form in aqueous solutions for more than a month.

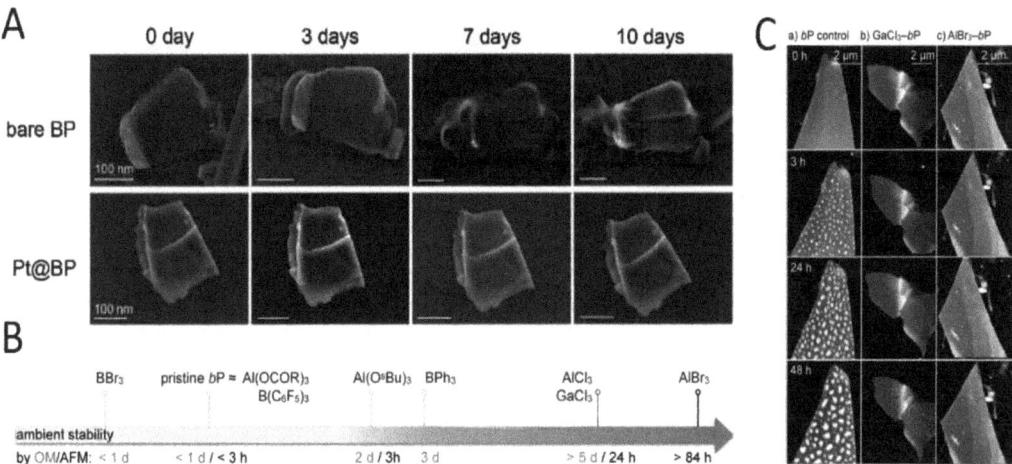

Figure 5. SEM images of bare BP and Pt@BP exposed in water for (**A**). Reproduced with permission [78]. Copyright 2019, ACS Publications. (**B**) Commercial group 13 Lewis acids investigated for the passivation of BP nanoflakes, ordered by their relative effectiveness according to optical and atomic force microscopy (R = C$_{17}$H$_{35}$); (**C**) AFM (**a**–**c**) analysis of control-BP, GaCl$_3$-BP, and AlBr$_3$-BP after 0 to 48 h of ambient exposure. Reproduced with permission [79]. Copyright 2021, Wiley.

In 2017, Fonsaca et al. [82] reported a simple and effective method to modify BP by preparing conductive polymer polyaniline (PANI) nanocomposite to passivate BP, using a liquid/liquid interface method to prepare highly stable, uniform, independent polyaniline covered BP film, which has high stability in environmental conditions. It was stable in environmental conditions for 60 days, while unmodified bare BP was degraded in only three days [90]. As shown in Figure 6C, it was found by optical image and Raman characterization that BP/Polyaniline nanocomposites could be stably maintained for about 20 days under environmental conditions. After that, due to the absorption of water, some physical changes occurred on the surface of the composites, such as volume expansion and irregular surface reflection hindering the focusing of the microscope [91,92]. After 30 days, although the morphology of the thin sheet changed, its Raman characteristics remained unchanged, indicating that only part of the BP on the surface was oxidized, which also indicated that the conductive properties of the material remained unchanged during this period of time. It was not until 60 days later that the BP/Polyaniline nanocomposites lost the Raman characteristics. This modification method is highly effective for the passivation of small layers of BP, increasing the service life of BP by about 600%. In 2020, Maria Caporali et al. [93] coated BP with colloidal nickel nanoparticles, and after Raman and XPS tests, the oxidation rate of BP modified by nickel nanoparticles decreased by more than three times.

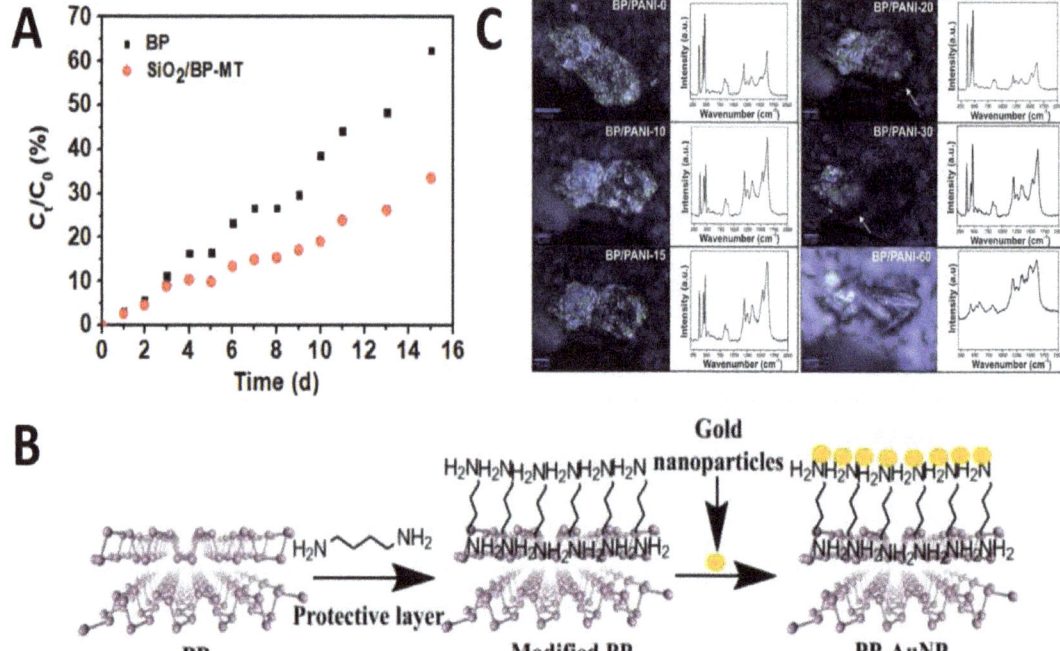

Figure 6. (**A**) The exposed BP and modified BP degraded over time. Reproduced with permission [87]. Copyright 2020, The Royal Society of Chemistry. (**B**) HMA is coated on BP and attached to the surface with gold for biosensors. Reproduced with permission [85]. Copyright 2018, Elsevier. (**C**) Optical images and Raman spectra (λ = 532 nm) of BP/PANI nanocomposite deposited over a glass substrate after 0, 10, 15, 20, 30, and 60 days of exposure to ambient conditions. Reproduced with permission [82]. Copyright 2017, Springer Nature Limited.

3.3. Doping

Doping has been proved to be a simple and effective method to adjust the intrinsic properties of 2D materials. For BP, by introducing heteroatoms, the 2D structure can be adjusted to improve its physical and chemical properties, including instability, and different doping types and ways can improve the performance of BP in some respects. In 2017, Xu et al. [94] developed a stable complementary metal oxide semiconductor (COMS) compatible electron doping method, which is realized with the strong field-induced effect from the K^+ center of the silicon nitride (Si_xN_y), as shown in Figure 7A. By doping Si_3N_4, BP remains stable for more than a month, and the electron transfer efficiency of BP can also be improved. This work provides a promising N-dope strategy for BP as well as other 2D semiconductors and paves the way toward high-performance BP-based complementary logic electronics, light-emitting diode, photovoltaic devices, and so on. Lv et al. [95] proposed a sulfur doping to inhibit the degradation of BP, and the S-doped BP field effect transistors (FETs) were more stable under environmental conditions, with carrier mobility reduced from 607 to 470 $cm^2\ V^{-1}\ s^{-1}$ (still 77.4%) after 21 days of exposure in air. Environmental stability is an important characteristic of semiconductor devices [96,97]. In order to study the stability of S-doped BP FETs, the exposure time was increased, and the transmission curve was tested at a relatively stable humidity of 45–50% and temperature of 25 °C, as shown in Figure 7B. Compared with undoped BP FETs, the drain-source current I_{ds} decreases much more slowly due to the gradual increase in resistance due to degradation.

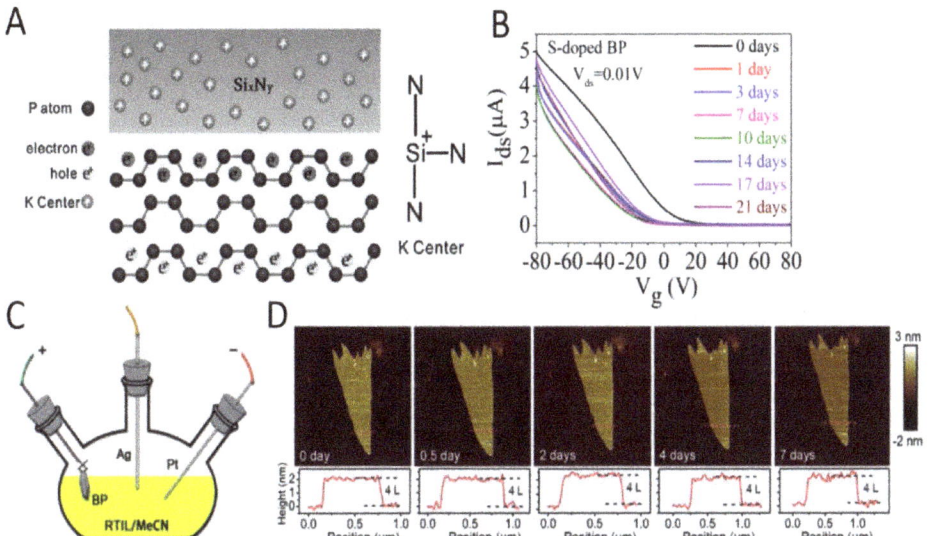

Figure 7. (**A**) Scheme of the field-induced n-doping of BP with Si_xN_y. Reproduced with permission [94]. Copyright 2017, Wiley. (**B**) Transfer curves of S-doped BP FETs at fixed drain−source voltage (V_{ds}) of 0.01 V after exposure times of 0, 1, 3, 7, 10, 14, 17, and 21 days. Reproduced with permission [95]. Copyright 2018, ACS Publications. (**C**) Schematic of the experimental setup. (**D**) AFM images showing evolution of the morphology of a fresh few-layer FP under persistent exposure to ambient conditions from 0 days (no exposure) to 7 days. Height variations of the FP along the positions indicated by the red dashed lines for different exposure periods are shown below the images. Reproduced with permission [98]. Copyright 2018, Wiley.

Wang et al. [99] studied the relationship between the electronic structure and magnetic properties of BP doped with Si and S by first-principle calculation. The results show that the stability of doped phosphorene could be improved by increasing the size of heteroatom or by decreasing the doping concentration. In the absence of spin polarization, the band structure of Si and S doped phosphorene always exhibits a metallic state, indicating that the band gap is not sensitive to the plane size and doping amount of the supercell. The significance of BP doping is revealed through theoretical research. On the basis of theoretical research, Zheng et al. [100] proposed a surface modification method for small layers of BP by high vacuum Al in-situ deposition, which significantly increased the electron mobility. The results show that the covalent bond is formed between BP surface layer and Al atom, and the electron mobility is greatly increased by more than six times. Because of its high electron and hole mobility, Al-doped BP can be used as high-performance logic devices or other functional electronic and optoelectronic devices. Wang et al. [101] studied the stability and electronic structure of al doped low layer BP prepared by ALD method. Al doping not only improved the stability of BP, but also improved the threshold voltage and electronic mobility of BP-based FET. Tang et al. [98] prepared large-scale F-doped phosphorusene (FP) using a simple one-step ionic liquid-assisted electrochemical stripping route. The experimental device is shown in Figure 7C. FP inherits the anti-oxidation and anti-hydration properties of fluorine atoms with high electronegativity, so that it has air-stable photothermal properties and can be stored for more than a week, as shown in Figure 7D.

This section summarized some passivation methods and their passivation effects of 2D BP in recent years, as shown in Table 2.

Table 2. Passivation method and effect comparison of 2D BP.

Passivation Technique	Coating Method	Oxidation Time	Ref.
10 nm Al_2O_3 thin film	ALD	>90 days	[102]
Ag^+ adsorption	The few layers of BP were transferred to Si/SiO_2 wafers with PDMS film as medium, and placed into the mixed solvent of NMP and $AgNO_3$	>5 days	[38]
PDDA hydrophilic ligand adsorption	PDDA and BP spontaneously and uniformly adsorbed by electrostatic action	>15 days	[73]
Cationic cisplatin	Reacted cisplatin, oxaliplatin, and cisplatin oxidized by H_2O_2 with BP	>10 days	[78]
Group 13 Lewis acids	Under inert gas, the few layers of BP were stripped onto Si/SiO_2 wafers and adsorbed with group 13 Lewis acid	>5 days	[79]
3-aminopropyl triethoxysilane and methyl triethyl silane	BP nanosheets were hydrolyzed and condensed with 3-aminopropyl triethoxysilane and methyl triethyl silane to form hydrophobic shells	>15 days	[87]
Hexamethylene-diamine	HMA monomolecular layer passivation of BP in chloroform	>30 days	[85]
PANI nanocomposites	A highly stable and uniform PANI covered BP film was prepared by liquid/liquid interface method	>20 days	[82]
Complementary metal oxide semiconductor (CMOS)	A strong field effect produced by the K^+ center of silicon nitride (Si_xN_y)	>30 days	[94]
Sulfur doping	BP was synthesized under high temperature and high pressure, and then mixed with 1wt% sulfur	>21 days	[95]
Fluoride ionic liquid	Electrochemical stripping and synchronous fluorination	>7 days	[98]

4. Applications of 2D BP based LIBs

LIBs work by converting electrical and chemical energy, or charging and discharging, between their anode and cathode [103–106]. Therefore, in theory, as long as the chemical structure of positive and negative active materials of LIBs is stable and the conversion cycle of electric energy and chemical energy is carried out, the battery can be used for a long time. The performance of LIBs mainly depends on the stability, conductivity, and conversion rate of anode and cathode materials [107–110]. At present, the cathode materials of LIBs are mainly lithium iron phosphate [111], nickel cobalt manganese ternary compound [112], and the cathode materials are mainly graphite and carbon [113,114]. Anode materials, especially active parts, play an important role in LIBs, so they are now the main focus. It is mainly because of the key technological breakthrough of LIBs anode material, which can greatly improve the energy density and power density of LIBs, and that plays an extremely important role in the performance of LIBs. Lithium example batteries typically consist of an anode, a cathode, and an electrolyte, where the charge flow is generated by the intercalation and delamination of lithium-ions between the anode and cathode. As is known to all, the selection of electrode materials plays an important role in the performance of LIBs.

The theoretical capacity of BP is 2596 mAh g^{-1}, which provides the basis for the capacity of LIBs. At the same time, BP has good electrical conductivity, which provides the possibility for rapid charge and discharge preparation of high-rate LIBs. These characteristics show great potential in the preparation of high-performance LIBs. On the other hand, BP's unique fold structure provides more space for the insertion of Li^+, and Li atoms can combine with P atoms to form strong bonds. Qiu et al. [115] believed that this process can be divided into BP-LiP-Li_2P-Li_3p, which provided a low diffusion energy barrier (0.08 eV) for the mobility of Li atoms [116,117]. Zhang et al. [118] calculated theoretically that the diffusion mobility of Li^+ in BP along ZZ direction was about 10^7–10^{11} times that of AC direction. This directional diffusion mobility is far superior to other 2D materials such as graphene and MoS_2, which allows BP to have ultra-fast charge/discharge characteristics. In addition, the 2D structure of BP has high reversibility in the process of lithium intercalation. In other words, during the charging and discharging process, the volume change of monolayer BP is only 0.2%, while that of bulk BP is 300% [119,120]. The thin layer BP becomes one of the most promising candidate materials for LIBs in the future due to its large storage space, extremely fast diffusion rate of Li$^+$, and stable and reversible structure.

On this basis, the experimental study of BP as anode material of LIBs was carried out. Castillo et al. [121] prepared a small layer BP with an average transverse size of 30 nm and an average thickness of 13 layers by liquid phase stripping method as the anode of LIBs. At the current density of 100 mA g^{-1}, the electrochemical performance of the prepared BP was tested with a half cell. The initial capacity was 1732 mAh g^{-1}. However, the electrode showed large capacity fading in the first 10 cycles. After 100 charge and discharge cycles, the specific capacity attenuates to 480 mAh g^{-1}. Zhang et al. [122] also prepared several layers of BP (5–12 layers) by liquid phase spalling method. However, the BP electrode with few layers exhibits the same rapid capacitance decay, resulting in a coulomb efficiency (CE) of only 11.4% and a reversible specific capacity of only 210 mAh g^{-1}, which are not satisfactory. They suggest that these poor properties may be due to the thicker layers of BP, resulting in large volume changes during the cycle. Therefore, many scholars proposed the construction of BP composite materials to overcome the volume expansion, low coulomb efficiency, and low reversible capacity of BP in LIBs.

4.1. BP/C Composite Materials

In 2014, Sun et al. [123] found that four different BP/C composites were prepared by high-energy mechanical milling using different carbon sources (graphite, graphite oxide, carbon black, fullerene). The effects of four different P-C bonds on chemical properties were studied.

The coherent P-C bond formed in BP-G composite material provides high capacity and good cycling capability for battery performance. Figure 8A evaluates the electrochemical lithium storage properties of RP, BP, and graphite mixtures (BP/G) without high energy mechanical milling, and BP-G composites chemically bonded by high energy mechanical milling by constant current charge–discharge measurements. CP and CP/G are the specific capacities calculated according to the weight of BP and BP-G composite materials, respectively. During the lithium process, when the potential drops from the initial 2V to 0.78, 0.63, and 0 V. The multistep reaction is attributed to the transformation of BP-LiP-Li$_2$P-Li$_3$P. Despite the initial CP charge of BP/G capacity electrode of 2479 mAh g^{-1}, the coulomb efficiency caused by the first cycle was only 58% resulting in the loss of electrical contact with G. BP and G (BP-G) through P-C bond, when circulating in the voltage range of 0.01 and 2.0 V, has an initial CP specific capacity of 2786 mAh g^{-1}. The reversible CP specific capacity is about 2382 mAh g^{-1}. The hysteresis (δ Ep) of the discharge and charging platform is reduced from 0.43 V to 0.31 V. The decrease of δ Ep indicates that PC bond improves coulomb efficiency due to better connection between particles in the process of delithium and lithium embedding with large volume variation. As shown in Figure 8B, in the constant current charge–discharge curve, it can be seen that BP-G has the highest specific capacity compared with four different P-C bonds. In 2017, Jiang et al. [124], for the first time, directly synthesized BP on conductive carbon paper (BP-CP) by efficient thermal evaporation deposition method to form black phospho-carbon composite material (BP-CP), which was successfully used as the anode material of LIBs, and evaluated the electrochemical lithium storage performance of RP, BP-CP, and bare CP by constant current charge–discharge measurement (see Figure 8C). The discharge/charge capacities of bare CP and RP are 31.9/318.9 and 1668.7/852.4 mAh g^{-1}, respectively, which are too small to be used as anode materials for LIBs. From the semiconductor point of view, BP has a low genetic conductivity. In order to improve the electrochemical performance of BP, BP samples grown directly on carbon paper may have better conductivity, which can promote the electrochemical behavior of BP in the discharge/charging reaction with lithium-ions. Compared with bare CP and RP, the charge and discharge capacities of BP-CP at 0.1 C current density increased to 2219.1 and 2168.8 mAh g^{-1}, respectively (Figure 8C).

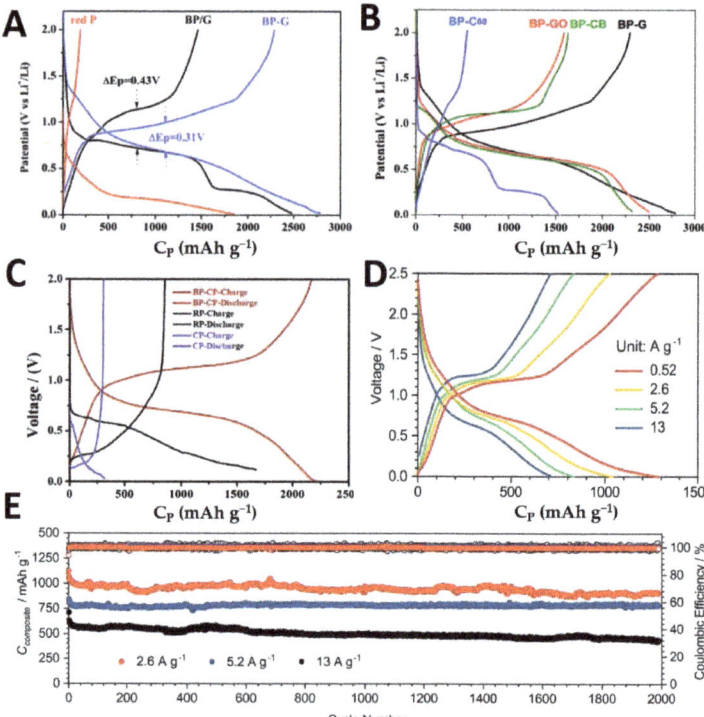

Figure 8. (**A**) First charge–discharge curves of Red P, BP/G, and BP-G at 0.01–2.0 V current density. (**B**) First charge–discharge curves of BP-C60, BP-GO, BP-CB, and BP-G at 0.01–2.0 V current density. Reproduced with permission [123]. Copyright 2014, ACS Publications. (**C**) The charge–discharge profiles of the initial cycle in the voltage range of 0–2 V. Reproduced with permission [124]. Copyright 2018, Elsevier. (**D**) Constant current charge and discharge and charge distribution under different current densities. (**E**) Cyclic performance of BP-G /PANI at current densities of 2.6, 5.2, and 13 A g^{-1}. Reproduced with permission [125]. Copyright 2020, Springer Nature Limited.

In 2020, Jin et al. [125] developed the preparation of polyaniline coated BP-graphite by in-situ polymerization of BP as an active anode for high-rate and high-capacity LIBs by ball milling a mixture of polyaniline and graphite. The formation of covalent bonds with graphite and carbon inhibits the boundary reconstruction of layered BP particles and ensures the rapid entry of Li$^+$ into the open edge. The covalently bonded BP-graphite particles were coated with expanded polyaniline to form a stable solid electrolyte interface phase, which inhibited the continuous growth of lithium fluoride and carbonate with poor conductivity, and greatly reduced the volume expansion problem of BP when used in LIBs. The expanded polyaniline induced the doping of Li$^+$ and proton in the polymer matrix, absorbing HYDROGEN fluoride and promoting the charge transfer between electrode and electrode electrolyte interface. According to the reversible capacity test at different current densities, the reversible capacity can reach 1300 mAh g^{-1} at 0.52 A g^{-1}, as shown in Figure 8D. Reversible capacity of 440 mAh g^{-1} at high current density of 13 A g^{-1} in 2000 cycles (Figure 8E). When the load mass is 3.6 mg cm^{-2}, the capacity is maintained at 750 mAh g^{-1} at 0.52 A g^{-1}.

4.2. Other BP Composite Materials

Due to the large volume change and low cycle life of BP when it is used as the anode of LIBs, Jin et al. [126] designed and prepared a BP nickel-cobalt alloy composite

(BP/NiCo MOF), which is composed of Ni^{2+} and Co^{2+} in benzenedicarboxylic acid (BDC). The hydroxyl group in BDC^{2-} can not only chelate with metal ions, but also bond with BP to form a stable 2D hybrid structure. BP/NiCo MOF electrode can provide abundant redox active sites to ensure the lithium storage capacity of the electrode. The nanostructures and porous structures after two days in ambient conditions not only retained a good charge transfer, but also buffered multi-loop volume expansion, so that the materials have good cyclic stability and excellent velocity performance, and ultimately improved the electrochemical performance. At the rate of 0.1 A g^{-1}, the capacity of the first cycle charge and discharge can reach 2483 mAh g^{-1}. After five cycles, the capacity of the cycle charge and discharge still remains at 1512 mAh g^{-1}, as shown in Figure 9A. After analysis, the irreversible capacity loss is due to the decomposition of electrolyte to form solid electrolyte interface (SEI) and the partial capture of Li^+ at the active site of hybrid structure. As shown in Figure 9B, by comparing the charge–discharge curves under different current densities, its cyclic charge–discharge capacity can still reach nearly 500 mAh g^{-1} under the condition of high current density 3 A g^{-1}. The composite material has high stability, a long cycle life, and good rate performance due to the synergistic effect of low layer BP and binary nickel-cobalt alloy structure. It provides a new way for the design and engineering of high-performance lithium storage systems.

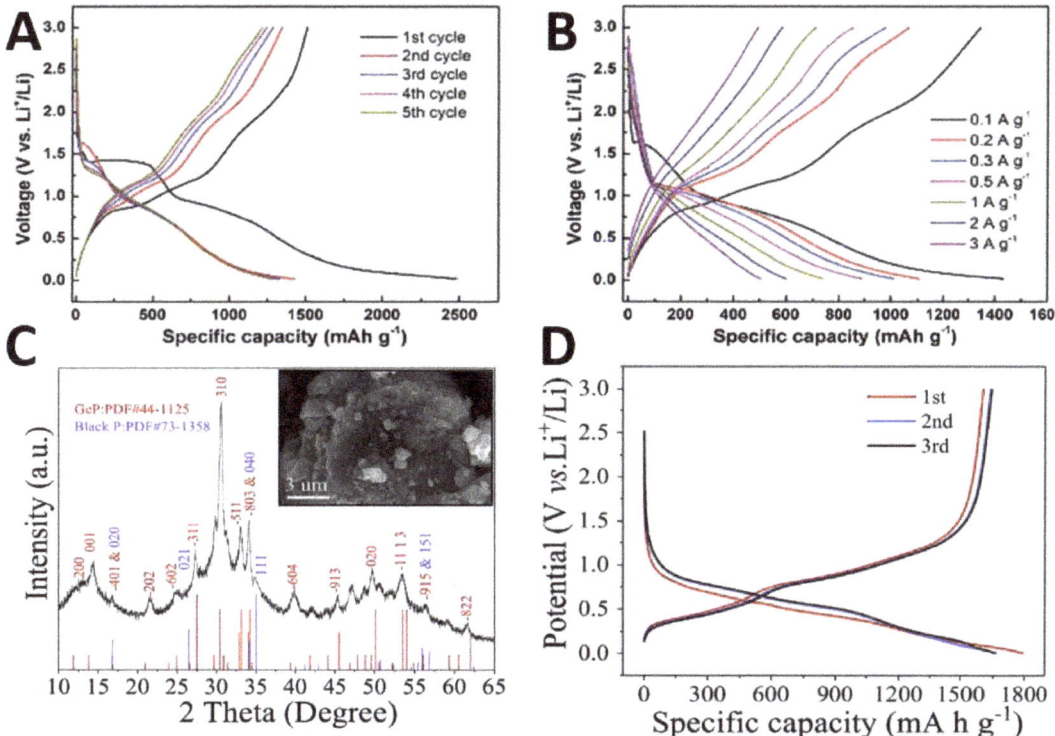

Figure 9. (**A**) Discharge/charge profiles of the first five cycles at 0.1 A g^{-1}. (**B**) Discharge/charge profiles at various rates. Reproduced with permission [126]. Copyright 2019, The Royal Society of Chemistry. (**C**) The XRD pattern and the inset shows the morphology. (**D**) The initial three discharge/charge profiles of the mixture of 2Ge and 3P milled at 7 h. Reproduced with permission [127]. Copyright 2019, Elsevier.

In 2019, Li et al. [127] designed and synthesized Ge_2P_3 composite material based on GeP_5's characteristics of metal conductivity, layered structure, large capacity, and high initial coulomb efficiency in the application of LIBs anode material, through the ball milling method to make a layered GeP and BP composite. The results showed that the layered GeP-BP composite has good lithium storage ability. It can be seen from Figure 9C that Ge_2P_3 powder is composed of layered GeP (JCPDS-44-1125) and layered P (JCPDS-73-1358). Due to the layered crystal structure, binary reaction components and rich heterogeneous interfaces of the composite electrode, the composite electrode has high performance of lithium storage. Figure 9D shows the three initial charge and discharge efficiencies of the layered GEP-BP composite electrode, which are 1795 mAh g^{-1} and 1610 mAh g^{-1}, respectively, and the initial coulomb efficiency is up to 89%. Three cyclic voltammetry curves show that the layered GEP-BP composite electrode has undergone three stages of lithium storage process: intercalation, conversion, and alloy [128].

5. Conclusions and Prospect

Since the successful preparation of 2D BP by the high-pressure method for the first time, researchers have made a lot of new progress in the structure, preparation and, reaction mechanism of 2D BP, especially as regards the emergence of catalytic method, which opened a new chapter in the preparation of 2D BP. However, in general, the following serious and urgent problems in the preparation of BP by catalytic method still remain: (1) The preparation of BP by the catalytic method is still an intermittent operation, which is not conducive to the large-scale preparation of 2D BP; (2) There are few studies on the preparation of 2D BP from cheap white phosphorus, which is not conducive to further reducing the preparation cost of 2D BP; (3) The formation mechanism of 2D BP is not clear, and the specific structure of the intermediate transition state during the conversion of red phosphorus to BP and the process of transition state to BP still need to be explored. To sum up, the preparation of BP with low-cost white phosphorus as raw material, in-depth study of the formation mechanism of 2D BP, and the realization of high-quality, low-cost, and large-scale preparation of 2D BP are the research directions that need to be followed in the future. In fact, the ultimate purpose of preparing BP is application, and the premise of application lies in the high quality and efficient preparation of nano-BP and its composite materials. In summary, ultrasonic stripping and electrochemical stripping are the two most commonly used methods for the preparation of nano-BP, and even effective means for the preparation of nano-BP matrix composites. However, both of these two methods have the problem of "uncontrollable" preparation of nano-BP, mainly that the size and thickness are not easy to control, which affects the application of nano-BP to a certain extent. Therefore, research on controllable preparation of nano-BP is of great significance to the application of nano-BP. Compared with the top-down method, the bottom-up solvothermal method and chemical vapor deposition method can produce nano-BP or even nano-BP matrix composites in one step. The preparation process is simple and efficient, especially the chemical vapor deposition method, which has the potential to prepare high quality nano-BP matrix composites. However, due to the immaturity of the preparation system and technology, many problems in the preparation process have not been solved, and industrial production is challenging to some extent.

At present, the coulombic efficiency, cycle performance, and rate performance of the battery have great room for improvement when nano-BP matrix composites are applied to the negative electrode of energy storage battery, and the construction of the composite material is particularly important. It is of great significance to promote the application of nano-BP in the field of energy storage using existing preparation methods or developing new methods to compound nanometer BP with existing materials or synthesis or finding new materials to construct nanometer BP matrix composites with excellent energy storage performance and develop efficient preparation methods. In addition, the theoretical calculation shows that black phosphorene also shows excellent hydrogen storage performance after metal doping. Putting the theory into practice will be an urgent research work in the next step. In summary, excellent performance depends on the structure–activity relationship between materials, and many challenges remain.

Funding: This work was financially supported by the Natural Science Foundation of Guangxi (No. 2021GXNSFAA220108, 2020GXNSFBA297122), Special Project for Guangxi Science and Technology Bases and Talents (No. AD20297134), National key research and development program (2021YFA0715404), Guangxi key research and development program (2021AB05083).

Institutional Review Board Statement: Not applicable.

Informed Consent Statement: Not applicable.

Data Availability Statement: The data presented in this study are available on request from the corresponding author.

Conflicts of Interest: The authors declare no competing financial interest.

References

1. Liu, K.; Liu, Y.; Lin, D.; Pei, A.; Cui, Y. Materials for lithium-ion battery safety. *Sci. Adv.* **2018**, *4*, eaas9820. [CrossRef] [PubMed]
2. Zubi, G.; Dufo-López, R.; Carvalho, M.; Pasaoglu, G. The lithium-ion battery: State of the art and future perspectives. *Renew. Sustain. Energy Rev.* **2018**, *89*, 292–308. [CrossRef]
3. Scrosati, B. Recent advances in lithium ion battery materials. *Electrochimica Acta* **2000**, *45*, 2461–2466. [CrossRef]
4. Lu, L.; Han, X.; Li, J.; Hua, J.; Ouyang, M. A review on the key issues for lithium-ion battery management in electric vehicles. *J. Power Sources* **2013**, *226*, 272–288. [CrossRef]
5. Barré, A.; Deguilhem, B.; Grolleau, S.; Gérard, M.; Suard, F.; Riu, D. A review on lithium-ion battery ageing mechanisms and estimations for automotive applications. *J. Power Sources* **2013**, *241*, 680–689. [CrossRef]
6. Maleki, H.; Al Hallaj, S.; Selman, J.R.; Dinwiddie, R.; Wang, H. Thermal Properties of Lithium-Ion Battery and Components. *J. Electrochem. Soc.* **1999**, *146*, 947–954. [CrossRef]
7. Schipper, F.; Erickson, E.M.; Erk, C.; Shin, J.; Chesneau, F.F.; Aurbach, D. Recent advances and remaining challenges for lithium ion battery cathodes. *J. Electrochem. Soc.* **2016**, *164*, A6220. [CrossRef]
8. Wang, Q.; Ping, P.; Zhao, X.; Chu, G.; Sun, J.; Chen, C. Thermal runaway caused fire and explosion of lithium ion battery. *J. Power Sources* **2012**, *208*, 210–224. [CrossRef]
9. Han, X.; Lu, L.; Zheng, Y.; Feng, X.; Li, Z.; Li, J.; Ouyang, M. A review on the key issues of the lithium ion battery degradation among the whole life cycle. *eTransportation* **2019**, *1*, 100005. [CrossRef]
10. Manthiram, A. A reflection on lithium-ion battery cathode chemistry. *Nat. Commun.* **2020**, *11*, 1550. [CrossRef] [PubMed]
11. Wrogemann, J.M.; Haneke, L.; Ramireddy, T.; Frerichs, J.E.; Sultana, I.; Chen, Y.I.; Brink, F.; Hansen, M.R.; Winter, M.; Glushenkov, A.M.; et al. Advanced Dual-Ion Batteries with High-Capacity Negative Electrodes Incorporating Black Phosphorus. *Adv. Sci.* **2022**, *9*, 2201116. [CrossRef] [PubMed]
12. Jin, H.; Huang, Y.; Wang, C.; Ji, H. Phosphorus-Based Anodes for Fast Charging Lithium-Ion Batteries: Challenges and Opportunities. *Small Sci.* **2022**, *2*, 2200015. [CrossRef]
13. Wang, H.; Liu, C.; Cao, Y.; Liu, S.; Zhang, B.; Hu, Z.; Sun, J. Two-Dimensional Layered Green Phosphorus as an Anode Material for Li-Ion Batteries. *ACS Appl. Energy Mater.* **2022**, *5*, 2184–2191. [CrossRef]
14. Li, H.; Li, C.; Tao, B.; Gu, S.; Xie, Y.; Wu, H.; Zhang, G.; Wang, G.; Zhang, W.; Chang, H. Two-Dimensional Metal Telluride Atomic Crystals: Preparation, Physical Properties, and Applications. *Adv. Funct. Mater.* **2021**, *31*, 2010901. [CrossRef]
15. Li, H.; Gu, S.; Sun, Z.; Guo, F.; Xie, Y.; Tao, B.; He, X.; Zhang, W.; Chang, H. The in-built bionic "MoFe cofactor" in Fe-doped two-dimensional MoTe 2 nanosheets for boosting the photocatalytic nitrogen reduction performance. *J. Mater. Chem. A* **2020**, *8*, 13038–13048. [CrossRef]
16. Guo, W.; Dong, Z.; Xu, Y.; Liu, C.; Wei, D.; Zhang, L.; Shi, X.; Guo, C.; Xu, H.; Chen, G.; et al. Sensitive Terahertz Detection and Imaging Driven by the Photothermoelectric Effect in Ultrashort-Channel Black Phosphorus Devices. *Adv. Sci.* **2020**, *7*, 1902699. [CrossRef] [PubMed]
17. Wang, Y.; He, M.; Ma, S.; Yang, C.; Yu, M.; Yin, G.; Zuo, P. Low-Temperature Solution Synthesis of Black Phosphorus from Red Phosphorus: Crystallization Mechanism and Lithium Ion Battery Applications. *J. Phys. Chem. Lett.* **2020**, *11*, 2708–2716. [CrossRef]
18. Liu, D.; Wang, J.; Bian, S.; Liu, Q.; Gao, Y.; Wang, X.; Chu, P.K.; Yu, X.-F. Photoelectrochemical synthesis of ammonia with black phosphorus. *Adv. Funct. Mater.* **2020**, *30*, 2002731. [CrossRef]
19. Lu, J.; Wu, J.; Carvalho, A.; Ziletti, A.; Liu, H.; Tan, J.; Chen, Y.; Neto, A.H.C.; Özyilmaz, B.; Sow, C.H. Bandgap Engineering of Phosphorene by Laser Oxidation toward Functional 2D Materials. *ACS Nano* **2015**, *9*, 10411–10421. [CrossRef]
20. Ma, W.; Lu, J.; Wan, B.; Peng, D.; Xu, Q.; Hu, G.; Peng, Y.; Pan, C.; Wang, Z.L. Piezoelectricity in Multilayer Black Phosphorus for Piezotronics and Nanogenerators. *Adv. Mater.* **2019**, *32*, e1905795. [CrossRef]
21. Bridgman, P. Two new modifications of phosphorus. *J. Am. Chem. Soc.* **1914**, *36*, 1344–1363. [CrossRef]
22. Liu, H.; Neal, A.T.; Zhu, Z.; Luo, Z.; Xu, X.; Tomanek, D.; Ye, P.D. Phosphorene: An Unexplored 2D Semiconductor with a High Hole Mobility. *ACS Nano* **2014**, *8*, 4033–4041. [CrossRef] [PubMed]

23. Lei, W.; Mi, Y.; Feng, R.; Liu, P.; Hu, S.; Yu, J.; Liu, X.; Rodriguez, J.A.; Wang, J.-O.; Zheng, L.; et al. Hybrid 0D–2D black phosphorus quantum dots–graphitic carbon nitride nanosheets for efficient hydrogen evolution. *Nano Energy* **2018**, *50*, 552–561. [CrossRef]
24. Liu, H.; Hu, K.; Yan, D.; Chen, R.; Zou, Y.; Liu, H.; Wang, S. Recent advances on black phosphorus for energy storage, catalysis, and sensor applications. *Adv. Mater.* **2018**, *30*, 1800295. [CrossRef]
25. Bridgman, P.W. The compression of 39 substances to 100, 000 kg/cm. *Proc. Am. Acad. Arts Sci.* **1948**, *76*, 55–70.
26. Maruyama, Y.; Suzuki, S.; Kobayashi, K.; Tanuma, S. Synthesis and some properties of black phosphorus single crystals. *Phys. B+C* **1981**, *105*, 99–102. [CrossRef]
27. Köpf, M.; Eckstein, N.; Pfister, D.; Grotz, C.; Krüger, I.; Greiwe, M.; Hansen, T.; Kohlmann, H.; Nilges, T. Access and in situ growth of phosphorene-precursor black phosphorus. *J. Cryst. Growth* **2014**, *405*, 6–10. [CrossRef]
28. Erande, M.B.; Suryawanshi, S.R.; More, M.A.; Late, D.J. Electrochemically Exfoliated Black Phosphorus Nanosheets—Prospective Field Emitters. *Eur. J. Inorg. Chem.* **2015**, *2015*, 3102–3107. [CrossRef]
29. Krebs, H.; Weitz, H.; Worms, K. Über die struktur und eigenschaften der halbmetalle. Viii. Die katalytische darstellung des schwarzen phosphors. *Z. Für Anorg. Und Allg. Chem.* **1955**, *280*, 119–133. [CrossRef]
30. Baba, M.; Izumida, F.; Takeda, Y.; Morita, A. Preparation of Black Phosphorus Single Crystals by a Completely Closed Bismuth-Flux Method and Their Crystal Morphology. *Jpn. J. Appl. Phys.* **1989**, *28*, 1019–1022. [CrossRef]
31. Dhanabalan, S.C.; Ponraj, J.S.; Guo, Z.; Qiaoliang, B.; Bao, Q.; Zhang, H. Emerging Trends in Phosphorene Fabrication towards Next Generation Devices. *Adv. Sci.* **2017**, *4*, 1600305. [CrossRef] [PubMed]
32. Li, L.; Yu, Y.; Ye, G.J.; Ge, Q.; Ou, X.; Wu, H.; Feng, D.; Chen, X.H.; Zhang, Y. Black phosphorus field-effect transistors. *Nat. Nanotechnol.* **2014**, *9*, 372–377. [CrossRef] [PubMed]
33. Chen, Y.; Jiang, G.; Chen, S.; Guo, Z.; Yu, X.; Zhao, C.; Zhang, H.; Bao, Q.; Wen, S.; Tang, D.; et al. Mechanically exfoliated black phosphorus as a new saturable absorber for both Q-switching and Mode-locking laser operation. *Opt. Express* **2015**, *23*, 12823–12833. [CrossRef] [PubMed]
34. Guan, L.; Xing, B.; Niu, X.; Wang, D.; Yu, Y.; Zhang, S.; Yan, X.; Wang, Y.; Sha, J. Metal-assisted exfoliation of few-layer black phosphorus with high yield. *Chem. Commun.* **2017**, *54*, 595–598. [CrossRef] [PubMed]
35. Mu, Y.; Si, M.S. The mechanical exfoliation mechanism of black phosphorus to phosphorene: A first-principles study. *Eur. Lett.* **2015**, *112*, 37003. [CrossRef]
36. Island, J.; Steele, G.A.; van der Zant, H.S.J.; Castellanos-Gomez, A. Environmental instability of few-layer black phosphorus. *2D Mater.* **2015**, *2*, 011002. [CrossRef]
37. Luo, Z.; Maassen, J.; Deng, Y.; Du, Y.; Garrelts, R.P.; Lundstrom, M.S.; Ye, P.D.; Xu, X. Anisotropic in-plane thermal conductivity observed in few-layer black phosphorus. *Nat. Commun.* **2015**, *6*, 8572. [CrossRef]
38. Guo, Z.; Chen, S.; Wang, Z.; Yang, Z.; Liu, F.; Xue-Feng, Y.; Wang, J.; Yi, Y.; Zhang, H.; Liao, L.; et al. Metal-Ion-Modified Black Phosphorus with Enhanced Stability and Transistor Performance. *Adv. Mater.* **2017**, *29*, 1703811. [CrossRef]
39. Brent, J.R.; Savjani, N.; Lewis, E.A.; Haigh, S.J.; Lewis, D.J.; O'Brien, P. Production of few-layer phosphorene by liquid exfoliation of black phosphorus. *Chem. Commun.* **2014**, *50*, 13338–13341. [CrossRef]
40. Hanlon, D.; Backes, C.; Doherty, E.; Cucinotta, C.; Berner, N.C.; Boland, C.S.; Lee, K.; Harvey, A.; Lynch, P.; Gholamvand, Z.; et al. Liquid exfoliation of solvent-stabilized few-layer black phosphorus for applications beyond electronics. *Nat. Commun.* **2015**, *6*, 8563. [CrossRef]
41. Late, D.J. Liquid exfoliation of black phosphorus nanosheets and its application as humidity sensor. *Microporous Mesoporous Mater.* **2016**, *225*, 494–503. [CrossRef]
42. Zhao, W.; Xue, Z.; Wang, J.; Jiang, J.; Zhao, X.; Mu, T. Large-Scale, Highly Efficient, and Green Liquid-Exfoliation of Black Phosphorus in Ionic Liquids. *ACS Appl. Mater. Interfaces* **2015**, *7*, 27608–27612. [CrossRef] [PubMed]
43. Sciannamea, V.; Jérôme, R.; Detrembleur, C. In-Situ Nitroxide-Mediated Radical Polymerization (NMP) Processes: Their Understanding and Optimization. *Chem. Rev.* **2008**, *108*, 1104–1126. [CrossRef] [PubMed]
44. Su, S.; Xu, B.; Ding, J.; Yu, H. Large-yield exfoliation of few-layer black phosphorus nanosheets in liquid. *N. J. Chem.* **2019**, *43*, 19365–19371. [CrossRef]
45. Yasaei, P.; Kumar, B.; Foroozan, T.; Wang, C.; Asadi, M.; Tuschel, D.; Indacochea, J.E.; Klie, R.F.; Salehi-Khojin, A. High-Quality Black Phosphorus Atomic Layers by Liquid-Phase Exfoliation. *Adv. Mater.* **2015**, *27*, 1887–1892. [CrossRef]
46. Xiao, H.; Zhao, M.; Zhang, J.; Ma, X.; Zhang, J.; Hu, T.; Tang, T.; Jia, J.; Wu, H. Electrochemical cathode exfoliation of bulky black phosphorus into few-layer phosphorene nanosheets. *Electrochem. Commun.* **2018**, *89*, 10–13. [CrossRef]
47. Li, J.; Chen, C.; Liu, S.; Lu, J.; Goh, W.P.; Fang, H.; Qiu, Z.; Tian, B.; Chen, Z.; Yao, C.; et al. Ultrafast Electrochemical Expansion of Black Phosphorus toward High-Yield Synthesis of Few-Layer Phosphorene. *Chem. Mater.* **2018**, *30*, 2742–2749. [CrossRef]
48. Lin, Z.; Liu, Y.; Halim, U.; Ding, M.; Liu, Y.; Wang, Y.; Jia, C.; Chen, P.; Duan, X.; Wang, C.; et al. Solution-processable 2D semiconductors for high-performance large-area electronics. *Nature* **2018**, *562*, 254–258. [CrossRef]
49. Zu, L.; Gao, X.; Lian, H.; Li, C.; Liang, Q.; Liang, Y.; Cui, X.; Liu, Y.; Wang, X.; Cui, X. Electrochemical prepared phosphorene as a cathode for supercapacitors. *J. Alloy. Compd.* **2018**, *770*, 26–34. [CrossRef]
50. Ambrosi, A.; Sofer, Z.; Pumera, M. Electrochemical Exfoliation of Layered Black Phosphorus into Phosphorene. *Angew. Chem.* **2017**, *129*, 10579–10581. [CrossRef]

51. Duan, X.D.; Wang, C.; Pan, A.L.; Yu, R.Q.; Duan, X.F. Two-dimensional transition metal dichalcogenides as atomically thin semiconductors: Opportunities and challenges. *Chem. Soc. Rev.* **2015**, *44*, 8859–8876. [CrossRef] [PubMed]
52. Li, H.; Deng, H.; Gu, S.; Li, C.; Tao, B.; Chen, S.; He, X.; Wang, G.; Zhang, W.; Chang, H. Engineering of bionic Fe/Mo bimetallene for boosting the photocatalytic nitrogen reduction performance. *J. Colloid Interface Sci.* **2021**, *607*, 1625–1632. [CrossRef] [PubMed]
53. Smith, J.B.; Hagaman, D.; Ji, H.-F. Growth of 2D black phosphorus film from chemical vapor deposition. *Nanotechnology* **2016**, *27*, 215602. [CrossRef]
54. Xu, Y.; Shi, X.; Zhang, Y.; Zhang, H.; Zhang, Q.; Huang, Z.; Xu, X.; Guo, J.; Zhang, H.; Sun, L.; et al. Epitaxial nucleation and lateral growth of high-crystalline black phosphorus films on silicon. *Nat. Commun.* **2020**, *11*, 1330. [CrossRef] [PubMed]
55. Fei, R.; Yang, L. Strain-Engineering the Anisotropic Electrical Conductance of Few-Layer Black Phosphorus. *Nano Lett.* **2014**, *14*, 2884–2889. [CrossRef]
56. Liu, H.; Du, Y.; Deng, Y.; Ye, P.D. Semiconducting black phosphorus: Synthesis, transport properties and electronic applications. *Chem. Soc. Rev.* **2014**, *44*, 2732–2743. [CrossRef]
57. Ge, S.; Li, C.; Zhang, Z.; Zhang, C.; Zhang, Y.; Qiu, J.; Wang, Q.; Liu, J.; Jia, S.; Feng, J.; et al. Dynamical Evolution of Anisotropic Response in Black Phosphorus under Ultrafast Photoexcitation. *Nano Lett.* **2015**, *15*, 4650–4656. [CrossRef]
58. Liu, M.; Feng, S.; Hou, Y.; Zhao, S.; Tang, L.; Liu, J.; Wang, F.; Liu, B. High yield growth and doping of black phosphorus with tunable electronic properties. *Mater. Today* **2020**, *36*, 91–101. [CrossRef]
59. Lange, S.; Schmidt, P.; Nilges, T. Au3SnP7@Black Phosphorus: An Easy Access to Black Phosphorus. *Inorg. Chem.* **2007**, *46*, 4028–4035. [CrossRef]
60. Tian, B.; Tian, B.; Smith, B.; Liu, Y. Facile bottom-up synthesis of partially oxidized black phosphorus nanosheets as metal-free photocatalyst for hydrogen evolution. *Proc. Natl. Acad. Sci. USA* **2018**, *115*, 4345–4350. [CrossRef] [PubMed]
61. Lee, S.; Yang, F.; Suh, J.; Yang, S.; Lee, Y.; Li, G.; Choe, H.S.; Suslu, A.; Chen, Y.; Ko, C.; et al. Anisotropic in-plane thermal conductivity of black phosphorus nanoribbons at temperatures higher than 100 K. *Nat. Commun.* **2015**, *6*, 8573. [CrossRef]
62. Zhao, G.; Wang, T.; Shao, Y.; Wu, Y.; Huang, B.; Hao, X. A Novel Mild Phase-Transition to Prepare Black Phosphorus Nanosheets with Excellent Energy Applications. *Small* **2016**, *13*, 1602243. [CrossRef] [PubMed]
63. Zhang, Y.; Rui, X.; Tang, Y.; Liu, Y.; Wei, J.; Chen, S.; Leow, W.R.; Li, W.; Liu, Y.; Deng, J.; et al. Wet-Chemical Processing of Phosphorus Composite Nanosheets for High-Rate and High-Capacity Lithium-Ion Batteries. *Adv. Energy Mater.* **2016**, *6*, 1502409. [CrossRef]
64. Yang, Z.; Hao, J.; Yuan, S.; Lin, S.; Yau, H.M.; Dai, J.; Lau, S.P. Field-Effect Transistors Based on Amorphous Black Phosphorus Ultrathin Films by Pulsed Laser Deposition. *Adv. Mater.* **2015**, *27*, 3748–3754. [CrossRef]
65. Kovalska, E.; Luxa, J.; Hartman, T.; Antonatos, N.; Shaban, P.; Oparin, E.; Zhukova, M.; Sofer, Z. Non-aqueous solution-processed phosphorene by controlled low-potential electrochemical exfoliation and thin film preparation. *Nanoscale* **2019**, *12*, 2638–2647. [CrossRef]
66. Sansone, G.; Maschio, L.; Usvyat, D.; Schütz, M.; Karttunen, A. Toward an Accurate Estimate of the Exfoliation Energy of Black Phosphorus: A Periodic Quantum Chemical Approach. *J. Phys. Chem. Lett.* **2015**, *7*, 131–136. [CrossRef]
67. Zhang, Y.; Wang, H.; Luo, Z.-Z.; Tan, H.T.; Li, B.; Sun, S.; Li, Z.; Zong, Y.; Xu, Z.; Yang, Y.; et al. An Air-Stable Densely Packed Phosphorene-Graphene Composite Toward Advanced Lithium Storage Properties. *Adv. Energy Mater.* **2016**, *6*, 1600453. [CrossRef]
68. Erande, M.B.; Pawar, M.S.; Late, D.J. Humidity Sensing and Photodetection Behavior of Electrochemically Exfoliated Atomically Thin-Layered Black Phosphorus Nanosheets. *ACS Appl. Mater. Interfaces* **2016**, *8*, 11548–11556. [CrossRef] [PubMed]
69. Zhu, X.; Zhang, T.; Sun, Z.; Chen, H.; Guan, J.; Chen, X.; Ji, H.; Du, P.; Yang, S. Black Phosphorus Revisited: A Missing Metal-Free Elemental Photocatalyst for Visible Light Hydrogen Evolution. *Adv. Mater.* **2017**, *29*, 1605776. [CrossRef]
70. Nine, J.; Cole, M.A.; Tran, D.N.H.; Losic, D. Graphene: A multipurpose material for protective coatings. *J. Mater. Chem. A* **2015**, *3*, 12580–12602. [CrossRef]
71. Huang, Y.; Qiao, J.; He, K.; Bliznakov, S.; Sutter, E.; Chen, X.; Luo, D.; Meng, F.; Su, D.; Decker, J.; et al. Interaction of Black Phosphorus with Oxygen and Water. *Chem. Mater.* **2016**, *28*, 8330–8339. [CrossRef]
72. Favron, A.; Gaufrès, E.; Fossard, F.; Phaneuf-L'Heureux, A.-L.; Tang, N.Y.-W.; Levesque, P.; Loiseau, A.; Leonelli, R.; Francoeur, S.; Martel, R. Photooxidation and quantum confinement effects in exfoliated black phosphorus. *Nat. Mater.* **2015**, *14*, 826–832. [CrossRef] [PubMed]
73. Lei, W.; Zhang, T.; Liu, P.; Rodriguez, J.A.; Liu, G.; Liu, M. Bandgap- and Local Field-Dependent Photoactivity of Ag/Black Phosphorus Nanohybrids. *ACS Catal.* **2016**, *6*, 8009–8020. [CrossRef]
74. Zhu, W.; Yogeesh, M.N.; Yang, S.; Aldave, S.H.; Kim, J.-S.; Sonde, S.; Tao, L.; Lu, N.; Akinwande, D. Flexible Black Phosphorus Ambipolar Transistors, Circuits and AM Demodulator. *Nano Lett.* **2015**, *15*, 1883–1890. [CrossRef] [PubMed]
75. Feng, Q.; Liu, H.; Zhu, M.; Shang, J.; Liu, D.; Cui, X.; Shen, D.; Kou, L.; Mao, D.; Zheng, J.; et al. Electrostatic Functionalization and Passivation of Water-Exfoliated Few-Layer Black Phosphorus by Poly Dimethyldiallyl Ammonium Chloride and Its Ultrafast Laser Application. *ACS Appl. Mater. Interfaces* **2018**, *10*, 9679–9687. [CrossRef] [PubMed]
76. Wu, W.; Wang, L.; Yuan, J.; Zhang, Z.; Zhang, X.; Dong, S.; Hao, J. Formation and Degradation Tracking of a Composite Hydrogel Based on UCNPs@PDA. *Macromolecules* **2020**, *53*, 2430–2440. [CrossRef]
77. Feng, J.; Li, X.; Zhu, G.; Wang, Q.J. Emerging High-Performance SnS/CdS Nanoflowers Heterojunction for Ultrafast Photonics. *ACS Appl. Mater. Interfaces* **2020**, *12*, 43098–43105. [CrossRef]

78. Zhang, J.; Ma, Y.; Hu, K.; Feng, Y.; Chen, S.; Yang, X.; Loo, J.F.-C.; Zhang, H.; Yin, F.; Li, Z. Surface Coordination of Black Phosphorus with Modified Cisplatin. *Bioconjugate Chem.* **2019**, *30*, 1658–1664. [CrossRef]
79. Tofan, D.; Sakazaki, Y.; Mitra, K.L.W.; Peng, R.; Lee, S.; Li, M.; Velian, A. Surface Modification of Black Phosphorus with Group 13 Lewis Acids for Ambient Protection and Electronic Tuning. *Angew. Chem.* **2021**, *133*, 8410–8417. [CrossRef]
80. Bayne, J.M.; Stephan, D.W. Phosphorus Lewis acids: Emerging reactivity and applications in catalysis. *Chem. Soc. Rev.* **2015**, *45*, 765–774. [CrossRef]
81. Abellán, G.; Wild, S.; Lloret, V.; Scheuschner, N.; Gillen, R.; Mundloch, U.; Maultzsch, J.; Varela, M.; Hauke, F.; Hirsch, A. Fundamental Insights into the Degradation and Stabilization of Thin Layer Black Phosphorus. *J. Am. Chem. Soc.* **2017**, *139*, 10432–10440. [CrossRef] [PubMed]
82. Fonsaca, J.; Domingues, S.H.; Orth, E.S.; Zarbin, A.J.G. Air stable black phosphorous in polyaniline-based nanocomposite. *Sci. Rep.* **2017**, *7*, 10165. [CrossRef] [PubMed]
83. Guo, R.; Zheng, Y.; Ma, Z.; Lian, X.; Sun, H.; Han, C.; Ding, H.; Xu, Q.; Yu, X.; Zhu, J.; et al. Surface passivation of black phosphorus via van der Waals stacked PTCDA. *Appl. Surf. Sci.* **2019**, *496*, 143688. [CrossRef]
84. Artel, V.; Guo, Q.; Cohen, H.; Gasper, R.; Ramasubramaniam, A.; Xia, F.; Naveh, D. Erratum: Protective molecular passivation of black phosphorus. *NPJ 2D Mater. Appl.* **2017**, *1*, 27. [CrossRef]
85. Liang, S.; Wu, L.; Liu, H.; Li, J.; Chen, M.; Zhang, M. Organic molecular passivation of phosphorene: An aptamer-based biosensing platform. *Biosens. Bioelectron.* **2018**, *126*, 30–35. [CrossRef]
86. Wan, B.; Yang, B.; Wang, Y.; Zhang, J.; Zeng, Z.; Liu, Z.; Wang, W. Enhanced stability of black phosphorus field-effect transistors with SiO_2 passivation. *Nanotechnology* **2015**, *26*, 435702. [CrossRef]
87. Li, M.; Zhao, Q.; Zhang, S.; Li, D.; Li, H.; Zhang, X.; Xing, B. Facile passivation of black phosphorus nanosheets via silica coating for stable and efficient solar desalination. *Environ. Sci. Nano* **2019**, *7*, 414–423. [CrossRef]
88. Galceran, R.; Gaufres, E.; Loiseau, A.; Piquemal-Banci, M.; Godel, F.; Vecchiola1, A.; Bezencenet, O.; Martin, M.-B.; Servet, B.; Petroff, F.; et al. Stabilizing ultra-thin black phosphorus with in-situ-grown 1 nm-Al_2O_3 barrier. *Appl. Phys. Lett.* **2017**, *111*, 243101. [CrossRef]
89. Li, K.; Ang, K.-W.; Lv, Y.; Liu, X. Effects of Al_2O_3 capping layers on the thermal properties of thin black phosphorus. *Appl. Phys. Lett.* **2016**, *109*, 261901. [CrossRef]
90. Kuriakose, S.; Ahmed, T.; Balendhran, S.; Bansal, V.; Sriram, S.; Bhaskaran, M.; Walia, S. Black phosphorus: Ambient degradation and strategies for protection. *2D Mater.* **2018**, *5*, 032001. [CrossRef]
91. Wang, C.-X.; Zhang, C.; Jiang, J.-W.; Rabczuk, T. The effects of vacancy and oxidation on black phosphorus nanoresonators. *Nanotechnology* **2017**, *28*, 135202. [CrossRef] [PubMed]
92. Luo, W.; Zemlyanov, D.Y.; Milligan, C.A.; Du, Y.; Yang, L.; Wu, Y.; Ye, P.D. Surface chemistry of black phosphorus under a controlled oxidative environment. *Nanotechnology* **2016**, *27*, 434002. [CrossRef] [PubMed]
93. Caporali, M.; Serrano-Ruiz, M.; Telesio, F.; Heun, S.; Verdini, A.; Cossaro, A.; Dalmiglio, M.; Goldoni, A.; Peruzzini, M. Enhanced ambient stability of exfoliated black phosphorus by passivation with nickel nanoparticles. *Nanotechnology* **2020**, *31*, 275708. [CrossRef] [PubMed]
94. Xu, Y.; Yuan, J.; Zhang, K.; Hou, Y.; Sun, Q.; Yao, Y.; Li, S.; Bao, Q.; Zhang, H.; Zhang, Y. Field-Induced n-Doping of Black Phosphorus for CMOS Compatible 2D Logic Electronics with High Electron Mobility. *Adv. Funct. Mater.* **2017**, *27*, 1702211. [CrossRef]
95. Lv, W.; Yang, B.; Wang, B.; Wan, W.; Ge, Y.; Yang, R.; Hao, C.; Xiang, J.; Zhang, B.; Zeng, Z.; et al. Sulfur-Doped Black Phosphorus Field-Effect Transistors with Enhanced Stability. *ACS Appl. Mater. Interfaces* **2018**, *10*, 9663–9668. [CrossRef] [PubMed]
96. Hu, X.; Li, G.; Yu, J. Design, Fabrication, and Modification of Nanostructured Semiconductor Materials for Environmental and Energy Applications. *Langmuir* **2009**, *26*, 3031–3039. [CrossRef] [PubMed]
97. Werner, M.; Fahrner, W. Review on materials, microsensors, systems and devices for high-temperature and harsh-environment applications. *IEEE Trans. Ind. Electron.* **2001**, *48*, 249–257. [CrossRef]
98. Tang, X.; Liang, W.; Zhao, J.; Li, Z.; Qiu, M.; Fan, T.; Luo, C.S.; Zhou, Y.; Li, Y.; Guo, Z.; et al. Fluorinated Phosphorene: Electrochemical Synthesis, Atomistic Fluorination, and Enhanced Stability. *Small* **2017**, *13*, 1702739. [CrossRef] [PubMed]
99. Wang, K.; Wang, H.; Zhang, M.; Liu, Y.; Zhao, W. Electronic and magnetic properties of doped black phosphorene with concentration dependence. *Beilstein J. Nanotechnol.* **2019**, *10*, 993–1001. [CrossRef]
100. Zheng, Y.; Hu, Z.; Han, C.; Guo, R.; Xiang, D.; Lei, B.; Wang, Y.; He, J.; Lai, M.; Chen, W. Black phosphorus inverter devices enabled by in-situ aluminum surface modification. *Nano Res.* **2018**, *12*, 531–536. [CrossRef]
101. Wang, Z.; Lu, J.; Wang, J.; Li, J.; Du, Z.; Wu, H.; Liao, L.; Chu, P.K.; Yu, X.-F. Air-stable n-doped black phosphorus transistor by thermal deposition of metal adatoms. *Nanotechnology* **2019**, *30*, 135201. [CrossRef] [PubMed]
102. Gamage, S.; Li, Z.; Yakovlev, V.S.; Lewis, C.; Wang, H.; Cronin, S.B.; Abate, Y. Nanoscopy of black phosphorus degradation. *Adv. Mater. Interfaces* **2016**, *3*, 1600121. [CrossRef]
103. Li, Z.; Huang, J.; Liaw, B.Y.; Metzler, V.; Zhang, J. A review of lithium deposition in lithium-ion and lithium metal secondary batteries. *J. Power Sources* **2014**, *254*, 168–182. [CrossRef]
104. Brain, M.; Bryant, C.W.; Pumphrey, C. How Batteries Work. Battery Arrangement Power 14. 2015. Available online: http://mdriscoll.pbworks.com/w/file/fetch/103781211/Lesson%203%20How%20Batteries%20Work%20-%20HowStuffWorks.pdf (accessed on 14 December 2021).

105. Kim, T.; Song, W.; Son, D.-Y.; Ono, L.K.; Qi, Y. Lithium-ion batteries: Outlook on present, future, and hybridized technologies. *J. Mater. Chem. A* **2019**, *7*, 2942–2964. [CrossRef]
106. Schalkwijk, W.V.; Scrosati, B. *Advances in Lithium-Ion Batteries 1–5*; Springer: Berlin/Heidelberg, Germany, 2002.
107. Santhanagopalan, S.; Guo, Q.; Ramadass, P.; White, R.E. Review of models for predicting the cycling performance of lithium ion batteries. *J. Power Sources* **2006**, *156*, 620–628. [CrossRef]
108. Yamada, M.; Watanabe, T.; Gunji, T.; Wu, J.; Matsumoto, F. Review of the Design of Current Collectors for Improving the Battery Performance in Lithium-Ion and Post-Lithium-Ion Batteries. *Electrochem* **2020**, *1*, 124–159. [CrossRef]
109. Jiang, X.; Chen, Y.; Meng, X.; Cao, W.; Liu, C.; Huang, Q.; Naik, N.; Murugadoss, V.; Huang, M.; Guo, Z. The impact of electrode with carbon materials on safety performance of lithiumion batteries: A review. *Carbon* **2022**, *191*, 448–470. [CrossRef]
110. Ma, X.; Chen, M.; Zheng, Z.; Bullen, D.; Wang, J.; Harrison, C.; Gratz, E.; Lin, Y.; Yang, Z.; Zhang, Y.; et al. Recycled cathode materials enabled superior performance for lithium-ion batteries. *Joule* **2021**, *5*, 2955–2970. [CrossRef]
111. Yang, X.-G.; Liu, T.; Wang, C.-Y. Thermally modulated lithium iron phosphate batteries for mass-market electric vehicles. *Nat. Energy* **2021**, *6*, 176–185. [CrossRef]
112. Song, L.; Du, J.; Xiao, Z.; Jiang, P.; Cao, Z.; Zhu, H. Research Progress on the Surface of High-Nickel Nickel–Cobalt–Manganese Ternary Cathode Materials: A Mini Review. *Front. Chem.* **2020**, *8*, 761. [CrossRef]
113. Chen, K.H.; Goel, V.; Namkoong, M.J.; Wied, M.; Müller, S.; Wood, V.; Sakamoto, J.; Thornton, K.; Dasgupta, N.P. Enabling 6C Fast Charging of Li-Ion Batteries with Graphite/Hard Carbon Hybrid Anodes. *Adv. Energy Mater.* **2020**, *11*, 2003336. [CrossRef]
114. Liu, T.; Han, X.; Zhang, Z.; Chen, Z.; Wang, P.; Han, P.; Ding, N.; Cui, G. A high concentration electrolyte enables superior cycleability and rate capability for high voltage dual graphite battery. *J. Power Sources* **2019**, *437*, 226942. [CrossRef]
115. Qiu, M.; Sun, Z.T.; Sang, D.K.; Han, X.G.; Zhang, H.; Niu, C.M. Current progress in black phosphorus materials and their applications in electrochemical energy storage. *Nanoscale* **2017**, *9*, 13384–13403. [CrossRef] [PubMed]
116. Guo, S.; Hu, X.; Zhou, W.; Liu, X.; Gao, Y.; Zhang, S.; Zhang, K.; Zhu, Z.; Zeng, H. Mechanistic Understanding of Two-Dimensional Phosphorus, Arsenic, and Antimony High-Capacity Anodes for Fast-Charging Lithium/Sodium Ion Batteries. *J. Phys. Chem. C* **2018**, *122*, 29559–29566. [CrossRef]
117. Pang, J.; Bachmatiuk, A.; Yin, Y.; Trzebicka, B.; Zhao, L.; Fu, L.; Mendes, R.G.; Gemming, T.; Liu, Z.; Rummeli, M.H. Applications of Phosphorene and Black Phosphorus in Energy Conversion and Storage Devices. *Adv. Energy Mater.* **2017**, *8*, 1702093. [CrossRef]
118. Zhang, C.; Yu, M.; Anderson, G.; Dharmasena, R.R.; Sumanasekera, G. The prospects of phosphorene as an anode material for high-performance lithium-ion batteries: A fundamental study. *Nanotechnology* **2017**, *28*, 075401. [CrossRef]
119. Zhao, S.; Kang, W.; Xue, J. The potential application of phosphorene as an anode material in Li-ion batteries. *J. Mater. Chem. A* **2014**, *2*, 19046–19052. [CrossRef]
120. Kou, L.; Chen, C.; Smith, S.C. Phosphorene: Fabrication, properties, and applications. *J. Phys. Chem. Lett.* **2015**, *6*, 2794–2805. [CrossRef]
121. Castillo, A.E.D.R.; Pellegrini, V.; Sun, H.; Buha, J.; Dinh, D.A.; Lago, E.; Ansaldo, A.; Capasso, A.; Manna, L.; Bonaccorso, F. Exfoliation of Few-Layer Black Phosphorus in Low-Boiling-Point Solvents and Its Application in Li-Ion Batteries. *Chem. Mater.* **2018**, *30*, 506–516. [CrossRef]
122. Zhang, Y.; Sun, W.; Luo, Z.-Z.; Zheng, Y.; Yu, Z.; Zhang, D.; Yang, J.; Tan, H.T.; Zhu, J.; Wang, X.; et al. Functionalized few-layer black phosphorus with super-wettability towards enhanced reaction kinetics for rechargeable batteries. *Nano Energy* **2017**, *40*, 576–586. [CrossRef]
123. Sun, J.; Zheng, G.; Lee, H.-W.; Liu, N.; Wang, H.; Yao, H.; Yang, W.; Cui, Y. Formation of Stable Phosphorus–Carbon Bond for Enhanced Performance in Black Phosphorus Nanoparticle–Graphite Composite Battery Anodes. *Nano Lett.* **2014**, *14*, 4573–4580. [CrossRef] [PubMed]
124. Jiang, Q.; Li, J.; Yuan, N.; Wu, Z.; Tang, J. Black phosphorus with superior lithium ion batteries performance directly synthesized by the efficient thermal-vaporization method. *Electrochim. Acta* **2018**, *263*, 272–276. [CrossRef]
125. Jin, H.; Xin, S.; Chuang, C.; Li, W.; Wang, H.; Zhu, J.; Xie, H.; Zhang, T.; Wan, Y.; Qi, Z.; et al. Black phosphorus composites with engineered interfaces for high-rate high-capacity lithium storage. *Science* **2020**, *370*, 192–197. [CrossRef] [PubMed]
126. Jin, J.; Zheng, Y.; Huang, S.-Z.; Sun, P.-P.; Srikanth, N.; Kong, L.B.; Yan, Q.; Zhou, K. Directly anchoring 2D NiCo metal–organic frameworks on few-layer black phosphorus for advanced lithium-ion batteries. *J. Mater. Chem. A* **2018**, *7*, 783–790. [CrossRef]
127. Li, X.; Li, W.; Shen, P.; Yang, L.; Lia, Y.; Shi, Z.; Zhang, H. Layered GeP-black P (Ge2P3): An advanced binary-phase anode for Li/Na-storage. *Ceram. Int.* **2019**, *45*, 15711–15714. [CrossRef]
128. Li, W.; Li, X.; Yu, J.; Liao, J.; Zhao, B.; Huang, L.; Abdelhafiz, A.; Zhang, H.; Wang, J.-H.; Guo, Z.; et al. A self-healing layered GeP anode for high-performance Li-ion batteries enabled by low formation energy. *Nano Energy* **2019**, *61*, 594–603. [CrossRef]